IP Networking over Next-Generation Satellite Systems

International Workshop, Budapest, July 2007

T0182003

Linghang Fan • Haitham Cruickshank •
Zhili Sun
Editors

IP Networking over Next-Generation Satellite Systems

International Workshop,
Budapest, July 2007

 Springer

Linghang Fan
Centre for Communication
    Systems Research
University of Surrey
Guildford, Surrey, GU2 7HX
United Kingdom
L.Fan@surrey.ac.uk

Haitham Cruickshank
Centre for Communication
    Systems Research
University of Surrey
Guildford, Surrey, GU2 7HX
United Kingdom
H.Cruickshank@surrey.ac.uk

Zhili Sun
Centre for Communication
    Systems Research
University of Surrey
Guildford, Surrey, GU2 7HX
United Kingdom
Z.Sun@surrey.ac.uk

ISBN-13: 978-1-4614-9883-4   ISBN-13: 978-0-387-75428-4 (eBook)

Printed on acid-free paper.

9 8 7 6 5 4 3 2 1

springer.com

Preface

With the boom of Internet, IP-based applications, such as WWW and multimedia, have been an essential part of our life, and there is an ever-increasing demand for accessing high-speed Internet services anywhere and anytime. This trend unavoidably has huge impact on the design of the next-generation satellite systems. In addition, with its broadcasting nature and global coverage, satellite systems also can play an important role in the next-generation Internet. For example, satellite systems can be a good driver for the deployment of IPv6 in the Internet and can provide a fast way to reach end-users because they do not rely on construction of a high-speed terrestrial networks. Thus satellites have the potential to bridge significant gaps in global connectivity issues.

To the naïve observer IP over satellite problem has been solved in the past and does not have any new challenges. However, recent satellite research in several EU projects show that there are still many unresolved issues; such as efficient deployment of IPv6 over satellites, interworking with other access technologies such as WLAN and WiMax, QoS provisioning over multi-segment networks (including satellites), security and On-Board Processing satellites usage for challenging applications such as IP multicast over satellites.

The papers in this book were selected from the 'International Workshop on IP Networking over Next-generation Satellite Systems (INNSS'07)', which was held on July 5, 2007 in Budapest, Hungary as a part of the 16th IST Mobile and Wireless Communications Summit conference.

This workshop focuses on the IP networking issues in the next-generation satellite systems and the papers cover the following topics:

- IPv6 over satellites
- Architecture of the next-generation satellite systems
- Satellite and terrestrial network integration
- Quality of Service (QoS) and Resource management
- Mobility management
- Multicast
- Security and privacy
- Communication protocols

- Network monitoring and measurement
- Performance enhancement

It is hoped that this book, which is the proceeding of this workshop, will be found useful as a reference work for engineers and researchers.

We sincerely thank IST projects: Satellite-based Communications Systems within IPv6 (SATSIX), IST European Satellite Communications Network of Excellence (SATNEX) and IST Integrated Multi-layer Optimization in broadband DVB-S.2 Satellite Networks (IMOSAN) for sponsoring the publication of this book.

We also thank the members of advisory board and technical committees for their support (listed below).

Advisory Board

- Michel Mazzella (Alcatel Alenia Space, France)
- Anton Donner (DLR, Germany)
- Avi Gal (Gilat, Israel)
- Tasos Kourtis (Demokritos Research Institute, Greece)
- Catherine Morlet (ESA)
- Istvan Frigyes (Budapest Univ. of Technology and Economics, Hungary)
- Isabelle Buret (Alcatel Alenia Space, France)
- Barry Evans (University of Surrey, United Kingdom)
- Erich Lutz (DLR, Germany)
- Zhili Sun (University of Surrey, United Kingdom)
- Carlo Caini (University of Bologna, Italy)

Technical Program Committee

- José Antonio Guerra Expósito (Hispasat SA, Spain)
- Cedric Baudoin (Alcatel Alenia Space, France)
- Pascal Berthou (CNRS/LAAS, France)
- Hermann Bischl (DLR, Germany)
- Ricardo Castellot (Telefonica I+D, Spain)
- Bernhard Collini-Nocker (University of Salzburg, Austria)
- Tomaso de Cola (CNIT, Italy)
- Borja de la Cuesta (University of Valladolid, Spain)
- Gorry Fairhurst (University of Aberdeen, United Kingdom)
- Georgios Gardikis (Demokritos Research Institute, Greece)
- Thierry Gayraud (CNRS/LAAS, France)
- Giovanni Giambene (University of Siena, Italy)
- Sunil Iyengar (University of Surrey, United Kingdom)
- Lei Liang (University of Surrey, United Kingdom)

- Mario Marchese (University of Genoa, Italy)
- Inge Melhus (SINTEF, Norway)
- Robert Mort (Systek, United Kingdom)
- Niovi Pavlidou (Aristotle University, Greece)
- Antonio Pietrabissa (University of Rome, Italy)
- Filippo Rodriguez (Telespazio, Italy)
- Robert Rumeau (CNES, France)
- Arjuna Sathiaseelan (University of Aberdeen, United Kingdom)
- Sandro Scalise (DLR, Germany)
- Alessandro Vanelli-Coralli (University of Bologna, Italy)
- Andreas Voigt (Fokus Fraunhofer, Germany)
- Lloyd Wood (Cisco)
- Ana Yun Garcia (Alcatel Alenia Space, Spain)

Linghang Fan
Haitham Cruickshank
Zhili Sun

Contents

Contributors

S. Abdellatif
LAAS-CNRS, University of Toulouse, France

J. Aguiar
University of Valladolid, Calle del Cementerio s/n, 47011 Valladolid, Spain

F. Arnal
Thales Alenia Space, 26 Avenue J-F Champollion, 31037 Toulouse Cedex 1,
France

Y.H. Aoul
University of Versailles, CNRS-PRiSM Lab, 45 Av.des Etats Unis, 78035
Versailles, France

D. Barvaux
B2i, 1 avenue de l'Europe, 31400 Toulouse, France

C. Baudoin
Thales Alenia Space, 26 Avenue J-F Champollion, 31037 Toulouse Cedex 1,
France

P. Berthou
University of Toulouse, LAAS-CNRS, 7 avenue du Colonel Roche, 31077
Toulouse Cedex 4, France

A. Burdena
Centre for Technological Research of Crete, Estavromenos,
Heraklion 71500, Greece

E. Callejo
Thales Alenia Space Espana, 7 calle Einstein (PTM), 28760 Tres Cantos
(Madrid), Spain

B. Carro
University of Valladolid, Calle del Cementerio s/n47011 Valladolid, Spain

M. Catalán
Hispasat S.A., C/ Gobelas 41, 28023 Madrid, Spain

H. Cruickshank
University of Surrey, CCSR, Guildford GU2 7XH, Surrey, UK

B. de la Cuesta
University of Valladolid, Calle del Cementerio s/n47011 Valladolid, Spain

M. Dervin
Thales Alenia Space, 26 Avenue J-F Champollion, BP 33787, 31037 Toulouse
Cedex 1, France

G. Dimitriadis
Aristotle University of Thessaloniki, Greece

A. Donner
Deutsches Zentrum für Luft- und Raumfahrt e.V. (DLR), Institute of
Communications and Navigation, Germany

L. Duquerroy
Thales Alenia Space, 26 Avenue J-F Champollion, 31037 Toulouse Cedex 1,
France

D. Elkouss
Thales Alenia Space, 7 calle Einstein (PTM), 28760 Tres Cantos (Madrid), Spain

À.V. Estrem
TriaGnoSys GmbH, Wessling, Germany

G. Fairhurst
Department of Engineering, University of Aberdeen, Fraser Noble Building,
Kings College, Aberdeen AB24 3UE, UK

L. Fan
University of Surrey, Guildford GU2 7XH, Surrey, UK

T. Gayraud
CNRS/LAAS, Toulouse University of Science, 7 avenue du Colonel Roche,
31077 Toulouse Cedex 4, France

R. Giménez
Telefonica Investigación y Desarrollo, C/ Emilio Vargas 6, 28043 Madrid, Spain

M. Gineste
University of Toulouse/LAAS-CNRS, France

J.A. Guerra
Hispasat S.A., C/Gobelas 41, 28023, Madrid, Spain

K. Han
Radio Broadcasting Research Division Broadband Wireless Multimedia Research
Team, Electronics and Telecommunications Research Institute (ETRI), Korea

M. Han
School of Electronics, Telecommunications and Computer Engineering, Korea
Aerospace University, Korea

Y. Hecht
Gilat Satellite Networks, Petah Tikva, Israel

N. Hennion
Thales Alenia Space, 100 bd du midi – BP 99, 06156 Cannes la bocca, France

S. Iyengar
University of Surrey (CCSR), Guildford GU2 7XH, Surrey, UK

B. Jacquemin
CNRS/LAAS, Toulouse University of Science, France

A. Jahn
TriaGnoSys GmbH, Wessling, Germany

S. Karapantazis
Aristotle University of Thessaloniki, Greece

D. Lee
School of Electronics, Telecommunications and Computer Engineering, Korea
Aerospace University, Korea

N. Lee
Radio Broadcasting Research Division Broadband Wireless Multimedia Research
Team, Electronics and Telecommunications Research Institute (ETRI), Korea

L. Liang
University of Surrey (CCSR), Guildford GU2 7XH, Surrey, UK

G. Mastorakis
Department of Information and Communication Systems Engineering, University
of the Aegean, Samos 83200, Greece

A. Mehaoua
University of Versailles, CNRS-PRiSM Lab, 45, Av. des Etats Unis, 78035
Versailles, France

I. Melhus
SINTEF ICT – Communication Systems, O.S Bragstads plass 2C, Norway

M. Mezzalla
Thales Alenia Space, 26 Avenue J-F Champollion, 31037 Toulouse Cedex 1,
France

C. Miguel
Universidad Politécnica de Madrid, Escuela Técnica Superior de Ingenieros de
Telecomunicación, Avda. Complutense s/n, Ciudad Universitaria, 28040 Madrid,
Spain

I. Moreno
Thales Alenia Space, 7 calle Einstein (PTM), 28760 Tres Cantos (Madrid), Spain

R. J. Mort
Systek Consulting Ltd., Havant, UK

R. Muñoz
Hispasat, Gobelas 41, 28023 Madrid, Spain

J. Nicol
B2i, 1 avenue de l'Europe, 31400 Toulouse, France

F. Nivor
University of Toulouse France/LAAS-CNRS, 7 avenue du Colonel Roche, 31077
Toulouse Cedex 4, France

E. Pallis
Department of Applied Informatics and Multimedia, TEI of Crete, Estavromenos,
Heraklion 71500, Greece

F.-N. Pavlidou
Aristotle University of Thessaloniki, Greece

D. Pérez
Telefonica I + D, Parque Tecnológico de Boecillo 47151 BoecilloValladolid, Spain

A. Pietrabissa
Dipartimento di Informatica e Sistemistica (DIS), Università degli Studi di Roma
"La Sapienza", Via Ariosto 25, 00184 Rome, Italy

A. Ramos
Telefonica I + D, Spain, Parque Tecnológico de Boecillo, 47151 Boecillo,
Valladolid, Spain

F. Rodriguez
Telespazio S. P .A., Via Tiburtina 965, 00156 Rome, Italy

G. Santoro
Dipartimento di Informatica e Sistemistica (DIS), Università degli Studi di Roma
"La Sapienza", Via Ariosto 25, 00184 Rome, Italy

A. Sathiaseelan
Department of Engineering, University of Aberdeen, Fraser Noble Building,
Kings College, Aberdeen AB24 3UE, UK

Z. Sun
University of Surrey (CCSR), Guildford GU2 7XH, Surrey, UK

J.A. Torrijos
Telefonica Investigación y Desarrollo, C/ Emilio Vargas 6, 28043 Madrid,
Spain

F. Vallejo
Thales Alenia Space, 7 calle Einstein (PTM), 28760 Tres Cantos (Madrid),
Spain

A. Yun
Thales Alenia Space, 7 calle Einstein (PTM), 28760 Tres Cantos (Madrid),
Spain

P. Zautasvili
Hungaro Digitel Plc, Komp str. 2., H-2310 Szigetszentmiklós-Lakihegy,
Hungary

Chapter 1
New Architecture for Next Generation Broadband Satellite Systems: The SATSIX Approach

C. Baudoin, L. Fan, E. Callejo, A. Pietrabissa, F. Rodriguez, A. Ramos, G. Fairhurst, F. Arnal, and G. Santoro

Abstract The EU-funded IST FP6 Project *Satellite-based communications systems within IPv6 (SATSIX)* is implementing and validating innovative concepts and cost-effective solutions for broadband satellite systems that build upon the DVB-RCS/S2 standards and services. It promotes the introduction of IPv6 into satellite-based communications systems and the development of hybrid networks that combine the use of satellite with wireless access technologies. The main objective is to lower the cost of broadband satellite access, through the development of new satellite access techniques and the integration of wireless local loops (WiFi and WiMax). Simulations, test beds and trial networks will show how satellite broadband access can be integrated into Next Generation Networks, based on IPv6, to support new multimedia applications.

Introduction

The SATSIX project focuses on the use of satellite systems to offer attractive solutions for the access networks in three key scenarios, targeting the corporate, the collective and residential users. New application scenarios have been defined for both corporate and consumer markets (including 3-play and TV over DVB-S2).

The corresponding user requirements and services definition define the overall SATSIX network, satellite access and payload architectures, where key techniques are investigated. At the network layer, WLL with a satellite access using WiFi or WiMax will enhance network services including mobility, multicast, security, and QoS. At the access layer, performance will be enhanced by use of an adaptive physical layer and the associated Radio Resource Management (RRM) and Connection Acceptance Control (CAC), optimized encapsulation and header compression, cross-layer optimization and state-of-the-art multimedia transport protocols. The introduction of a convergent payload will improve both the performance and flexibility of the satellite segment.

These key techniques are being validated by simulation and tested by developing an end-to-end emulation test bed allowing evaluation of IPv6 multimedia applications (voice, data, video) using an emulation of next generation satellite access. Live

L. Fan et al. (eds.), *IP Networking over Next-Generation Satellite Systems*
© Springer 2008

experimentation will also be performed to evaluate user requirements with key technologies including IPv6 over DVB-RCS and WiFi.

Scenarios and Services

Three different scenarios have been analysed in the Satsix project, Corporate, Collective, and Residential. In the following sections the main characteristics will be shown together with an overview of the services for each scenario.

Corporate Scenario

The Corporate scenario (see Fig. 1.1) is based on a complete and integrated solution for the provision of applications for companies that have distant offices, warehouses, manufactures, subsidiaries or individual workers located in rural or hardly reachable areas by wired backbone.

The high-level architecture includes support for tunnelling between distant offices located beyond the satellite access on the local loop and the Headquarters located on the terrestrial backbone, forming an end-to-end Virtual Private Network, VPN.

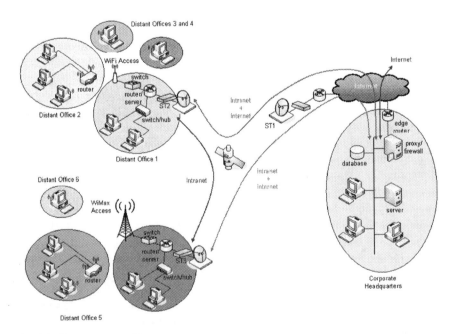

Fig. 1.1 Corporate scenario

This kind of scenario allows distant offices to have broadband connection with their headquarters without deployment of wired terrestrial infrastructure. The connection to the local loop may be Wireless Local Loop, WLL, or wired; whatever their local loop topology, the corporate VPN's need total integrity of their data, and an appropriate QoS level for the type of information they carry.

In all cases, the distant offices have direct access to the Internet through the Headquarters. The distant offices which are not located in the same spot can however 'talk' to each other without passing through the Headquarters, provided they are using their Intranet facilities only.

This figure represents a configuration with both WiFi and WiMax accesses.

The principal services that will be provided within this type of scenario are internal LAN-to-LAN connectivity, access to the internet and extranet networking, assuring secured connections with business customers. Beyond the interconnection services, VoIP and Video conferencing are allowed, supported by an ad hoc QoS architecture, multicasting and broadcasting capabilities (intrinsically provided by the satellite architecture) and applications including video surveillance and resource sharing.

Collective Scenario

The collective scenario aims to provide a high level of connectivity to citizens and small companies, belonging to rural communities that are not able to currently be connected via ADSL. This solution is based on a hybrid network combining a satellite segment and terrestrial wiring as well as wireless access such as WiFi, or WiMAX.

Three types of interconnections may be provided for the local loop and may be used simultaneously:

- Wired access: the Satellite Terminal (ST) interfaces with Ethernet wiring, providing access in a building.
- WiFi: the ST interfaces with WiFi Access Point, which distributes and collects the traffic over a local area. PCs are equipped with WiFi cards that allow them to move within or at a short distance from the premises.
- WiMax: the ST interfaces with a WiMax Base station, which distributes and collects the traffic over a local area. WiMAX may also be used as a backhaul for WiFi, where a 802.11 network is connected to the 802.16 Subscriber Station (SS). These solutions can be put in place for medium-sized local areas, (e.g. a dock area).

In this scenario, the link between the ST and the Satellite Access Provider's Gateway is seen as a LAN. The local loop is a sub-LAN and no direct Internet connectivity is possible between two STs connected to a local loop.

The connection to the satellite on the WLL side is a single link for the whole community that shares the satellite segment.

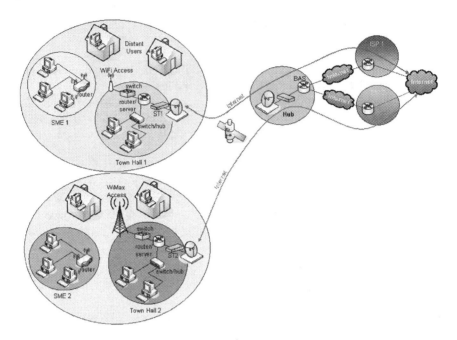

Fig. 1.2 Collective scenario, WLL access example

Fig. 1.2 shows an example of collective terminal scenario with WLL access. The collective scenario is represented by public places in which people can access the service without specific requirements but requiring a subscription for access to the broadband wireless network and applications.

Examples of this scenario are: a hotel, public places, a mountain hut, islands, harbour areas, places in emergency (war or natural disasters).

The applications in this kind of scenario, include: e-Learning, tele-Medicine, e-Payment, games and any other services, may improve the comfort of users located far from large urban areas or that are not reached by broadband wired networks.

Different type of services may be provided within this architecture, from web browsing and email to instant messaging, support for the peer to peer, interactive gaming, movie/music streaming and VoIP. Emergency messages e.g. alert messages may be broadcast locally to users in potentially dangerous situations.

Residential Scenario

Finally, a residential user scenario is directly related to the 'Triple Play' concept, which should is a key goal to achieve competition with other solutions like ADSL or Cable in low-density population areas. The principle of the 'Triple Play' concept is to provide the three basic access services (Television, Internet access and Telephony) through one single satellite communication network.

This contrasts with the 'terrestrial' communications offered by telecom operators today using 'Dual play', which provides both voice telephony and ADSL internet access over the same network (via traditional copper telephone cables). A 'Triple Play' satellite service will offer the three access services together to remote communities, at an affordable cost.

The architecture will make use of both transparent and regenerative satellite systems. The optimal solution is a hybrid satellite system, which includes both transparent and regenerative satellite systems, each used depending on the service. Users may either benefit from the on-board processing of the regenerative satellite system (to reduce delay in real time traffic for example) or may instead use instead the transparent transponder.

A graphical view of the scenario is depicted in Fig. 1.3. In the transparent mode (red), users make use of the Hub & NOC station to communicate one to each other, whereas a direct communication will be achieved if a regenerative mode is used (green).

Applications include Digital TV, in particular Television over IP (TVoIP) on DVB-S2 and video broadcasting (SDTV and HDTV) directly over DVB-S2.

Television is mainly broadcasted digitally and therefore can be combined with different services to make it highly interactive. New experiences will be offered to users through a wide range of applications varying from video on demand (VoD),

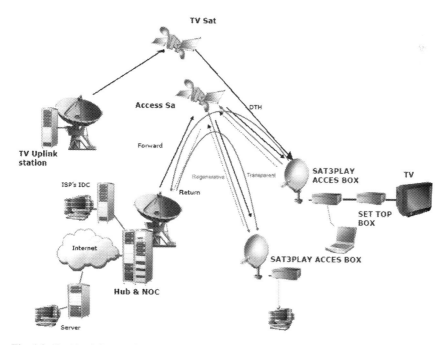

Fig. 1.3 Residential scenario

time shifting or multiple audio streaming to the complete thematic personalization of the video images and scenes.

Other applications include general Internet access (web browsing), VoIP, telephony, videoconference, and peer to peer applications.

SATSIX Network Architecture

The SATSIX network architecture may be applied to two network scenarios: the transparent star and the regenerative mesh/star. The main difference between these two is the support for a regenerative mesh topology, using satellite has the on-board processing (OBP), allowing a Return Channel Satellite Terminal (RCST) to not only communicate with the gateway/NCC via the satellite, but also to directly communicate with other RCSTs (in a single satellite hop) as depicted in Fig. 1.4.

The SATSIX project also addresses an approach for the convergence of these two satellite network scenarios that would provide advantages due to the exploitation of regenerative or transparent connectivity only for those services or applications that really require each of them (i.e. depending on the type of service and application).

Both WiFi and WiMAX terrestrial access networks may be connected to the RCSTs to provide shared access to the traffic forwarded on the satellite link.

The RCSTs for a next generation communications satellite system must be low cost and high-performance satellite terminals, and support IPv6. All terminals will use adaptive DVB-RCS air interface for transmission and DVB-S/S2 for reception

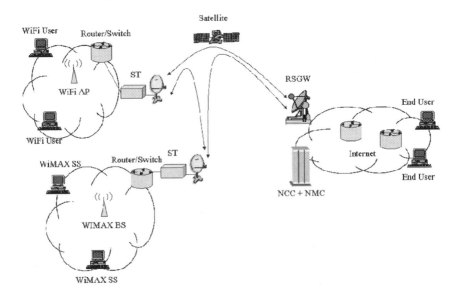

Fig. 1.4 Network architecture for the Regenerative Mesh Topology

(DVB-S has to be maintained for backwards compatibility). At present the terminals transmission is based on the non-adaptive DVB-RCS specification and the reception on DVB-S.

A reference functional architecture has been proposed for each of the two topologies, respectively. These two architectures integrate QoS, multicast, security, mobility and transport functions. The highlights are:

- The satellite segment can interwork with Internet DiffServ QoS to provide end-to-end QoS at the network level. This architecture supports end-to-end QoS, and also dynamic QoS according to application and user needs.
- The architecture provides multicast management (i.e. MLDv2 [1]) for IPv6.
- The architecture supports key management protocols (i.e. SATIPSec [2] and GSAKMP), key distribution systems e.g. LKH and Layer 2 security enhancing security level provided by Layer 3 solutions such as SatIPSec.
- The architecture has enhanced standard IPv6 mobility (i.e. HMIPv6 [3] and FMIPv6 [4], to enable Mobile IPv6 in a satellite system.
- The architecture provides an enhanced TCP PEP in order to efficiently manage TCP-based connections.

Figure 1.5 describes the reference network architecture:

The network and functional architectures may be applied to three different scenarios: a corporate application scenario, a residential application scenario and a collective access terminal scenario.

Satellite Access Architecture

The SATSIX satellite access architecture (see Fig. 1.6) supports both transparent and regenerative/hybrid payload based on the features provided by the connection-control protocol, C2P.

SATSIX brings innovative access techniques to this architecture:

- Adaptive physical layer support: a complete definition of the adaptive return and forward physical layer has been defined that meets the scenarios and system requirements, and supports an optimised RRM and CAC using innovative control-theoretic and operations research methodologies.
- Integrated QoS at MAC layer: interactions between IP QoS and RRM have been defined, enabling the layer 2 to operate as an efficient and integrated global QoS framework. An IP/MAC back pressure feedback has been introduced to improve the overall scheduling and reduce the MAC queues sizes, and cross layer optimization has been defined to modify allocation provided by the DAMA according to SIP information.
- Layer 2 security framework: this architecture aims to ensure a very high security level to satellite applications by protecting the data and the control plane at MAC layer. It relies on GSE [5] or ULE [6] security extension header [7, 8] and DVB-

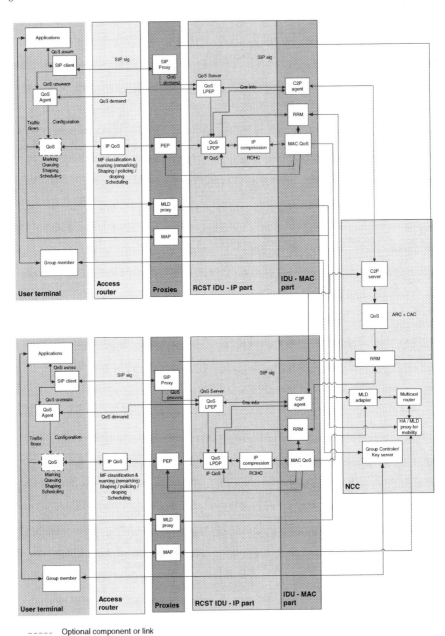

- - - - - Optional component or link

Fig. 1.5 SATSIX reference network architecture for the Regenerative Mesh Topology

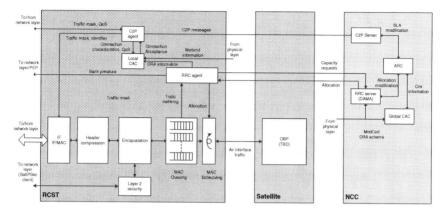

Fig. 1.6 SATSIX reference satellite access architecture for the Regenerative Mesh Topology

RCS optional security mechanisms for AAL5 and uses SATIPSec or GSAKMP key management protocols adapted to the DVB-RCS context.

- IPv6-friendly encapsulation schemes: The ULE and GSE encapsulations provide an efficient way to transport IPv4 and IPv6 traffic over satellite systems with GSE also providing a scheduling algorithm that can optimize the filling of DVB-S2 frames based on QoS constraints.
- Efficient header compression: header compression has been introduced to reduce the transmission overhead.
- The satellite access architecture is compliant with the BSM SI-SAP definition with an evolution to focus on QoS, RRM, security and IPv6.
- C2P adaptation: an extension provides services for access and network functionalities such as IPv6, DVB-S2 support, full dynamic multicast, ATM and MPEG profiles support and DVB-RCS security in a star and mesh.

Evaluation and Validations

The SATSIX project relies on different methods to evaluate and validate access and network features. Simulations validate specific algorithms in a convenient way and are the more appropriate tools when no implementation is available or when conditions cannot be met in a test bed or trial context (load for example). Emulation is the key tool to test innovative access mechanisms that cannot be implemented in real equipment within the frame of the project. It also offers an ideal environment to evaluate network functions with real applications. Finally, trials are the right way to demonstrate how features defined within the frame of SATSIX can be handled in deployed satellite systems.

Simulations

The first validation and evaluations of the SATSIX architecture and algorithms and procedures are being performed using simulation. The simulations take as inputs the system and user requirements, the architecture and the algorithms defined within the project, and identify requirements for the emulation test bed and system trials. The simulations include: adaptive physical layer characterization; Radio Resource Management (RRM) procedures (Connection Admission Control (CAC) and Bandwidth-on-Demand (BoD) algorithms); Internet multimedia transport protocols; multicast protocols.

The physical layer simulations seek to provide input data to the RRM simulations. These are performed for different system scenarios and take into account Fade Mitigation Techniques (FMT) on the return and forward links. The FMT mechanisms consist of a feedback loop between the satellite terminal and the gateway, designed to track and mitigate the channel attenuation on the physical link: Based on regular estimations of the signal to noise ratio (SNR), the FMT loop allows to control the modulation type, the code rate and the symbol rate, to optimize the spectral efficiency on the transmission. A realistic time-series (including worst case rain attenuation) has been considered for the simulations of the FMT loop. These simulations allowed to generate the required combinations of {modulation/coding/symbol rate} for a set of users located in the same spot affected by rain. These data are provided as an input to the RRM simulations.

Connection Acceptance Control (CAC) simulations evaluate the research-based algorithm in terms of scalability with respect to the number of classes, fairness among traffic classes (in terms of blocking probabilities), and is capable of supporting adaptive traffic. The algorithm integrates the adaptive physical layer model.

Bandwidth on Demand (BoD) simulations test the control theoretic-based algorithms under different network loads and with different traffic mixes (VoIP, Video Conference, Multimedia Streaming, Data). The algorithms seek to optimize the satellite access delay. These results provide data to the transport layer simulations.

The transport layer simulations analyse the approaches and techniques used by applications to enable effective congestion control for multimedia when the network path includes a DVB-RCS satellite system. The following protocols are simulated: TCP, UDP with TFRC, Datagram Congestion Control Protocol (DCCP) with Faster Restart (FR) and Quick Start (QS). The considered multimedia traffic sources are voice over IP (real-time audio streaming), audio (non-real-time audio streaming), video (including TV streaming over IP) and data using TCP.

A set of multicast simulations evaluate the two standardized multicast service models, i.e., Any Source Multicast and Source Specific Multicast, over the architecture. The results will show the mean end-to-end delay under different network conditions.

A Wireless Local Loop (WLL) scenario consisting of WiFi and WiMAX networks connected to a satellite terminal is defined, and simulation scenarios for QoS and mobility. These simulations will test protocols and algorithm development in

SATSIX: protocols to handle micro- and macro-mobility (like MIPv6, HMIPv6 and FMIPv6), FMT techniques in the forward link scenarios, scheduling algorithms in the satellite terminals, frame constitution and DAMA Controller algorithms for adaptive coding.

Emulation

The SATSIX emulation test bed (PLATINE) will demonstrate network and application service integration with the possibility to interoperate with terrestrial networks. PLATINE emulates DVB-RCS and DVB-S2 transparent or regenerative satellite systems and offers full native IPv4 and IPv6 support. It includes ULE/MPEG2-TS and AAL5/ATM stacks together with the adaptive physical layer simulation and the associated RRM. A complete QoS architecture mixing SIP proxies, QoS Agent /QoS Server architecture, enhanced IP/MAC scheduling and cross layer techniques is available. Finally, a layer 2 security framework will be implemented. Various network features, such as IPv6 mobility, dynamic multicast and its interaction with mobility (using an MLD proxy) are conducted in the frame of SATSIX, allowing testing of new schemes, protocols and services for next generation satellite networks.

The PLATINE architecture is distributed over a LAN and relies on services offered by a synchronization and bloc library. This allows a modular design (see Fig. 1.7) and preserves room for future evolution.

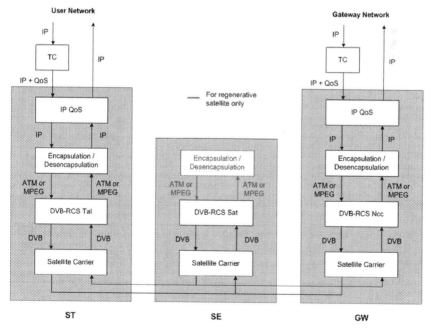

Fig. 1.7 PLATINE architecture

PLATINE also offers a convenient way to collect results, where various measurement points, as well events or errors, are handled by a distributed framework and displayed in real time using a dedicated user interface.

Trials

Live demonstrators in SATSIX will prove the capability of the satellite to provide affordable services to a large range of customers. The SATSIX trials will validate this concept, demonstrating its feasibility in demonstration scenarios. The tests will be performed over four real Satellite Systems: the Regenerative AMERHIS system of HSA and the Transparent DVB-RCS systems of AASF, HDT and HSA, to validate key SATSIX features selected, the subsequent recording of results and the methods of evaluating the results, as well as the implementation plan and overall integration tests. All these tasks will be performed for each of the following trial scenarios:

- Scenario 1: Corporate applications using the AASF DVB-RCS transparent access platform will assess and validate the IPv6 application scenarios over a DVB-RCS star network and operational architecture. In the frame of this trial, native IPv6 prototypes will be developed, based on DVB-RCS/S2 Thales Alenia Space Hub and terminal. In addition, an IPv6 PEP will be implemented, as well as dynamic QoS support features with the integration of an Access Resource Controller linked with a SIP proxy. These features will be evaluated using innovative collaborative working applications.
- Scenario 2: Collective Access Terminal on DVB-RCS transparent access platform. This scenario is developed and evaluated at HDT premises with the participation of HDT for the Collective Access Terminal. HDT's DVB-RCS Hub will be used. This combination of access technologies can reduce the costs for users and allowing sharing the cost of the satellite connection using WLL. This live trial is able to assess the real user's behaviour and help evaluate not only the measured parameters but also the user experience. Moreover, IPv6 mobility and multicast will be demonstrated to
- Scenario 3: Corporate applications on DVB-RCS regenerative access platform, developed and evaluated on HSA's pilot AMERHIS system. This trial will benefit from improvements brought by the SATLIFE IST project. It will assess and validate the IPv6 application scenarios defined in the project over a DVB-RCS mesh network and operational architecture. The objective of this scenario is to demonstrate that the regenerative platform (the OBP) is able to work with IPv6 and ULE. This will require the integration of an IPv6/ULE encapsulator with an IPv6/ULE receiver, which can be seen as a first step towards a future move to GSE integration.
- Scenario 4: Residential applications on the DVB-RCS transparent access platform. The applications are developed and evaluated using HSA's pilot DVB-RCS

system. The principal residential application of satellite networks is Digital TV services (audio, video and data broadcasting to end users). This trial will focus on IPTV over DVB-S2 with IP QoS functionalities.

For the different scenarios the most innovative IPv6 network features that will be demonstrated inside SATSIX are: WLL inter-working, dynamic multicast, dynamic QoS, application/Network security, mobility and PEP/RRM interaction.

Conclusions

The SATSIX project is studying key techniques to decrease the cost of satellite access and to improve the functionalities offered by satellite systems. It will promote the introduction of IPv6 in this context and demonstrate the relevance of these solutions with regards to real users requirements by performing trials involving real users and equipments.

Acknowledgements This work is supported by the IST FP6 SATSIX project, funded by European Commission (EC). The financial contribution of the EC towards this project is greatly appreciated.

References

[1] R. Vida, L. Costa, Multicast Listener Discovery Version 2 (MLDv2) for IPv6, IETF RFC 3810, June 2004.
[2] L. Duquerroy, S. Josset, O. Alphand, P. Berthou, T. Gayraud, SatIPSec: An Optimized Solution for Securing Multicast and Unicast Satellite Transmissions, 22nd AIAA International Communications Satellite Systems Conference (ICSSC), Monterey (USA), 9–12 May 2004.
[3] H. Soliman, C. Castelluccia, K. El Malki, L. Bellier, Hierarchical Mobile IPv6 Mobility Management (HMIPv6), IETF RFC 4140, August 2005.
[4] R. Koodli, Fast Handovers for Mobile IPv6, IETF RFC 4068, July 2005.
[5] ETSI, Digital Video Broadcasting (DVB); Generic Stream Encapsulation (GSE) Protocol, DVB Project A116, May 2007.
[6] G. Fairhurst, B. Collini-Nocker, Unidirectional Lightweight Encapsulation (ULE) for Transmission of IP Datagrams over an MPEG-2 Transport Stream (TS), IETF RFC 4326, December 2005.
[7] H. Cruickshank, P. Pillai, S. Iyengar, L. Duquerroy, Security Extension for Unidirectional Lightweight Encapsulation Protocol, IETF Work-in-Progress, draft-cruickshank-ipdvb-sec-02.txt, June 2006.
[8] H. Cruickshank, L. Duquerroy, P. Pillai, Security Requirements for the Unidirectional Lightweight Encapsulation (ULE) Protocol, IETF Work-in-Progress, draft-ietf-ipdvb-sec-req-02.txt, 2007.

Chapter 2
Satlife: A Big Step into the Enhancement of the Regenerative Satellite Generation

Miriam Catalán de Domingo, Isaac Moreno Asenjo, Ana Yun García, Fernando Vallejo Lázaro, and José Antonio Guerra Expósito

Abstract This paper offers an overview in the new developments carried out and services offered thanks to SATLIFE project [1]. Based on the AmerHis system [2], SATLIFE [3] has been working in the improvement of the new AmerHis capabilities, looking for a decrease in costs and the provision of new services not yet offered over a regenerative DVB-RCS platform. Satlife concept is introduced, and an analysis of Amerhis/Ibis service enhancements is carried out, together with a deep overview of the new services and functionalities developed. In the phase of real trials, Satlife is testing all the features developed over these two last years.

Satlife Concept

'Satellite Access Technologies: Leading Improvements for Europe', Satlife, has been the first project to bring technological innovations and solutions in the DVB-RCS regenerative systems through the AmerHis improvement. Today, after two years, most of the objectives have been reached: SATLIFE has become a key project in the innovation area for positioning and adoption of DVB-RCS regenerative systems, promoting the development of the Information Society.

Hispasat, together with the group of leading companies participants in the project (Thales Alenia Space España, Telefónica I + D, Thales Alenia Space Francia, Nera, EMS, Shiron, Indra Espacio, Thales, University of Surrey, Universidad Politécnica de Madrid and Telefónica Pesquisas e Desenvolvimento), face the challenge of real trials. Therefore, all the prototypes and developments design for each company have been integrated and tested in the laboratory.

L. Fan et al. (eds.), *IP Networking over Next-Generation Satellite Systems*
© Springer 2008

Amerhis/Ibis Service Enhancements

Digital TV

One of the main uses of satellite networks today is the provision of Digital TV services. Audio, video and data are broadcasted to final users by satellite in a one-way communication, but two-way communication is needed for interactive applications. Thus, broadcast is ensured in one or several spot beams through on-board switching and duplication. The Video Broadcasting service is improved in Satlife. A specific terminal is introduced to transmit the video from the Service Provider to the Users.

As it is shown in Fig. 2.1, when having several VSPs some signalling information has to be assembled in the NCC (Network Control Center) and redistributed to the receivers.

The Video Service Provider is based on the RCST, having its functionality and in addition being able to transmit MEPG2 video and data.

Thus, it is necessary to add a Video Streaming Unit to the RCST, which must feed the RCST with the MPEG-2 video stream to be broadcasted. The main difference of this Service Provider RCST is that it implements an interface between the streaming unit with the purpose of receiving the video in MPEG-2 format to be broadcasted. This interface consists of video input and a clock output. The clock is required to synchronize the Video Unit with the RCST (Fig. 2.2).

As shown in Fig. 2.3, this SP-RCST removes the IP, UDP and RTP headers to obtain MPEG2 original packets transmitted to the satellite. Thus, all the MPEG2

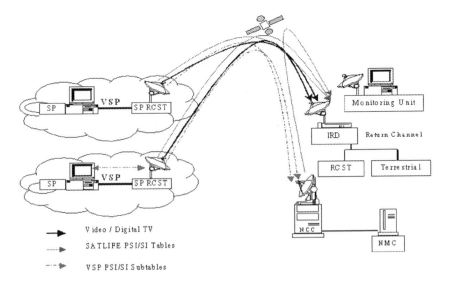

Fig. 2.1 Video broadcast service

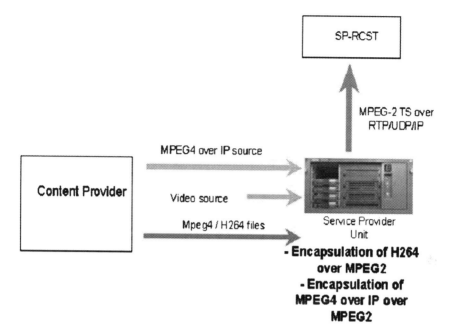

Fig. 2.2 Satlife video service provisioning

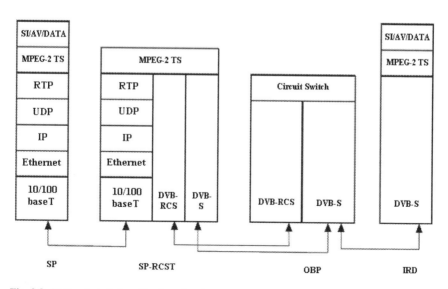

Fig. 2.3 Protocol stack for video broadcasting

contributions from each station are multiplexed on board of the satellite and transmitted in a DVB-S saturated carrier to be received by an IRD, who recovers the MPEG2 signal sent by the Service Provider.

The commercial advantages of this system are very clear, since it removes the need to have a centralized installation in the uplink to give DVB-S services as it is currently happening in the existing digital TV platforms. Now, small service providers may distribute them in different places and have their own uplinks from their premises.

Multiconference

This service allows to establish multimedia communication between two or more users. It provides the same communication capabilities than POTS with new added value services provided by the multimedia technology. Thanks to the mesh regenerative scheme, QoS can be improved, taking advantage of a single satellite hop. In particular, the system allows the use of multicast connections, session and conference control establishment, security and adaptation of the different audio/video topologies depending on the kind of users.

Multiconference functionalities that have been improved specifically in the project are:

1. SIP introduction and compatibility and new H.323 aspects, allowing a new basic interaction between both protocols, avoiding that H.323 clients can only establish multiconference sessions with other H.323 clients and similar with SIP clients. This task is carried out by the regenerative gateway. Besides, H.323 final users with IP private addresses can be registered with the RSGW Gatekeeper while they are carrying out calls towards H.323 users in the Internet network. Thus, now there is no need to remove from the Gatekeeper register as it was necessary with AmerHis.

Between novelties introduced by Satlife, there is the audio/video communication based on SIP protocol; thus, Satlife regenerative gateway supports the following SIP services using a SIP proxy:

- SIP voice calls with ISDN/PSTN terminals: the Satlife RSGW allows a voice call to be established from a Satlife subscriber with a SIP user agent registered in the RSGW SIP proxy to an ISDN/PSTN terminal. Thus, the RSGW is in charge of translating both traffic and signalling from ISDN protocol to/from SIP protocol. For this internetworking service, calls can be initiated from both ends, ISDN or Satlife, with independence of the RSGW subscriber address type (public or private) (Fig. 2.4).
- SIP voice/video with SIP terminals located at the Internet: a Satlife subscriber SIP user agent is able to establish a communication (voice and video) to an external SIP terminal located at the Internet. Both of them will communicate using SIP protocol through the Access Router located at the RSGW (Fig. 2.5).

Fig. 2.4 SIP voice call to ISDN/PSTN

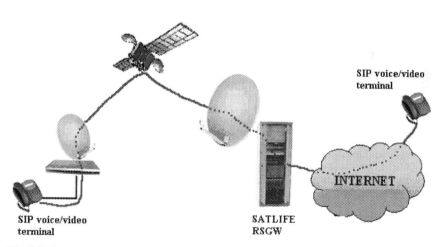

Fig. 2.5 SIP voice/video calls to Internet SIP endpoints

For this service it is necessary to differentiate based on the type of IP address assigned to the subscriber:

- ○ A SatLife UT with a public IP address is able to send and receive SIP voice/video calls being actively registered in the RSGW SIP Proxy. It means that the SIP proxy is in charge of finding the Internet user agent (in case of an outgoing call) or finding the SatLife user agent (in case of an incoming call).

ͻ A SatLife UT with a private IP address is able to make outgoing SIP voice/
video calls only if it is deregistered from the RSGW SIP proxy. In any case,
it could be registered to a third party SIP proxy located in the Internet.

- Voice/Video calls with SIP terminals within the same VSN (Virtual Satellite
Network): A Satlife subscriber with a SIP user agent registered in the RSGW SIP
proxy is able to establish a communication (voice and video) to another Satlife
subscriber located in the same VSN and registered at the same or a different
RSGW. Thus, the RSGW will be in charge on the signalling communication in
order to allow both parties to start the communication, but after this phase a direct
communication from one user to the other (without going through the RSGW)
will be performed. In that way only one satellite hop will be performed as the
OBP will be in charge of routing the traffic. Calls can be set-up with independence of the RSGW subscriber address type (public or private) (Fig. 2.6).

2. Call termination optimization for ISDN/PSTN voice/video calls

The Satlife RSGW allows optimizing the termination of the H.323 voice/video
service by not terminating an ISDN/PSTN call initiated by a Satlife UT necessarily
in the subscriber's RSGW, but by using a collaborative ITSP node. As a collaborating
ITSP node we understand an external ITSP, accessible through Internet/Intranet and

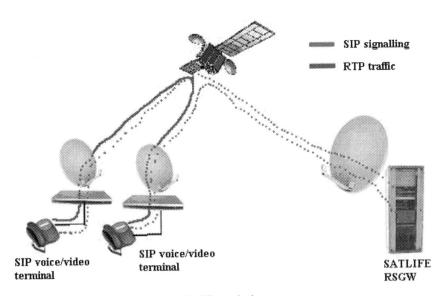

Fig. 2.6 Voice/video calls between satlife SIP terminals

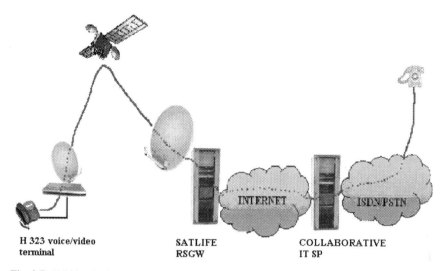

Fig. 2.7 H.323 voice/video call termination optimization

with an adequate network infrastructure, or a secondary RSGW belonging to the same VSN.

If the RSGW service provider reaches adequate agreements, it may be more convenient for cost reasons to let these ITSPs terminate certain calls instead of doing so locally. The RSGW service provider decides for each collaborating ITSP the conditions under which the routing of a call is going to be performed. These conditions are based on call destination and/or type of service (voice/video) and may be defined taking into account the best rates offered by each ITSP in each case (Fig. 2.7).

3. H.323 service improvements

- Internet calls improvements: thanks to improved NAT traversal functionality, a SatLife H.323 endpoint with private IP address can be actively registered with the RSGW Gatekeeper while placing calls towards H.323 users located in the Internet. It is no longer necessary to de-register first from the Gatekeeper, as required for AmerHis.
- NATed endpoints support: the Satlife RSGW supports NAT traversal, allowing a H.323 terminal, which is NATed at the subscriber RCST, to make voice/video calls towards ISDN/PSTN, Internet H.323 endpoints and non-NATed H.323 endpoints within the same VSN. Thus a NATed H.323 terminal is able to perform H.323 voice/video calls even if the subscriber RCST NAT does not support a H.323 ALG.

Internet/Intranet Access improved Using PEP Protocol in Mesh and Star Configurations

The purpose of the TCP/IP enhancement functionality is to speed up TCP connections over a large latency satellite link. The functionality can be used for both TCP connections between terminals and TCP connection between terminals and Internet hosts. Typical popular applications that use TCP as transport protocol are FTP and HTTP (for web browsing).

Conditions particular to geostationary satellites severely constrict the performance of TCP and reduce the end users experience of accessing the Internet over satellite. Large latency, high bit error rates and asymmetric bandwidth which are characteristic for satellite networks makes TCP less suitable when used over satellite. Some performance improvements can be achieved by simply tuning the TCP parameters, but the effectiveness of this is limited (Fig. 2.8).

To overcome the TCP problems over satellite, several commercial products (commonly known as TCP accelerators) have been developed which optimize the TCP performance over satellite. These products have in common that they all use mechanisms described in RFC 3135, TCP Performance Enhancing Proxy (TCP PEP).

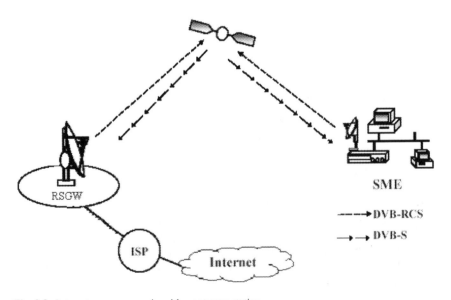

Fig. 2.8 Internet access scenario with a gateway station

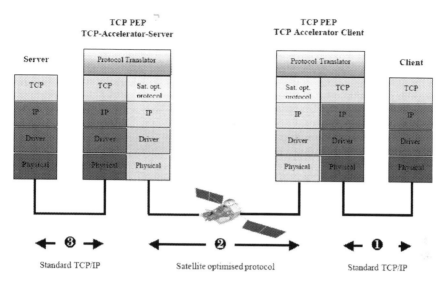

Fig. 2.9 TCP Acceleration over satellite

The TCP PEP implementations make use of the split-connection principle. A split connection terminates the TCP connection from an end system (client) and establishes a corresponding TCP connection to the other end system (server). This is done in order to use a third connection between two TCP PEPs which is optimized for the satellite link (see Fig. 2.9). Both connection 1 and 3 in Fig. 2.9 are complete separate TCP connections, while connection 2 is normally a proprietary satellite optimized protocol or a modified TCP protocol optimized for the satellite link. Data streams are forwarded from one connection to another, and extensive buffering mechanisms are required. Other traffic than TCP shall simply be forwarded untouched through the TCP PEPs.

This architecture makes it possible to achieve vastly improved performance while remaining entirely transparent to the end user and fully compatible with the Internet infrastructure. No changes are required to the client and server, and all applications continue to function without modifications. TCP congestion avoidance mechanisms remain in place over the terrestrial connections and also maintain full TCP reliability and end-to-end flow control.

Multicast Service

Thanks to Satlife project, Multicast service combines for the first time star and mesh configurations simultaneously in a satellite system, and allows the definition of multicast routes at the NCC (Fig. 2.10).

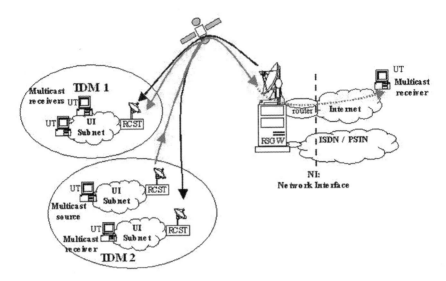

Fig. 2.10 Satlife multicast scenario

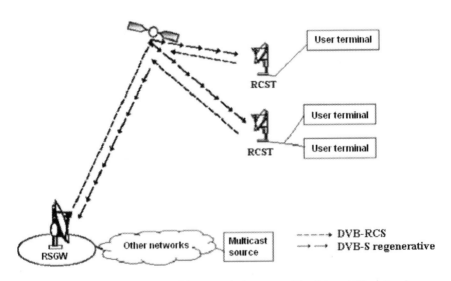

Fig. 2.11 Multicast scenario with multicast terminal users outside of the satellite network

The multicast service implies that a data packet can be delivered to a group of different receivers. Here, three architectural scenarios may be considered: 'star' multicast, 'mesh' multicast, and a final one which combines properties from both of them. The main different is the involvement of a RSGW in the first one, because the multicast source is outside the satellite network; in the mesh setup, the multicast server is a RCST receiving an IP stream, which is multicasted to several users. The third case also implies the use of an RSGW, but the source is inside the satellite.

In Fig. 2.11, the multicast source is inside the satellite network, but multicast terminals are either inside or outside the network. In fact, this is similar to the star configuration, but implies that the gateway must support bidirectional multicast.

New Services and Functionalities

E-Learning

There are several kinds of E-learning applications, each one having different service requirements.

Two main groups can be established:

- Real-time conversational e-learning service (Synchronous E-learning).
- Low-interactivity e-learning service (Asynchronous E-learning) (Fig. 2.12).

The scenario for e-learning is similar to the multiconference scenarios. The conference model suitable for SATLIFE imposes the usage of multicast with only one satellite hop for the media flows. Both 'star' and 'mesh' multicast might be used depending on the involved users and the place where the teacher or teachers are located. The scenario also shows some users that might have unicast access to the network. They need some way to access the multicast lecture using MCUs.

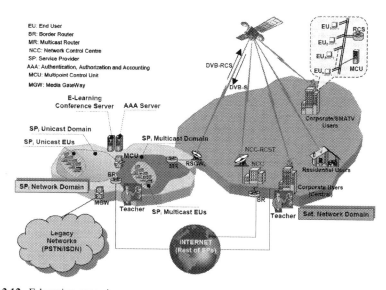

Fig. 2.12 E-learning scenario

Video on Demand

Video On Demand (VoD) is an audiovisual service where a streaming server is remotely controlled by the user, so that it emulates the functionality of a home video player. The interactivity provided by the service implies both content selection and streaming control.

This VoD service uses mesh connections, allowing a DVB-RCST to be a video service provider, in this case, for VoD services. The video client is a commercial IP set-top-box that supports IP unicast connections using RTSP connected to the satellite network through another DVB-RCST (Fig. 2.13).

The previous picture shows the VoD server connected to the DVB-RCST, so the video and audio transported is encapsulated in IP. The upper layers include UDP for audio and video real-time transmission and TCP for the session control. The DVB-RCST in the client side de-encapsulates IP traffic and distributes it to the IP set-top-boxes, which in turn extract and decode video.

Software Download

A satellite data service targeting PCs that provides value-added services at high speed. The broadcast service is similar to TV point-to-multipoint transmissions. Basically, the contents are transmitted in a carousel, which enables users to access

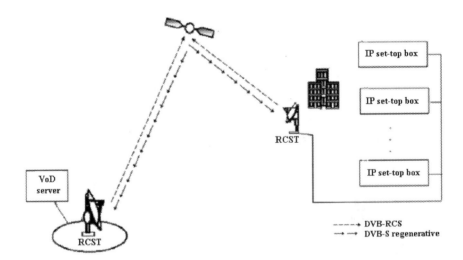

Fig. 2.13 Video on demand service conceptual architecture

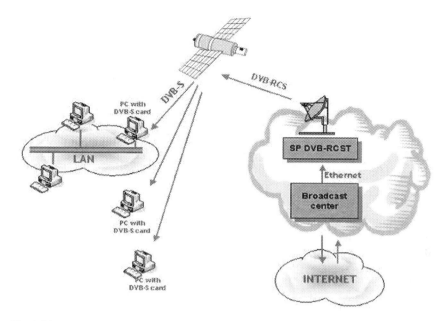

Fig. 2.14 Satellite connection scheme

relevant Internet contents without being on-line since the contents are received continuously through the satellite link.

Through this service, users will access a huge diversity of contents in a near-on-demand fashion. Data will be supplementary available not only in broadcast but also in the more versatile multicast mode (Fig. 2.14).

In the previous figure, the user terminal is a PC with a DVB-S card, or some kind of multimedia set-top box with a hard disk. The idea is to use a transparent DVB-S and thus encapsulate software into MPEG2-TS; otherwise, DVB-RCS and IP could be used, which is regenerated in the satellite and received as DVB-S in the user terminal.

Nomadic Access

The nomadic access to the satellite system is based on a fixed mobile solution with an automatic scanner polarizer and beam positioning system for a auto folding two-way satellite antenna. This service is aimed for the nomad users who want to have high speed access in remote locations where cable and DSL do not exist, or for mobile terminals.

Figure 2.15 below shows an architecture sample from a scenario where all agents and application providers are inside the satellite network.

Figure 2.16 shows the architectural scenario with the application provider outside the satellite network.

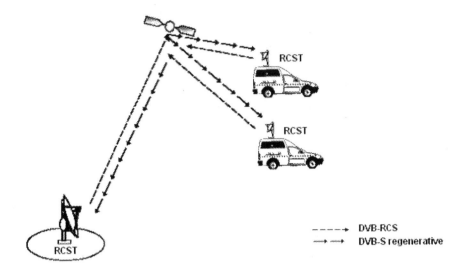

Fig. 2.15 Nomadic access scenario. Application provider inside the satellite network

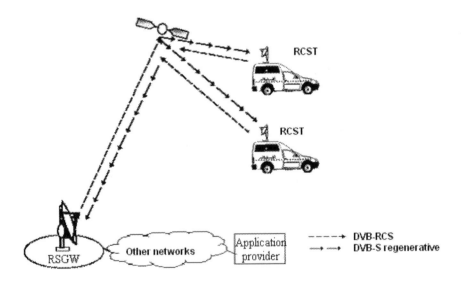

Fig. 2.16 Nomadic access scenario. Application provider outside the satellite network

SNMP Based DVB-RCS Authentication Protocol

The authentication protocol to be used by SNMP follows the Quick Key Exchange as mentioned in the DVB-RCS document, thus making the design conformant with the DVBRCS specifications. This Key Exchange protocol uses the cookie value

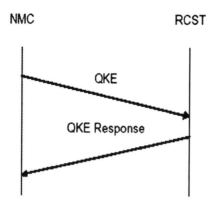

Fig. 2.17 Flow Diagram of the QKE Authentication protocol

(secret static value shared between the RCST and NMC) to authenticate the RCST to the NCC via NMC. The flow diagram and the byte structure of these messages are shown in Fig. 2.17.

The QKE and QKE Response messages are encapsulated via SNMP get and the response to SNMP Get messages. SNMP is used for transport of the authentication messages. The NMC knows the OID of each client which is defined in the MIB of the RCST.

Thus, security at RCST level is improved in the project.

Conclusions

As explained in the paper, Satlife has become a key project in the way to enhance the regenerative systems. As a first step in the AmerHis improvements, Satlife has carried out new developments in order to improve AmerHis features and being able to offer new innovative services. The new on-board processor concept introduced by AmerHis has been a clear revolution in the satellite arena. Thus, Satlife means a first step in order to evolve this new concept towards the optimization of the system to be able to decrease costs and increase efficiency and QoS between others.

References

[1] Deliverable DE422—SATLIFE.
[2] ESA-AMERHIS PROJECT 3.
[3] IST-1-507675. SATLIFE PROJECT.

Chapter 3
SATSIX Satellite System and Network

E. Callejo, A. Yun, C. Baudoin, F. Rodriguez, and J.A. Guerra

Abstract Satellite platforms that already provide broadband access for Internet/ Intranet (IPv4) and PSTN/ ISDN terrestrial networks are evolving towards the convergence with other network access technologies as WiFi and WiMax searching for mobility, QoS and IPv6 protocol evolution. This paper presents the benefits of the Next Generation Broadband Satellite Multimedia Systems and Network architecture future enhancements in the frame of EC-funded IST FP6 SATSIX (Satellite-based communications systems within IPv6) project.

Introduction

Satellite communication is becoming an established part of the institutional communications, both for civilian and military use. Broadband Satellite Networks do help to reduce the Digital Divide in regions with little accessibility to terrestrial network access or where it is uneconomic to install cable or fixed wireless. These regions with low density might be low, but does not mean communication is less important. Consumers and business customers sparse in remote areas may benefit from broadband satellite networks and enjoy broadband services. This will surely, enhance the quality of life and comfort of users located far from large urban areas.

The consortium of the SATSIX project wants to go a step beyond in satellite communication. The objectives identified have been: IPv6, mobility and new generation of satellite payloads based on ACM/DVB-S2. For the complete design of the satellite architecture, three different service scenarios with threes kinds of users have been identified. These services scenarios are: Residential, Collective and Corporate. Each scenario has several service profiles associated, including e-learning, telemedicine, e-payment, games, VoIP, VoD, interactive TV, e-mail, Web browsing or file sharing among others.

In the frame of SATSIX project, the satellite network architecture presented must accomplish the needs of the services scenarios defined. This paper will introduce this architecture based on the different criteria followed in its definition. It will also point out the NGN Satellite Systems requirements that will allow introducing

L. Fan et al. (eds.), *IP Networking over Next-Generation Satellite Systems*
© Springer 2008

concepts such as IPv6, mobility or DVB-S2 into already existing Broadband Satellite Multimedia (BSM) networks. Then, the main criteria for the evolution of BSM network within the frame of SATSIX project will be summarized. The next section presents SATSIX communication satellite network reference architectures, protocol stack and services supported. Finally, the paper deals with each one of main aspects analyzed for this evolution towards a New Generation Satellite Network, that will improve satellite communications in terms of efficiency (i.e. DVB-S2, GSE), IPv6 support (ULE), mobility (i.e. MIP), QoS (i.e. SIP, C2P) and interworking with other networks (i.e. interface with WiFi/WiMAX access networks).

SATSIX Satellite Architecture

The aim of SATSIX project is to define the future IPv6 satellite network architecture taking into account previous lines of evolution [8]. For this design, SATSIX satellite network is built taking as baseline already existing DVB-RCS/DVB-S networks [1, 5] and tries to enhance them in terms of flexibility, QoS support, efficient multicast, and integration with terrestrial wired and wireless networks (i.e. WiFi and WiMAX). SatLife (FP6 project) is the reference considered for the regenerative satellite network scenario. SATIP6 (FP6 project) as a preliminary study for IPv6 support for DVB-RCS [2, 7] systems. Finally, ESA-SatLabs compliant star transparent DVB-RCS network as the reference transparent satellite network scenario.

Standards

SATSIX architecture is based on European standards, DVB-S(2) [3, 4] for the forward link and DVB-RCS for the return link. DVB-S2 standard is an evolution from DVB-S. The main benefits of DVB-S2 are:

– 30% greater efficiency than DVB-S.
– An increased range of applications by combining the functionality of DVB-S (for direct-to-home applications), and DVB-DSNG (for professional applications).
– Techniques such as adaptive coding to maximize the usage of value satellite transponder resources.
– Four modulation modes: QPSK, 8PSK, 16APSK and 32APSK.
– Additional roll-off factors for tighter bandwidth.
– Powerful FEC system that results in a performance which is at times only 0.7 dB from the Shannon limit.

The forward link waveform uses a TDM access and is based on DVB-S2 standard in ACM mode. The possible MODCODs (value that identifies the modulation and FEC code) to be used within SATSIX framework is a subset of all possible DVB-S2 MODCODs proposed by the standard. The selection of the best MODCOD (value that identifies the modulation and FEC code) will be determined by the level of robustness and protection required by the information to be transmitted.

DVB-RCS was reopened last January. Its main objective is to create a new version of the standard, backward compatible only adding whatever is required to support DVB-RCS mobile and C2P (Connection Control Protocol) for mesh DVB-RCS communications, plus small clarifications and corrections.

In the scope of SATSIX, the influence of standards have been considered in the migration towards DVB-S2 and the use of an adaptive return link: intended as the signaling to transfer MODCOD, modulation and coding formats among space and ground segment elements.

Mobility is being considered in the sense of a seamless integration of DVB-RCS satellite networks with terrestrial WiFi and WiMAX networks. WiMAX / WiFi users will communicate with the WiFi / WiMAX base station, and the base station is connected to the DVB-RCS system via a Return Channel Satellite Terminal (RCST). In this context WiFi / WiMAX are used as wireless local loops integrated with the satellite network.

Payload

The reference broadband satellite network architectures addressed in this paper are based on geosynchronous orbit satellites. These satellites are classified based on the type of payload.

- Transparent: Transparent payload refers both to a pure transparent satellite transponder (bent-pipe) and to a satellite transponder which does not perform any demodulation but is able to perform some form of physical layer switching, e.g.:
  - by switching uplink carriers in a beam to different downlink beams by means of IMUX (Input Multiplexer) and OMUX (Output Multiplexer); or
  - by performing ADC (Analogue-to-Digital Conversion) operation on the uplink carriers, digital switching, and DAC (Digital-to-Analogue Conversion) operation.

Its evolution is moving towards a new concept of flexible payloads. This is a way to search for innovative solutions to reduce cost and production cycle, to add value to services and develop new applications. Flexible payloads should provide tunable performances and allow the satellite operator a mechanism to adapt to new business. Different domains of flexibility may be considered: orbit location, coverage, frequency plan, allocation, routing (e.g. RF switching) and waveform.

- Regenerative based on Adaptive Switching OBP: The regenerative payload, the On-Board Processor (OBP), provides regeneration functions that decouple the uplink and downlink air interface formats (modulation, coding, framing, etc.).

A DVB-RCS air interface is used on the return link (uplink) while a DVB-S/S2 air interface is used on the forward link (downlink). The OBP capability of the satellite transponder, allows both star and mesh (single-hop) communications.

The definition of a new generation satellite payload compatible with ACM and DVB-S2, the Adaptive Switching OBP architecture had been considered. This Adaptive Switching OBP should be capable to route traffic from any or several U/L to any or several D/L, based on different criteria and uplink encapsulation methods (e.g. ATM, MPEG, GSE). Special care is taken in order to analyze the impact of GSE over DVB-S2/ACM OBP.

A third type of payload has also been considered, a hybrid payload. This type of payload is built from the concept of dual transponders. Both transparent and regenerative transponders are mixed in the satellite, and they are used with a certain intelligence based on the service requested. This hybrid payload provides advantages to the system in terms of system capacity, system coverage and system services, but with an increment in complexity due to the flexible use of the transponders.

NGN

Next Generation Networking technologies are impacting current and future Satellite-based networks. SATSIX satellite architecture has open the pave to new ways to enhance QoS based on the usage of SIP proxies. SIP proxies allow the support of dynamic QoS control that reduces delay and jitter, and ensures a more efficient usage of the satellite resources.

IPv6

Internet is running out of IP addresses. Terrestrial networks are already experiencing the move towards IPv6. IPv6 does not only solves the IP addresses limitation, but brings solutions for better routing mechanisms, improvements in IP mobility, changes in IP multicasting and QoS. Satellite network manufacturers are not yet convinced of the need to change to IPv6, but the satellite world must be prepared for the IPv4 to IPv6 transition.

Networks, Topologies, Connectivity Types and Services

The reference broadband transparent and regenerative satellite network architectures addressed in this paper are classified in two different families depending on the network topology:

1. Star/mesh transparent (no processing on board): "Traditional" VSAT network with a single Hub that can broadcast information to all terminals within the satellite beam. One transmission from the hub can reach all terminals in the coverage area, and multicasting from terminals is solved with a double-hop communication through the hub. This Hub or gateway allows the interconnectivity with external terrestrial networks.
2. Star/mesh regenerative (based on OBP, On Board Processing technologies). One single Hub or NCC is used for control and management. Multiple gateways/feeders are envisaged to provide distributed interconnectivity with external terrestrial networks. The new generation satellite based on OBP allows direct connectivity between two terminals (only one hop), cross full connectivity between different beams, and replications on board, allowing an efficient multicast service.

The end user will have access to services like high speed Internet, VoIP, ISDN/PSTN low delay audio/video conferencing thanks to star connectivity. On the other hand, mesh connectivity will allow services like VPN or Lan to Lan interconnectivity.

Regenerative Platform

SATSIX baseline for the definition of a new generation regenerative platform has been AmerHis System. The AmerHis System represents the first regenerative DVB-S/DVB-RCS satellite platform. It was designed as a response to cover the growing demand in multimedia broadband services and the adaptation of real time services to the satellite world. This kind of system has been standardized under ETSI SES BSM families as RSM-B (Regenerative Satellite Multimedia) Family B 0.

The regenerative platform is composed of the OBP (On Board Processor), a fleet of standard RCS (Return Channel Satellite) Terminals, a MS (Management Station) acting as Network Control Center and Network Management Center, several RSGWs (Regenerative Satellite Gateways) for interconnection with terrestrial networks and a group of VSPs (Video Service Providers) for the transmission of video.

The core of the system is the OBP, a multibeam switch in the sky. The regenerative platform provides mesh and star topology, both cases in a single hop, allowing real time and multimedia connectivity (Fig. 3.1).

A DVB-RCS Terminal behaves as an IP router. It can transmit up to 4 Mbps and receive up to 8–16 Mbps allowing the deployment of any IP application (e.g. Internet/Intranet Access and Virtual Private Networks between terminals) and being interoperable with the terrestrial networks. The RCSTs supports different customer service profiles, given a correspondence between traffic profiles and SLAs defined by the operator.

The elements of the regenerative platform are introduced next:

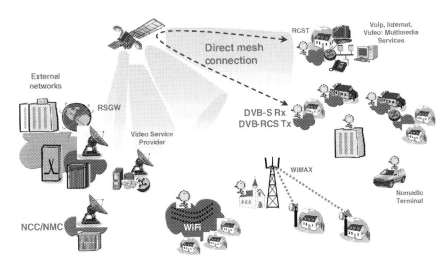

Fig. 3.1 Different services on DVB-S regenerative access platform

The Management Station (MS) manages all the elements of the system. It also controls the sessions, resources and connections of the ground terminals. It is composed by:

- Network Control Center (NCC), which controls the Interactive Network, provides session control, routing and resource access to the subscriber regenerative RCSTs and manages the OBP configuration and DVB-S/DVB-RCS tables.
- NCC-RCST, the satellite terminal of the MS, supporting modulation and demodulation functions to access to the satellite.
- Network Management Center (NMC), in charge of the management of all the system elements. The AmerHis NMC is split in two systems:
- Element Manager (EM): It is responsible for the management of the redundant NCC (including NCC-RCST) and of the GWs.
- Network and Service Manager (NSM): It is responsible for the management of the VSNs, Service Providers, RCSTs, and telecom services and NCC-initiated connections.
- The Regenerative Satellite Gateway (RSGW) provides functions similar to those offered by the GW in the transparent networks (TSGW). It provides interconnection with terrestrial networks (ISDN/POTS, Internet, and Intranet). At the same time, it manages all its subscribers, guaranteeing their Service Level Agreement (SLA). It also establish point to multipoint connections to provide a dynamic Star Multicast Service.

Within one Interactive Network, it may be possible to define a "two-level" architecture by introducing the concept of Virtual Satellite Network (VSN). The VSN is a group of RCSTs with a certain addressing plan assigned and with a GW assigned to it also. Each VSN may be considered a specific Administrative Domain, which receives a set of satellite physical resources, a set of logical resources etc. If included in Interactive Network, the VSNs do not modify the typologies of communication supported between network elements.

The VSN (Virtual Satellite Network) concept is introduced to allow the definition of isolated networks within the whole Satellite Interactive Network. One VSN represents a group of RCSTs and RSGWs, an amount of traffic capacity and an independent IP addressing plan. The Network Management System, part of the MS, allows provisioning different VSNs, each one of them may be assigned to a different service provider. The VSN concept represents a more efficient and flexible way to manage and distribute the satellite resources.

Transparent Platform

The network has a typical star topology. The central node or Hub contains the NMS, the satellite front-end, IP infrastructure and terrestrial networks interfaces (Fig. 3.2).

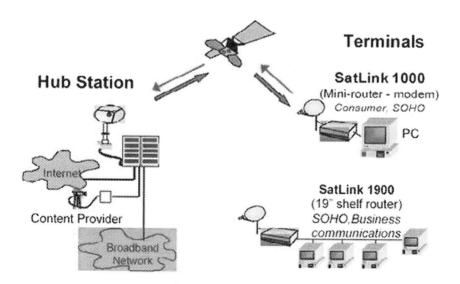

Fig. 3.2 Transparent platform equipment

The Hub or Gateway is mainly composed of the following subsystems:

Network Management System (NMS): Interfaces between ISP/operator and Hub. It manages the users' traffic and services provision. NMS also performs the control and supervision of HUB equipment, network and terminals.

Forward Link Subsystem (FLS): Encapsulates IP packets in MPEG-2 frames and transmits these frames on a TDM carrier, base-band modulated. In the RF equipment, the modulated bitstream is upconverted and sent to the antenna, which transmits it up to the satellite at frequencies in the Ku band.

Return Link Subsystem (RLS): The Gateway antenna receives the downlink signal at Ku band and downconvertes it to the 950–1450 MHz IF band. Thereafter, RLS filters the different return channels that compose a superframe, demodulates the signal, decodes the TDMA bursts form the terminal and extracts the frames containing the IP packets and sends them to the Hub station Ethernet.

Reference and Synchronization Subsystem (REFS): Receives time signals from GPS and provides synchronization and timing for the different Gateway subsystems.

IP infrastructure: It makes possible the establishment of the management and traffic networks as well as the routing between Internet and the different subsystems.

This platform provides star connectivity Fig. 3.3.

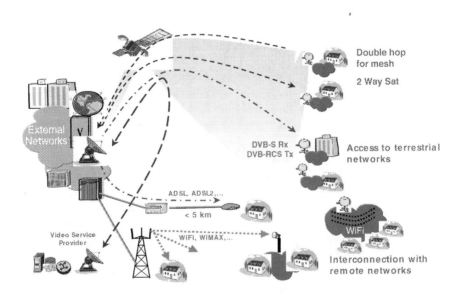

Fig. 3.3 Different services on DVB-S transparent access platform

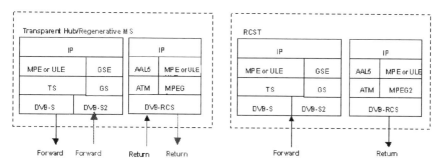

Fig. 3.4 SATSIX user plane protocol stack

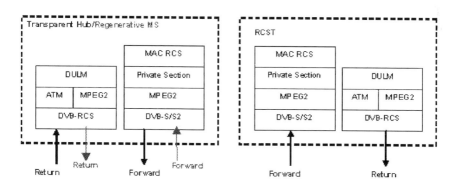

Fig. 3.5 SATSIX control plane protocol stack

Protocol Stack

Taking into account the impact of IPv6, DVB-S2 and new encapsulation techniques the protocol stack proposed within the frame of SATSIX project is the following one (Figs. 3.4 and 3.5).

Services over the Network Reference Scenarios

To cover the support of corporate, residential and collaborative services scenarios defined in SATSIX framework [6], one of the task for system engineering include the mapping between the services and the network scenarios and topologies and connection types, taking into account both regenerative and transparent platforms.

In the star transparent platform we can divide the IP connectivity in two groups:

– Star connectivity between an RCST and external networks. The RCST sends the data to the NCC/TSGW and it routes it to the Internet through an ISP network.
– Star connectivity between two RCSTs, all the data is received in the NCC/TSGW and routed to the other RCST. This means that the connection between RCSTs is indirect (double hop), the RCSTs only have direct connectivity to the GW.

In both cases all the data goes through the NCC/TSGW, the first one needs two satellite hops while the second one requires a satellite hop to reach the ISP.

This is applied both for unicast and multicast communications, but for multicast in multibeam system the flow has to be replicated by the NCC/TSGW as many times as the destination beams, so if there are four destination beams the multicast flow is transmitted four times in the Return Link.

The star/mesh Regenerative scenario allows the following kinds of connections:

– Mesh communications between RCSTs in any user beam with RCSTs in the same beam or any other user beam in one hop. The RCSTs can be connected to an isolated LAN or to the Internet; in the case that one of them has Internet access and the other don't have it the first one can act as gateway of the second. This connection can be unicast or multicast. Mesh multicast service consists on the ability to statically transmit multicast flows receivable by the rest of the terminals in the VSN (see 0), which can be in the same or different beams.
– Star communication between an RCST and a RSGW for Internet access. This connection is also in a single hop. Star multicast connectivity allows the RCSTs

Service	Corporate scenario	Residential scenario	Collective scenario
Internet/intranet access	X	X	X
Web browsing	X	X	X
FTP	X	X	X
E-mail	X	X	X
Peer-to-Peer		X	X
E-services (e-medicine, e-commerce, e-government, e-learning)			X
VoIP and video conferencing	X	X	X
Video Broadcast Service		X	X
Audio and video on demand (streaming)		X	X
Alert messages	X	X	X
Media content download (Store and Forward)	X	X	X
Interactive gaming		X	X
Software downloads	X	X	X
Remote control of applications	X		
Shared applications	X	X	X
Video surveillance	X	X	X
Web conference	X		X

to statically or dynamically receive a multicast flow coming from the RSGW. "Dynamically" means that the star flow will start only when a user "joins" a certain multicast session and it will stop when the RSGW Access Router does not receive membership reports for the group.

Next table presents the association between the presented services and the User Scenarios they are applicable to:

Depending on the network architecture the same service can be provided with mesh or star connectivity.

Both payload scenarios have to be able to provide the previous list of services, although depending on the features of the service the performance will be different in the two payload scenarios.

Functional Layers Overview

The next step in SATSIX System Engineering consisted in the definition of the System functional requirements.

From the System point of view both star transparent and star/mesh regenerative network scenarios were analyzed to tackle with SATSIX main objectives:

– NGN QoS, SIP proxy
– Connections support
– IPv6 support, ULE/GSE encapsulation
– DVB-S2/ACM
– WiFi/WiMAX interconnectivity

Connections Support, Connection Control Protocol (C2P)

The Connection Control Protocol (C2P) provides the interaction between RCSTs/ RSGWs and NCC/Hub to support set-up, modification and release of connections and channel bandwidth modification.

C2P implementation was done for AmerHis/SatLife, a star/mesh regenerative scenario following Annex J from DVB-RCS Guidelines and the C2P specification given in ETSI SES BSM 102 429-3, as the perfect solution to solve mesh connectivity.

Anyhow, C2P can be extended to other DVB-RCS scenarios. Within the frame of SATSIX, C2P definition is studied to accomplish a star transparent DVB-RCS scenario, IPv6, DVB-S2 support and full dynamic multicast group management (in star and mesh scenarios). The aim of C2P is to enhance the control plane of DVB-RCS systems by means of the following:

– Dynamic control of the set of communication parties;
– Quality of service driven dynamic allocation of bandwidth resources to satellite connections, closely linked to NGN SIP signaling;
– Configuration of the logical channel identifiers for ATM and MPEG profiles;
– Identification of calling and called addresses in either MAC or IPv4/IPv6 syntaxes, performing embedded ARP functionality or dynamic IP routing;
– DVB-S2 parameters dynamic assignment;
– Provide dynamic control of the set of communicating parties in DVB-RCS systems for both mesh and star topologies, in transparent and regenerative, mono and multi beam scenarios.
– Dynamic multicast group management, both for star and mesh topologies.

In a Next Generation Satellite System the NCC and the terminals shall support the C2P protocol, being capable of triggering C2P connections.

NGN QoS, SIP Proxy

In order to meet users requirements defined in the corporate, collective and residential scenarios, the following QoS components shall be used in both transparent and regenerative system architectures [11]:

• SIP proxy, in charge of the QoS aware applications. It mostly in charge of the SIP signalization capture to feed the other QoS components;
• QoS Agent, in charge of the non QoS aware applications. It sends the wanted QoS level for a given traffic flow to the QoS Server;
• QoS Server, in charge of collecting the QoS information on traffic flows and to configure accordingly the IP and MAC layers;
• PEP, in charge of the TCP and HTTP acceleration in order to limit the delay for the corresponding applications;
• IP Compression, in charge of the header compression in order to decrease the overhead introduced by the successive stacks and thus decrease the load and the delay;
• IP QoS, in charge of the classification, the marking; the policing/shaping/dropping and the scheduling at IP layer;
• MAC QoS, in charge of the queuing, dropping and scheduling at MAC layer;
• C2P agent, in charge of the MAC connection handling at the terminal side;
• C2P server, in charge of the MAC connection handling at the NCC side;
• Access Resources Controller (ARC), in charge of the management of MAC resources;
• MAC QoS, in charge of the queuing, dropping and scheduling at MAC layer;
• RRC agent, in charge of the capacity request emission and the extraction of the uplink packets function of the allocation carried in the TBTP;

- RRC server, in charge of the capacity requests gathering and the allocation processing

Two approaches can be envisaged for the dynamic QoS architecture.

- The first one is based on an "IP-oriented" approach using SIP proxies or specific signalization (QoS agent / QoS server / Access Resource Controller) between the ST and the GW. Proxies are in charge of configuring IP and MAC components directly, without any MAC signaling path.
- The second one relies on the concept of access connection in a "MAC-oriented" approach. The dynamic QoS depends on information provided by proxies (SIP, QoS server, …) but the communication between the ST and the GW is based on the C2P.

Mobility

In the frame of SATSIX it has been defined the functional architecture to be deployed in order to support application layer mobility scenarios.

Different definitions of mobility are emerging, and in SATSIX three mobility scenarios are defined [9], namely discrete mobility, continuous mobility and seamless mobility. Next, two mobility contexts; macromobility and micromobility, according to network hierarchy have been considered. To improve the mobility management, and reducing signaling overhead and handover delay when using MIPv6, the enhancement protocols HMIPV6, FMIPv6 and the combination FHMIPv6 are investigated.

When needed (for QoS purposes for instance), upper layer mobility in SATSIX provides SIP applications with QoS support, maybe combined with MIP6 architecture.

The localization of the access router, separated or integrated, with the satellite terminal/base station determines the transport format over the satellite (tunneling or native IPv6 transport).

With an integrated access router more than one subnet can exist behind the satellite terminal/base station. In this case, optimized techniques such as hierarchical mobile IP can improve the mobility process if moving to another subnet, since micro mobility inside the visited network is managed locally without involving the home agent. This case also includes Layer 2 mobility and handover in the WiMAX/WiFi part of the local network [10], where handover (fast handover) has to be combined with Mobile IPv6 solutions (MIPv6).

With an access router separated from the satellite terminal/base station, the transport over the satellite will be at Layer 2. The Layer 3 mobility will be anchored in the core network (behind the satellite gateway for a transparent satellite). Two alternatives of the Mobile IP(v4 or v6) protocol are then supported in the WiMAX NWG specifications:

- Client MIP (CMIP), where CMIP is an IETF compliant Mobile IP solution based on a Mobile IP enabled MS (Mobile Node)
- Proxy MIP (PMIP), where PMIP is an embodiment of the standard Mobile IP framework, performing on behalf of a client that is not MIP-aware or MIP-capable. The PMIP should be placed at the access router.

To support SIP mobility, the architecture has to implement all the devices defined in SIP architecture such as SIP registrars,...These elements are to be available once in the satellite access network. If the network topology is meshed, its location depends on the choice of satellite network operator. In a star topology, these elements will be located behind the gateway.

Security

The SATSIX system has to provide security mechanism to meet users' security requirements for the corporate, collective and residential scenarios:

- IPSec for end-user to end-user and site-to-site communications protection shall be provided (notice: incompatibility with PEPs and firewalls).
- IPSec/SATIPSEC for satellite/WLL segment protection shall be provided.
- Layer 2 security mechanisms of WIFI/Wimax standards shall be provided
- Layer 2 SATIPSEC and ULE /AAL5 security dataplane shall be provided
- DVB-S/RCS signaling protection mechanisms may be provided

The compliance with these requirements is critical for users as militaries, business organizations or Government customers.

DVB-S2

Both regenerative and transparent platforms shall be fully compliant with DVB-S2. Once analyzed the different topologies and connectivity types, the MODCODs selection procedure will not only be based on the type of traffic to be transmitted or the weather conditions, but will take into the source and destination of each communication. The MODCOD selected for a unicast connection in the DVB-S2 forward link will depend only on the source and destination terminals and the Hub/NCC. The MODCOD selected for a multicast connection in the DVB-S2 forward link will depend on the source terminal, all destination terminals receiving that flow and the Hub/NCC.

The new generation satellite payload should also be compatible with ACM and DVB-S2. The new Adaptive Switching OBP architecture, should be capable to route traffic from any or several U/L to any or several D/L based on a convergent architecture. The routing will done based on different criteria and uplink encapsulation methods (e.g. ATM, MPEG, GSE). Special care will be taken to analyze the impact of GSE over DVB-S2/ACM OBP.

ULE/GSE

Unidirectional Lightweight Encapsulation (ULE) mechanism permits the transport of IPv4 and IPv6 datagrams and other network protocol packets directly over the MPEG2 Transport Stream as TS Private Data. The introduction of Generic Streams by DVB-S2 makes necessary a new encapsulation, the Generic Stream Encapsulation, that relies in some fundamental design choices of ULE. Within the SATSIX project these two protocols are mainly considered for IPv6 encapsulation support. Their processing will impact not only on the OBP but also on the ground segment equipment.

The Next Generation Satellite System has to be designed to be as far as possible compatible with GSE/ULE. However, given that the vast majority of existing satellite systems use MPE, it has to be maintained during the transition to a long term solution.

Convergent Platform

The convergent or hybrid communication topology is indented as the combination of transparent and regenerative connections, in a way that could be observed by the end-user and INAP operators as a unique platform. The advantage is the exploitation of regenerative or transparent connectivity only for those services or applications that really require each of them (i.e. depending on the type of service and application).

The number of transponders available and its bandwidth determine the capacity of the system. When talking about the capacity of a convergent system, two different choices can be considered:

- To maintain separately the capacity of the transparent and the regenerative networks using dedicated transponders.
- To share transponders between both networks. A certain bandwidth of the transponder would be reserved to be used by each system. This option would fit better to the real case; transparent and regenerative services will coexist in the system, a percentage of the capacity would be dedicated to each type.

The first approach (vand the one considered in this paper) is to use dedicated transponders for the different parts, the payload will mot be integrated in the first step.

The convergent system should allow full connectivity between different beams. Depending on the level of integration of transparent and regenerative parts there are different modes of operation based on the level of integration of transparent and regenerative management systems:

Loose Integration: it only exists a common management system for configuring a terminal as transparent or regenerative. It is decided at provisioning time what system the terminal will use.

Medium integration: the terminal assignment as transparent or regenerative is changed by the NCC/HUB depending on the service. A logoff of the terminal is necessary to change DVB-S2 flow.

Tight integration: there is collaboration between the NCC / HUB to manage the available resources. If there are not resources in the first assigned platform the terminal will perform the communications with the second best configuration and no switching between platforms.

Full integration: terminals with two DVB-S2 receivers to handle both DVB-S2 streams. Depending on the type of service it will decide to use one path or the other.

SATSIX recommends the second/third option (medium integration) as a preliminary step in the development of convergent solutions, the other ones are much complex.

Conclusions

This paper presents the system and network architecture requirements of the Next Generation Satellite System. Three user scenarios (residential, corporate and collaborative) and two satellite network scenarios based on payloads transparent and regenerative have been considered as the framework for SATSIX system architecture specification. Even more the concept of a new approach, a hybrid system that integrates both types of payloads has been considered.

Finally, the main features considered within SATSIX frame such as standards evolution as DVB-S2/ACM techniques, adaptive return link, the enhancement of Connection Control Protocol (C2P) for DVB-RCS systems, the support of IPv6 and new encapsulation techniques (ULE and GSE), introduction of IP mobility (MIP and HMIP), the development of a dynamic QoS architecture (SIP proxies) and the interworking with WiFi/WiMAX have represented the main drivers for the complete definition of SATSIX new generation satellite network and have been described in this paper.

Acknowledgments The SATSIX consortium greatly appreciates the financial contribution and technical comments of the EC towards this project. This work is supported by the IST FP6 SATSIX project, funded by European Commission (EC).

References

[1] M. Wittig, and J. M. Casas, "A communications switchboard in the sky: AmerHis", ESA Bulletin, No. 115. Aug. 2003.

[2] ETSI EN 301 790, Digital Video Broadcasting (DVB); Interaction channel for satellite distribution systems, ver 1.4.1.

[3] ETSI EN 301 192, Digital Video Broadcasting (DVB); DVB specification for data broadcasting, ver 1.3.1.

[4] ETSI EN 302 307, Digital Video Broadcasting (DVB); Second generation framing structure, channel coding and modulation systems for Broadcasting, Interactive Services, News Gathering and other broadband satellite applications, ver 1.1.1.

[5] F. Vallejo, and A. Yun, "AmerHis: Triple Play over an OBP-based DVB-RCS Satellite Platform", 23rd AIAA International Communications Satellite Systems Conference (ICSSC-2005) and 11th Ka Band Broadband Communications Conference.

[6] E. Callejo, A. Yun, I. Jiménez, J. Prat, C. Baudoin, P. Loyer, L. Duquerroy, F. Arnal, M. Catalán, J. A. Guerra, R. Muñoz, F. Rodriguez, P. Zautasvili, and T. Gayraud, "Satellite network requirements", SATSIX-IST-026950.

[7] ETSI TS 102 429-1,2,3,4, "Satellite Earth Stations and Systems (SES); Broadband Satellite Multimedia; Regenerative Satellite Mesh – B (RSM-B); DVB-S/DVB-RCS family for regenerative satellites; Part 1 System Overview; Part 2 Satellite Access Layer; Part 3 Connection Control Protocol; Part 4 Specific Management Information Base".

[8] C. Baudoin, L. Fan, E. Callejo, A. Pietrabissa, F. Rodriguez, A. Ramos, G. Fairhurst, F. Arnal, and G. Santero, "New Architecture for Next Generation Broadband Satellite Systems: The SATSIX Approach". INNSS07.

[9] I. Melhus, F. Arnal, T. Gayraud, and B. Jacquemin, "SATSIX Mobility architecture and its performance evaluation". INNSS07.

[10] F. Rodríguez, L. Fan, and I. Melhus, "Interworking Strategy between DVB-RCS and WiMAX". INNSS07.

[11] A. Ramos, B. de la Cuesta, B. Carro, J. Aguiar, D. Pérez, C. Baudoin, T. Berthou, and T. Gayraud. "SATSIX QoS Architecture". INNSS07.

Chapter 4
Fast IP Handover Between Satellite Networks and Wireless LAN Networks for High-Speed Trains

Myunghee Han, Namkyung Lee, Kiseop Han, and Dongjun Lee

Abstract Internet services using satellites for high-speed trains are being widely studied these days. The major drawback of satellite networks is that antennas on the trains can not receive signals from the satellites in train stations. To solve this problem, IP handover to terrestrial wireless networks in the stations is considered. However, due to long disconnection time of layer 2 and layer 3, the performance of current IP handover schemes is not good in this handover case. In this paper, we propose a new IP handover scheme between a geo-satellite network and a terrestrial wireless LAN network for high-speed trains. Through simulations, we show that proposed handover scheme greatly reduces overall handover latency and increases TCP throughputs.

Introduction

There have been wide research efforts to provide internet service to high-speed mobiles such as fast trains. Particularly satellite communication systems have advantages of covering extensive area, which is incomparable with that of terrestrial wireless communication systems, and being less affected by fast fading. Previous satellite communication systems were mostly unidirectional multimedia broadcast systems using large bandwidth, however the needs of high rate data and bi-directional multimedia communication for satellite communication systems have been growing much recently. In this respect, ETSI (European Telecommunications Standards Institute) approved standards DVBS (Digital Video Broadcasting via Satellite) [1, 2] and DVBRCS (Digital Video Broadcasting and Return Channel via Satellite) [3, 4] for interactive broadband internet services using satellites.

For internet service to high-speed trains, satellite networks are being considered seriously due to less amount of handover compared to using terrestrial wireless networks. However, the major drawback of satellite networks is that trains' antennas cannot receive signals from the satellites in train stations or other shadow zones.

L. Fan et al. (eds.), *IP Networking over Next-Generation Satellite Systems*
© Springer 2008

To solve this problem, handover to a terrestrial wireless network in the stations is considered. However, due to inherent long disconnection time of layer 2 and layer 3 in this handover case, the performance of current IP handover scheme is not good. The major problem in handover between satellite and wireless network is that handover latency is too large. Therefore, it is necessary to reduce the handover latency. Although Fast Handover for Mobile IPv6 scheme [11] is proposed to reduce handover latency, the scheme does not show enough performance in this handover case. Also, as far as we know, there have been few researches regarding IP handover between terrestrial wireless networks and satellite networks. Therefore, we propose a new IP handover scheme for high-speed trains between satellite networks and terrestrial wireless networks.

This paper is organized as follows. The second section presents an overview of the operation of Fast Handover for Mobile IPv6 referenced from RFC 4068 [5]. In the third section, we examine latency of current fast handover scheme and propose a new IP handover scheme between satellite and wireless networks for high-speed trains. In the fourth section, we present simulation environments and results for the proposed handover scheme. The fifth section presents conclusions.

Fast Handover for Mobile IPv6

The material given below is from RFC 4068 [5]. The Fast Handover Protocol is an extension of Mobile IPv6 (FMIPv6) that anticipates the layer 3 handover based on layer 2 triggers. The layer 2 triggers could be utilized as in IEEE 802.21 [6]. Because trains move in fixed paths, FMIPv6 that reduces handover time by anticipating movements is easily applicable to trains.

Figure 4.1 describes signal flows of Fast MIPv6 between a MN (Mobile Node) and ARs (Access Routers).

The MN exchanges a RtSolPr (Router Solicitation for Proxy Advertisement) message and a PrRtAdv (Proxy Router Advertisement) message with the PAR (Previous Access Router). With the information provided in the PrRtAdv, the MN formulates a NCoA (New Care of Address) and sends a FBU (Fast Binding Update) message. The FBU authorizes PAR to bind PCoA (Previous Care of Address) to NCoA, so that newly coming packets can be forwarded to the current location of the MN.

The PAR starts buffering arriving packets with PCoA as destination and determines whether the NCoA is acceptable to the NAR through the exchange of HI (Handover Initiate) and Hack (Handover Acknowledgment) messages. After receiving Hack the PAR tunnels buffered and arriving packets to the NAR (New Access Router). When a link up trigger is delivered, the MN sends an FNA (Fast Neighbor Advertisement) message to the NAR to inform its existence. Until NAR receives FNA, it also buffers arriving packets from PAR.

The gap of receiving time between the FBU at the PAR and the FNA at the NAR causes disconnection time that means MN actually can't receive packets

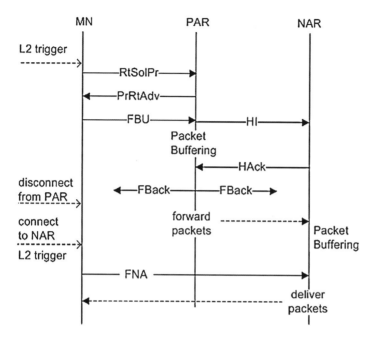

Fig. 4.1 Fast Handover for MIPv6 operation

from CN (Correspondent node). In the third section, we discuss about discon-nection time in detail.

Proposed Handover Scheme Between Satellite Networks and Wireless LAN Networks for High-Speed Trains

Satellite signals cannot be reached to trains in train stations. Consequently, when a high-speed train enters a train station being connected to a satellite network, it should handover from the satellite network to a terrestrial wireless network for seamless internet service. Afterwards, when the train leaves the train station, it handovers to the satellite network again.

Users in the train can communicate with CNs (Correspondent nodes) through MR (Mobile Router) in the train. NEMO BS (Network Mobility Basic Support), specified as RFC 3963 [7], adds a mobility function to IPv6 routers. It hides the address change of the mobile router and provides the nodes inside the mobile net-work with transparent access to the Internet. The users in the train can use static IPv6 addresses that do not change.

The Disconnection Time of Current FMIPv6 Scheme

In the current Fast Handover scheme, layer 2 access delays to new network are
included in disconnection time. Figure 4.2 describes signal flows for access to
WLAN (Wireless Local Area Network) and satellite network.

In addition, the current scheme has a layer 3 delay. After receiving the FBU, the
PAR forwards packets to the NAR. At this time, if NAR cannot receive FNA
quickly from the MR, the NAR cannot forward packets to the MR right away. If
the propagation return delay between MR and NAR is large such as in satellite networks,
the disconnection time is also large. The disconnection time can be calculated by
the equation below:

$$T_D = N_{SIG} * T_{MR-NAR} + T_{FNA} + T_P, \qquad (4.1)$$

where N_{SIG} is number of layer 2 access signals (entry signals), T_{MR-NAR} is the propa-
gation delay between MR and NAR, T_{FNA} is the propagation delay for FNA from
MR to NAR and T_P is the processing delay including scheduling delays at NCC
(Network Control Centre) or APs (Access Points).

In this paper, we assume that the propagation delays on WLAN and geo-satellite
network are 35 ms and 240 ms respectively. In consequence, disconnection times
are at least 245 ms and 1200 ms respectively by (4.1) and Fig. 4.2. This is the minimum calculated value that would take much longer in actual operating conditions.
Because of the long disconnection time, the current scheme cannot meet the
requirement of delay-critical service such as VoIP (Voice over Internet Protocol)
[9] or cause TCP performance deterioration. Hence we propose a new handover
scheme that can eliminate disconnection time T_D.

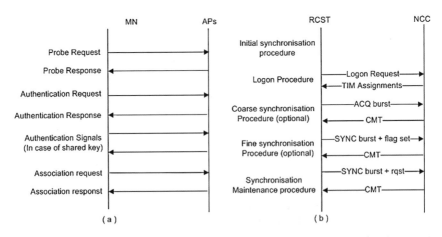

Fig. 4.2 Layer 2 signal flows (**a**) joining a network at IEEE 802.11 MAC layer [8] (**b**) example
of RCST (Return Channel Satellite Terminals) network entry signaling flow for DVB-RCS

The Proposed Handover Scheme

Typical fast handover schemes are based on the assumption of hard handover. In trains, simultaneous connection to different networks for a relatively long period is possible because the train is long so that antennas at its head and tail can be connected to different networks respectively. In IPv6, the MR can be assigned several IP care-of addresses. Hence it is able to connect to both satellite network and wireless network simultaneously. We use this feature in proposing a new handover scheme.

Figure 4.3 shows the proposed handover scheme used between satellite network and WLAN. The operation of the proposed scheme is as follows.

While an MR is connected to PAR, it completes layer 2 connection to the new network and sends FBU to PAR via NAR. After receiving the FBU, the PAR stops sending packets to the previous network and start sending packets to the MR via NAR. The MR cuts off the connection with PAR, after it receives a packet from NAR. The MIH Commands could be utilized as specified in IEEE 802.21 [6]. MIH Handover Complete command is a notification that handover has been completed, and newly coming packets may now be forwarded by NAR. It means that the MR can disconnect from PAR.

Comparing with the previous Fast Handover scheme, there are two differences.

1. Layer 2 connections are maintained through two networks simultaneously until the MR receives packets via NAR. The MIH commands can be used to inform MR to disconnect with previous network. Therefore, layer 2 access delay to new network does not affect TCP performance of the MR.
2. The MR sends a FBU via NAR after it attaches to NAR. This operation has some profits. In case of handover from WLAN to satellite network, in previous

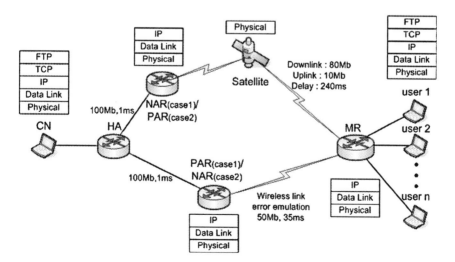

Fig. 4.3 Proposed handover scheme between satellite network and WLAN

Fast Handover scheme, after receiving the FBU, PAR starts sending packets to the MR via NAR. But NAR can not forward packets to the MR right away because NAR receives a FNA from the MR after 240 ms propagation delay in this handover case. This delay could be a major cause for TCP congestion control due to timeout. The proposed handover scheme can solve this problem. In case of handover from Satellite to WLAN, The FNA and the FBU are sent via faster paths (WLAN and wire-line between PAR and NAR), so that total delays of handover procedure can be reduced.

Simulation Environments and Results

In this paper, we assume that the propagation delays on WLAN and geo-satellite network are 35 ms and 240 ms respectively. In consequence, disconnection time are at least 245 ms and 1200 ms respectively by (4.1) and Fig. 4.2. We use a simulation network as shown in Fig. 4.4. The simulator is implemented using NS-2 [10]. We simulate two different cases that the train handovers from WLAN to satellite network (case1 on Fig. 4.4) and from satellite to WLAN network (case 2 on Fig. 4.4).

Users inside a train communicate with a CNs through the MR in the train. To simulate the movement of the MR, we turned on/off the link between the MR and ARs. Satellite network has bandwidths of 80 Mbps downlink and 10 Mbps uplink. The WLAN has bandwidths of 50 Mbps. We use "TCP-Reno" and "TCPSink" that embodied by NS-2. The TCP segment size is 1040 byte and receiver window size is 104 KB (rounded up to 100 segments). For an application, we use FTP application protocol. The file size is large enough so that mobile user can download a ftp

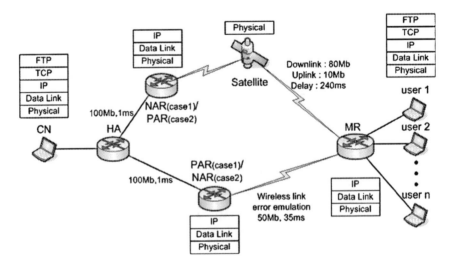

Fig. 4.4 Simulation configuration the network

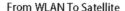

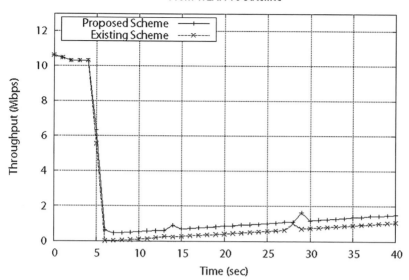

Fig. 4.5 TCP throughputs during handover from WLAN to satellite network

file from the CN infinitely. HA forwards the traffic through satellite network or WLAN. We added "MIPAgent class" to NS-2 for roles of deciding routing path and exchanging handover signals.

Figures 4.5 and 4.6 show TCP throughputs of downlink during handover from WLAN to satellite and from satellite to WLAN respectively. Vertical axis represents the TCP throughputs in Mbps, Horizontal axis represents time in seconds.

In Fig. 4.5, the throughput of previous scheme goes down to zero during handover due to disconnection time, while the throughput of proposed scheme goes down to 0.5 Mbps at the lowest. In the proposed scheme, we could not avoid falling throughput due to increase of RTT (Round-trip Time) because the MR moves to the slower satellite link. However it is still shown that the restoration of the throughput is advanced by around 15 s comparing with the previous scheme.

Figure 4.6 shows that the proposed scheme takes about 2 s to reach the maximum throughput after handover, while the previous scheme takes about 6 s. This is because there are no TCP congestion controls caused by disconnection time in case of the proposed scheme. The throughput decrease of the previous scheme results from TCP congestion control.

The simulation results show that the proposed scheme outperforms the previous Fast Handover scheme with respect to TCP throughput for both handover scenarios.

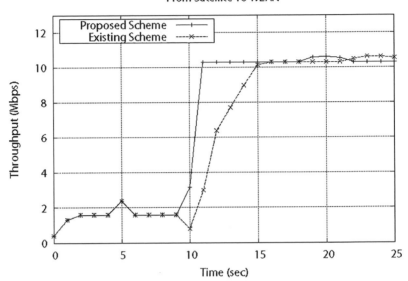

Fig. 4.6 TCP throughputs during handover from satellite network to WLAN

Conclusions

In this paper we proposed an IP handover mechanism between geo-satellite network and WLAN for high-speed trains. The main characteristics of the proposed scheme are that the MR maintains connections of previous network and new network simultaneously until packets arrive at the MR through new access router, and the MR sends an FBU via NAR after it attaches to NAR. This scheme can eliminate disconnection time due to handover. The proposed handover scheme outperforms the previous Fast Handover scheme with respect to TCP throughput.

References

[1] ETSI, EN 300 421, Framing structure, channel coding and modulation for the 11/12 GHz satellite services, v.1.1.2, Aug. 1997.
[2] ETSI, EN 301 192, DVB specification for data broadcasting, v.1.2.1, Jan. 1999.
[3] ETSI, EN 301 790, Interactive channel for satellite distribution systems, v.1.4.1, April 2005.
[4] ETSI, TR 101 790, Interactive channel for satellite distribution systems; guidelines for the use of EN 301 790, v.1.4.1, Sept. 2005.
[5] R. Koodli, et al., Fast handovers for mobile IPv6, RFC 4068, July 2005.
[6] IEEE 802.21 Working Group. http://www.ieee802.org/21/
[7] V. Devarapalli, R. Wakikawa, A. Petrescu, and P. Thubert, Network mobility (NEMO) basic support protocol, IETF, Tech. Rep. RFC3963, Jan. 2005.

[8] IEEE 802.11. http://www.ieee802.org/11/

[9] T.C. Schmidt and M. Wahlisch, Roaming real-time applications mobility services in Ipv6 networks, A technical presentation, 2003.

[10] NS2, Network simulator version 2.29, http://www.isi.edu/nsnam/ns/

[11] D. Johnson, et al., Mobility support in IPv6, RFC 3775, June 2004.

Chapter 5
SATSIX Mobility Architecture and Its Performance Evaluation

I. Melhus, F. Arnal, T. Gayraud, and B. Jacquemin

Abstract As wireless and satellite networks become widely used, new ways to think communication can be envisaged. In particular, a greater degree of connectivity is almost becoming a necessity for the users on the go, being connected whatever their locations are. Different definitions of mobility are emerging, and in SATSIX three mobility scenarios are defined, namely discrete mobility, continuous mobility and seamless mobility. Next, two mobility contexts; macro-mobility and micro-mobility, according to network hierarchy have been considered. To improve the mobility management, and reducing signaling overhead and handover delay when using MIPv6, the enhancement protocols HMIPV6, FMIPv6 and the combination FHMIPv6 are investigated. Application mobility is analyzed as a complement to network layer mobility. The project also addresses the impacts on several others key networking functions like multicast, QoS and PEP.

Introduction

As wireless and satellite networks become widely used, new ways to think communication can be envisaged. In particular, a greater degree of connectivity is almost becoming a necessity for the users on the go, being connected whatever their locations are. A mobile user wants to move freely from one access network to another or roam within a domain while keeping a session alive, and he may demand the same service or resource for the active session on the new routing path as he had on the old path. In principle, mobility can be implemented in various layers of the Internet architecture; link layer, network layer, transport layer and finally application layer. Link layer mobility is associated only with a specific access link technology while the mobile node (MN) or host is moving within the same subnet. Moreover, if mobility is implemented at the transport layer or the application layer, it would have to be implemented separately for each different transport and application layer protocol. In contrast, if mobility management is implemented at the IP layer, this layer is common to all link layer, transport layer and application layer protocols.

In SATSIX, mobility is specified at the network and application layers, where architecture and protocols focuses on terminal mobility, which is the ability of the mobile node or host to change its location. Mobility of the DVB-RCS terminal while moving is not investigated, mobility while moving is only defined for the user (subscriber) terminal connected to a satellite terminal.

The Mobile IP protocol is the standard mobility mechanism at the network layer. It makes sure that the moving node or host is reachable anywhere by its original address. In IPv6, the Mobile IPv6 protocol (MIPv6) [1] is part of the standard and uses features of the IPv6 protocol like Address Autoconfiguration and Neighbor Discovery. Two enhancement protocols are also defined: Hierarchical Mobile IPv6 (HMIPv6) [2] and Fast Handover Mobile IPv6 (FMIPv6) [3] or the combination of the two (FHMIPv6).

While the network mobility is managed by MIPv6 and its optimizations for TCP connections, the Session Initiation Protocol (SIP) [4] is used for real-time application mobility over UDP. SIP and application mobility makes a complement to the network layer mobility in a large range of applications like VoIP, instant messaging and multimedia conferencing. In SATSIX two scenarios are described: *nomadic mobility* and *mid-call mobility*.

Multicast is a central feature in SATSIX, so is also the combination of multicast and mobility. In the literature three approaches for multicast mobility are specified: *remote subscription, home subscription* and *MLD proxying*, the last one being a mix of the other two. The different approaches used in SATSIX scenarios are described.

Due to the high bandwidth* delay product, satellite links make use of accelerators or PEPs (Performance Enhancing Proxies). When the mobile node is visiting a foreign network, the combination of network mobility (Mobile IP) and PEPs causes various problems. Different issues and solutions needs to be considered and are thoroughly discussed and analyzed in the project.

Mobility gives a significant impact to the QoS management, and generates a new challenge for QoS provision as it will have to deal with terminals (nodes) changing their point of attachment to the network. For active application sessions on the mobile, the network should negotiate QoS along the new route as part of the handover procedure which could allow the mobile terminals with ongoing applications to keep or adapt the QoS in the visited networks.

Functional Requirements

Three mobility scenarios are defined:

- *Discrete mobility, i.e. nomadic (roaming) mobility.*
- *Continuous mobility, i.e. suspended service sessions.*
- *Seamless mobility, i.e. local (micro) mobility with handover and no interrupting.*
- Further, two mobility contexts according to the network hierarchy are considered:

- *Macro-mobility, which refers to inter-technology and/or inter-domain mobility*
- *Micro-mobility, which refers to intra-technology and intra-domain mobility.*

The functional requirements of mobility architecture in SATSIX are illustrated in Fig. 5.1.

(a) Nomadic: The user/teleworker, whose home network is located in a small town with satellite connection, is moving from the Home Network to his home (visited network/teleworker home) or to another other village offering public infrastructures or where his company is located (visited network/remote office), while keeping his/hers nomadic communication support.

(b) Continuous: The user/teleworker, is moving between different villages covered by a WiMAX BS linked to a satellite terminal while keeping the ongoing connections opened.

(c) Seamless: The user/teleworker is visiting and moving in a network linked with one satellite terminal, but composed of several separate local WiMAX/WiFi networks/hot zones (i.e. hospital, school, town hall,...).

Network Layer Mobility and MIPv6

The Mobile IP protocol enables a mobile host or node to maintain its IP address and transport layer connections while its point of attachment to the network changes. Each mobile node is always identified by its Home Address, regardless of its current point of attachment to the network. While away from home, MN is

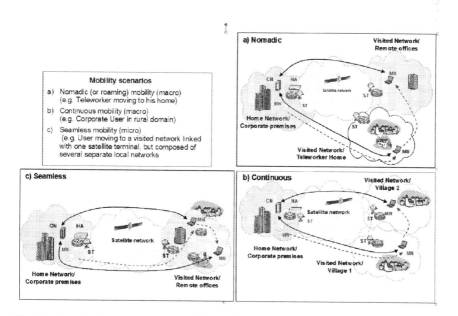

Fig. 5.1 Functional requirements of mobility architecture

associated with a Care-of Address (CoA), which provides information about the MN's current location.

The main difference between MIPv4 and MIPv6 is that there is no FA (Foreign Agent) in MIPv6. For packets delivery a MIPv6 mobile node uses its CoA as source address in the foreign link, in this way there will be no ingress filtering problem. The correspondence node (CN) uses IPv6 routing header rather than IP encapsulation, and therefore naturally supports Route Optimization (RO) to be used instead of Triangular Routing.

Mobile IPv6 Optimization Using HMIPv6

In order to reduce handover delay and signaling overhead over the satellite link for local mobility in the LAN (Local Area Network) usually connected to the satellite terminal, SATSIX defines the use of the two MIPv6 enhancement protocols, Hierarchical Mobile IP (HMIPv6) [2] and Fast Handover Mobile IPv6 (FMIPv6) [3] or the combination of the two FHMIPv6. Hierarchical Mobile IPv6 (HMIPv6) aims at reducing signaling messages during intra-domain and local movement by introducing a new network element called Mobility Anchor Point (MAP). FMIPv6 attempts to acquire information that is needed to join a new link before disconnecting communication at the old link and in that way minimizing handover latency.

In HMIPv6 the MAP provides a simplified signaling scheme as the mobile node sends Binding Updates to the local MAP rather than to the Home Agent (usually further away) and the CNs. Only one Binding Update message needs then to be transmitted by the MN before traffic from the HA and all CNs is rerouted to its new location. The use of the MAP mode is based on two other new entities, the Regional Care-of Address (RCoA) and the Local Care-of Address (LCoA). When a mobile node enters a foreign HMIPv6 domain for the first time, it has to configure two CoAs, one Local CoA that equals the MIPv6 CoA, and a Regional CoA. The RCoA is an IPv6 address obtained from the MAP subnet. Subsequent a successful registration with the MAP, a bidirectional tunnel is established between the mobile node and the MAP, and all packets sent by the mobile node are tunneled to the MAP and vice versa. Fig. 5.2 illustrates the HMIPv6 architecture and operation for macro- and micro- or local movement of the mobile node.

Mobile IPv6 Optimization Using FMIPv6

While Mobile IPv6 describes the protocol operations for a mobile node to maintain connectivity to the network during its handover from one access router to another, the combined handover latency is often sufficient to affect real-time

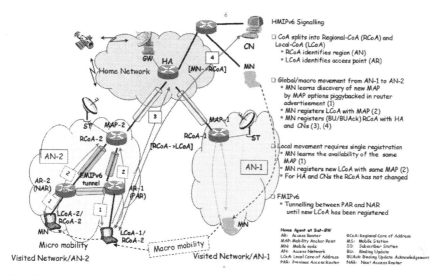

Fig. 5.2 Optimized MIPv6 with HMIPv6

applications. Throughput-sensitive applications can also benefit from reducing this latency. These handover operations involve movement detection, IP address configuration, and location update.

The problems to be solved are:

- How to allow a mobile node to send packets as soon as it detects a new subnet link?
- How to deliver packets to a mobile node as soon as its attachment is detected by the new access router?

The main idea in FMIPv6, see Fig. 5.3, is to reduce the Binding Update latency in MIPv6 by specifying a tunnel between the Previous CoA (PCoA) and the New CoA (NCoA). Fast Handover can be either *mobile-initiated* or *network-initiated*, depending on whether the network (Access Router) or Mobile Node initiate the handover. When mobile-initiated, the MN sends a FBU (Fast Binding Update) message to its Previous Access Router (PAR) to establish the tunnel. When feasible, the MN sends an FBU from PAR's link (predictive). Otherwise, it should be sent immediately after attachment to NAR (Next Access Router) has been detected (reactive). As a result, PAR begins tunneling packets arriving for PCoA to NCoA without waiting for the MIPv6 handover procedure to finish. Such a tunnel remains active until the MN completes the Binding Update with its correspondents. In the opposite direction, the MN reverses tunnel packets to PAR until it completes the Binding Update. PAR forward the inner packet in the tunnel to its destination (i.e. to the MN's correspondent). The reverse tunnel ensures that packets containing PCoA as a source IP address are not dropped due to ingress filtering.

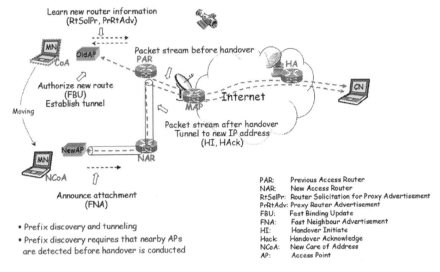

Fig. 5.3 Optimized MIPv6 with FMIPv6

Mobile IPv6 Optimization Using FHMIPv6

In the combined solution HFMIPv6, the tunnel for a handover is established between the MAP and the NAR, rather than between PAR and NAR. In this way the mobile node are doing the signaling for the handover with the MAP and not with PAR.

Application Layer Mobility

The Session Initiation Protocol (SIP) [4] is an application-layer signaling protocol for creating, modifying, and terminating sessions with one or more participants. SIP is based on a client-server architecture. It is used in a large range of applications like VoIP, instant messaging or multimedia conferencing and since November 2000, SIP has been accepted as a 3GPP signaling protocol and a permanent element of the IMS (IP Multimedia Subsystem) architecture. SIP endpoints are addressed by URL/URIs which looks like email addresses (sip:user@addressing_space.com) and SIP defines several logical entities.

- User Agents (UA) originate and terminate sessions.
- Localization Servers (LS) locate the user agents, mapping one public URI to a set of local URIs (sip:user@example.addressing_space.com) related to one terminal.

- Registrar Servers (RS) provide databases in which the location of the users is stored and modified when needed (if the user is moving for example).
- Proxy Servers (PS) relay requests to another server. They can be stateless or stateful.

Typically, a "SIP server" implements a localization server and a proxy server, with information provided by a built-in registrar. Depending on configuration and the specific request, the server acts as either a proxy or redirect server or a registrar.

SIP requests and responses consist of a text header and a MIME (Multipurpose Internet Mail Extensions) body, very similar to the format of HTTP requests. One major difference is that SIP requests can use any transport protocol, including UDP, with user agents and stateful proxies ensuring request reliability via retransmission for unreliable protocols, see Fig. 5.4.

Nomadic Mobility

SIP firstly allows managing the nomadic mobility, where the mobile node acquires a new address before having or receiving a call. In this case, the MN only registers his new address to his "home" Registrar Server. When a correspondent node wants to initiate a session with this mobile node, the call will be redirected to the new position of the MN, see Fig. 5.5.

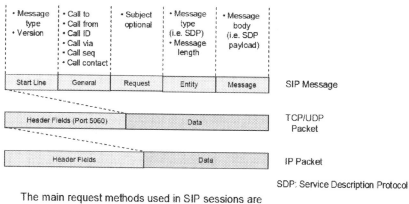

The main request methods used in SIP sessions are

- INVITE (Users to a session)
- ACK (Confirms a session establishment)
- BYE (Terminates a session)
- OPTIONS (Determines capabilities)
- CANCEL (Terminates a pending session)
- REGISTER (Binds a permanent address to current location)
- SUBSCRIBE (To Presence service)
- Notify (Of presence change)
- Message (Instant Message)

Fig. 5.4 SIP messages IETF RFC 3261

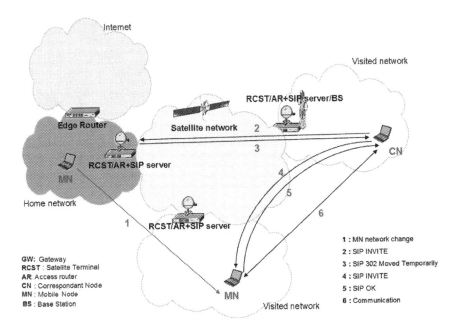

Fig. 5.5 Nomadic mobility with SIP

Mid-Call Mobility

The main mobility issue appears when the network change occurs during a communication (while a session is active). In this case, the main objective is to allow the user to carry on his communications in the new network. Firstly, we consider that, the mobile node is registered with a "SIP server" located in its home network. Each time the mobile node changes its location, it sends a new registration message to its home "SIP server", in the same way as the MN sends a registration message (Binding Update) to its home agent in Mobile IP. When the mobile node moves during an ongoing session, it must send a new INVITE message, called re-INVITE, to the correspondent node, using the same call identifier as in the original call.

The new IP address must be put in the *Contact* field of the SIP message, in order that the correspondent node knows where he has to send the future messages and in the *SDP* field to redirect the data traffic flow. However, the Home Address is used in the *From* field for identification. Finally, the mobile node updates its registration at its home "SIP server" in order that other calls can be correctly redirected, see Fig. 5.6.

In both case of mobility (nomadic and mid call), a significant issue is the ability to detect at the application layer when the IP address has changed. A simple solution is that the client polls the OS periodically, but the ability to have applications being aware of such changes are preferable.

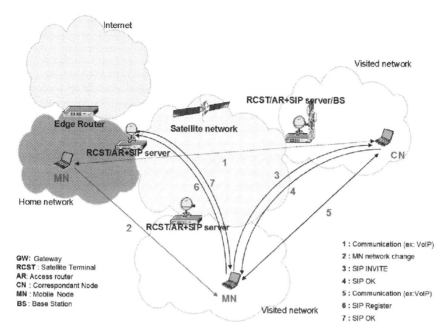

Fig. 5.6 Mid Call mobility with SIP

Multicast and Mobility

Multicasting is a main networking service in SATSIX, and there is a need for the mobile multicast protocol to enable a mobile node to take part in IPv6 multicast sessions irrespectively of its current point of attachment.

Different Approaches

The IETF draft [5] discusses three different approaches of mobility extensions to current layer multicast solutions (listener mobility):

(1) *Remote subscription* where the mobile node joins a multicast group via a local multicast router on the foreign link by using its Care-of-Address instead of its Home Address.

  - The advantages are: simple approach for multicast routing as no support is needed from the network, except the presence of multicast router in visited networks, and the absence of multicast filtering in the new visited network.
  - The disadvantages are: in case of frequent MN movement, the resulting frequent join/leaves to update the multicast routing tree would imply significant signaling traffic and possibly long handover latencies for the MN.

(2) *Home subscription* where the mobile node joins the group, send and receive packets from its Home Agent through the bidirectional tunneling.

- The advantages are: bidirectional tunneling does not require any reconstruction of the multicast tree while the multicast receiver is changing its location, as its mobility is transparent to the multicast tree. Furthermore, no other network support is required (in particular, multicast router on the visited networks!). MIPv6 extensions like HMIPv6 and FMIPv6 are natively maintained.
- The disadvantages are: triangle routing, long tunnels, large number of tunnels with overhead, tunnel congregating, packet latency.

(3) *MLD (Multicast Listener Discovery) proxying* is a mix of the previous ones where a multicast agent, near the Home Agent network, joins the multicast session on behalf of mobile node in the visiting network, it then forwards traffics to attached mobile nodes (as in unicast addressing). After the BU with the HA has been completed, the MN may even trigger a route optimization procedure with the multicast source.

   – The advantages are: best handover latency and static path delays between the two possible routes when route optimization is possible. Permanent overhead due to tunneling encapsulation may also be saved.
   – The disadvantages are: an unnecessary routing path may be maintained between the multicast source and the HA after the MN has started operating with an optimization route. Conversely, if the HA leaves the group after a certain delay, the MN could not always benefit from fast handovers.

These approaches are schematized in Figs. 5.7, 5.8 and 5.9.

Solution Analysis for the SATSIX Network Architecture

To analyze the previous solutions in the SATSIX network, it is worth reminding that limiting unnecessary signaling traffic over the satellite link is important to save

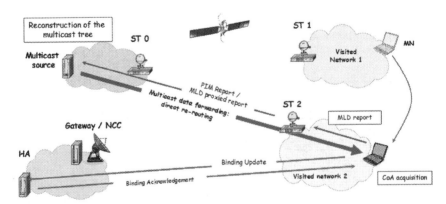

Fig. 5.7 Multicasting from remote subscription

2 - Bidirectional tunnelling with HA

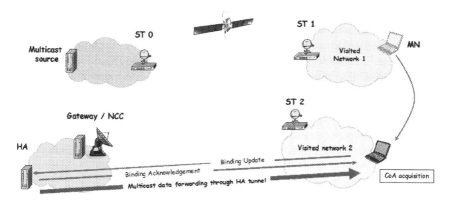

Fig. 5.8 Multicasting from bidirectional tunneling with HA

3 - MLD proxying at HA

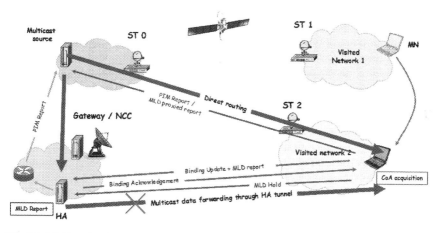

Fig. 5.9 Multicasting from MLD proxying at HA

radio resources, to optimize end-to-end packet delays, and to beneficiate from cheaper terrestrial networks resources. In order to perform the analysis, we will assume a dynamic multicast routing with mutlicast agent entities (routers or proxies) in all satellite terminals sub networks, and limiting the analysis to the case where MN is only a multicast listener.

Four criterions are proposed to assess the different approaches of mobility with multicast routing:

• Because the number of required satellite hops (1 or 2) to route packets from the multicast source to the MN have direct impact on the satellite network usage and on the end-to-end packet delays (roughly RTT/2 or RTT), how many hops are required in each solution ? The answer being linked to the system configurations, and we identify the cases for which a single hop is possible.

- Does a permanent overhead over the satellite links exists?
- How good are the handover performances in terms of delay?
- Additional issues being present?

The following conclusion is drawn from this Table 5.1.

Solution 1 may be quite complex with regards to the required protocol update which could induce routing signaling, and long delays to complete the handover (according to the protocol parameters tuning). Conversely, as an interesting point, mesh system configuration would allow an optimal routing independently from the relative location between the multicast source and the HA.

In Solution 2, the permanent overhead, the routing constraint, and the scalability issue at the HA (network congestion plus CPU load) make this approach having only limited interest.

Solution 3 finally appears as a best trade-off between the previous two solutions: the static routing is optimized (no overhead, best end-to-end delay) and the handover latency can be maintained low thanks to the HA membership. However, to avoid a permanent or too long HA membership could be difficult if the handover latency was simultaneously important to achieve. Trade-off on this leaving delay would require advanced mobility prediction models that do not exist yet.

PEP and Mobility

TCP Performance Enhancements Proxies (TCP-PEPs) are the general solution to overcome TCP limitations in wireless and especially in GEO satellite networking. Unfortunately, during active connections, they introduce new issues in the context

Table 5.1 Multicast mobility analysis in SATSIX

	Conditions required for single satellite hop	Permanent overhead	Responsiven ess to handover	Additional issues
1) Remote sub-scription	Mesh configuration or source and HA/MAP behind the same ST	No	Routing protocol dependent	
2) Bidirectional tunneling	Source and HA/MAP behind the same ST	Yes	Good	Scalability issue: static point of convergence at the HA
3) MLD proxying	Mesh configuration or source and HA behind the same ST	No	Best	Temporary duplicate routing Trade-off required to make the HA leave the group or not

of mobility managed with MIP6. In order to analyze them, we consider a worst-case study in which a MN moves from one satellite terminal subnet to another one (i.e. macro-mobility scenario) without coordination between the old and previous access routers. In this scenario, PEPs are implemented at the system gateway and at the satellite terminals in a symmetric and dual architecture.

Context Loss

Firstly, TCP-PEP connections require to be initialized as any classical TCP connection does. The most convenient way to do it for PEPs is to detect connection opening from the TCP SYN segments sent at their beginning; in order to initialize a new connection accordingly. Along the connection, data are buffered (added and removed) and ACKs are generated according to the current Tx and Rx window values. This information represents the context of a TCP connection.

After a MN has changed from the PEP1 to the PEP2 "attachment", PEP2 will suddenly receive segments or ACKs for which no connection context is normally known. In order to avoid discarding them (which should be the default behavior in current PEP implementation) finally leading to a forced TCP connection close, PEPs aware of mobility should implement a resynchronizing strategy in order to manage data and ACK during a transient period of uncertainty (e.g. optimistic or pessimistic approach). An alternate approach would be to transfer contexts between PEP1 and PEP2 (similarly to FMIP), but this would require good synchronization and heavy signaling between PEP1, PEP2, and MN, and could lead to much longer handover.

Moreover, in case this last approach would not be implemented, a deterministic sequence numbering should be implemented in the TCP-PEP connection segment with respect to the original segment.

Addressing

Considering the previous issue solved, the change of CoA must now be considered at the PEP level. Therefore, in order to guarantee that PEPs are always receiving incoming packets to relevant connection in any situation, only the *Home Address* (and not the CoA) plus the port numbers should be used as TCP flow identifiers. Therefore, PEP implementations should be carefully designed to detect the presence of Home Addresses in IPv6 datagrams.

Misrouting

Due to the latency to complete the binding update process (half a RTT at least) some data segment could be routed to the old PoA (i.e. to PEP1) whereas the MN is only reachable from PEP2 (note that this situation can only occur in dual-PEP architectures). Overcoming this problem is not straightforward because data segments could be ACKed to the gateway PEP after their reception from PEP1 without the possibility to retransmit them to PEP2 if they were suppressed from the gateway PEP buffer.

Consequently, in the absence of context transfers, gateway PEP should only be authorized to drop data segments after the corresponding distant PEP has received corresponding ACKs from the MN. Other approach could be to set a minimum delay before any segment suppression; in order PEP1 may report to the gateway PEP the absence of the MN on its sub network.

Security

Independently from mobility, end-to-end security (e.g. based on IPSec) is a main issue with PEP because IP payload data like TCP headers they have to inspect are encrypted. Consequently, when tunneling between HA and MN is used, the IPSec ESP option to protect MIP6 tunneled data seems to prohibit the enabling of PEP acceleration.

QoS and Mobility

Mobile IP ensures correct routing of packets to a mobile node as the mobile node changes its point of attachment to the Internet. However, it is also required to provide proper Quality of Service (QoS) forwarding treatment to the mobile node's packet stream at the intermediate nodes in the network, so that QoS-sensitive IP services can be supported over Mobile IP [6]. Currently, the mobile IP routing decision algorithms do not combine any information regarding the availability of the resources such as network bandwidth. When the mobile node changes its point of attachment to the network, the path across the access network will change. For active application sessions on the mobile, the network should negotiate QoS along the new route as part of the handover procedure. The QoS contract negotiated with the network should be maintained where possible, or renegotiated if a lower service level that is still suitable for the application can be achieved.

QoS-Architecture

In the following is presented an overview of one possible QoS-architecture which could allow mobile terminals with ongoing applications to keep or adapt their QoS in the visited networks. The QoS part of the architecture will be based on the QoS server, the QoS agent for non QoS-aware applications and the enhanced SIP Proxy for QoS-aware applications (using SIP). The mobility part of the architecture will be managed by Mobile IPv6 and its optimizations (HMIP, FMIP, FHMIP) for the TCP connections and by SIP-based mobility for real time communications over UDP. This can be achieved if we allow the mobile host to choose when to use its home address or care-of address. When sending RTP streams it will use the care-of address, and when establishing TCP connections, it will use the home address.

We consider here a mobile node with an ongoing SIP session (VoIP for example) and some TCP connections (ftp, web browsing ...) which moves from its home network to a visited network. It is considered, to simplify the problem, that all applications are exchanged with the same correspondent node. Moreover, a QoS agent is implemented on the MN and the CN and an enhanced SIP Proxy and a QoS server are implemented on each satellite terminal, see Fig. 5.10. This architecture allows the QoS to be properly configured for each application.

Procedure

When the mobile node is in the visited network, it has to register its new location with the home agent and the home SIP server. However, in order not to duplicate the registration messages (BU and SIP Register), some solutions must be found. To do that, it is possible to colocate the home agent and the SIP server in the same network device or to allow the home SIP server to query the home agent. At the same

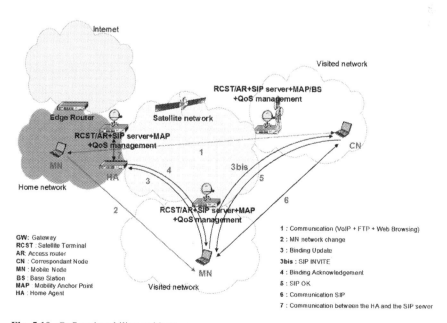

Fig. 5.10 QoS and mobility architecture

time as the MN sends a Binding Update to it's HA, it can simultaneously send an INVITE request to its correspondent node. After having received the BA and the SIP OK, the TCP connections can continue via the home agent or directly with the CN if the Route Optimization is activated and the SIP session directly between the MN and the CN. Concerning the SIP session, in order to keep or adapt the QoS previously established in the home network, the enhanced SIP Proxy implemented on the new access router intercept the SIP INVITE to manage QoS reservation with the QoS server. Concerning the other applications, if the new quality is not sufficient, the user could manually configure the QoS thanks to an enhanced QoS agent which could dynamically discover its new QoS Server.

Conclusions

The principles of the Mobile IPv6/IPv4 protocol which is the standard mobility mechanism at the network layer, is basically described. To improve the mobility management, and reducing signaling overhead and handover delay when using MIPv6 in SATSIX, the hierarchical (HMIPV6) and fast handovers (FMIPv6) protocols have been investigated and also the combination of the two FHMIPv6. HMIPv6 is an extension to MIPv6 that aims at reducing signaling messages during intra-domain and local movement and handover. By further introducing the SIP-protocol, SIP-mobility and application mobility is analyzed as a complement to network layer mobility.

Multicast is a focal feature in SATSIX, so is also the combination of multicast and mobility. Three approaches for multicast mobility have been described. Due to the high bandwidth*delay product, satellite links make use of accelerators or PEPs. When the mobile node is visiting a foreign network, the combination of network mobility and PEPs causes various problems. Different issues and solutions that need to be considered have been thoroughly discussed and analyzed. Mobility gives a significant impact to the QoS management, and generates a new challenge for QoS provision as it will have to deal with terminals changing their point of attachment to the network. An overview of a possible QoS architecture has been presented which could allow the mobile terminals with ongoing applications to keep or adapt their QoS in the visited networks.

Acknowledgments This work is supported by the IST FP6 SATSIX project, funded by European Commission (EC). The financial contribution of the EC towards this project is greatly appreciated.

References

[1] D. Johnson, C. Perkins, J. Arkko,"Mobility Support in Ipv6", IETF RFC 3775, June 2004.
[2] H. Soliman, C. Castelluccia, K. El Malki "Hierarchical Mobile IPv6 Mobility Management (HMIPv6)", IETF RFC 4140 (Experimental), August 2005.

[3] P. McCann, "Mobile IPv6 Fast Handovers for 802.11 Networks", IETF RFC 4068 (Informative), November 2005.

[4] J. Rosenberg "SIP Session Initiation Protocol", IETF RFC 3261, June 2002.

[5] T. C. Schmidt, M. Waehlisch "Multicast Mobility in MIPv6: Problem Statement", IETF, draft-schmidt-mobopts-mcastv6-ps-02.txt, March 2007.

[6] H. Chaskar, "Requirements of a Quality of Service (QoS) Solution for Mobile IP", IETF RFC 3583 (Informative), September 2003.

Chapter 6
Cross-Layer Anticipation of Resource Allocation for Multimedia Applications Based on SIP Signaling over DVB-RCS Satellite System

F. Nivor, M. Gineste, C. Baudoin, P. Berthou, and T. Gayraud

Abstract This paper introduces a cross-layer approach for improving QoS guaranties to interactive multimedia applications over an efficient satellite access assignment scheme (on-demand). It particularly focuses on the communication opening which represents the weakness of on-demand capacity allocation in the satellite context (due to significant delays). It finally presents experimental results of the various proposed enhancements.

Introduction

Geostationary satellite offers practical and easy-to-deploy way for Internet communications over the world. Nevertheless, satellite communication's characteristics, such as large delay and resource variability require particular treatments in terms of resource management and optimization. In this context, the introduction of on-demand Quality of Service (QoS) support over satellite communications is very important for interactive multimedia applications. This is particularly accurate for the *return link* that can be considered as a bottleneck due to the large set of distributed end-users accessing a limited uplink resource.

QoS *guarantee* on the return link has been often implemented using *static resource reservation* (e.g. for TV journalists who work where terrestrial infrastructure is not available). This technique avoids complex resource management and allows strong QoS guaranties. However, multimedia application customers requiring interactivity, such as audio-video conferencing, cannot use static reservations due to the high cost.

In order to allow at the same time a more efficient use of the resource and reduce the cost from their perspective, the DVB-RCS proposes a *dynamic approach* through its DAMA access scheme (Demand Assignment Multiple Access). DAMA is a classical technique to offer such dynamicity thanks to a distributed mechanism. It offers various types of services to the satellite operator.

In spite of these advances, QoS-enabled satellite services still have to be used efficiently for multimedia communications, particularly regarding the delay accumulation issue that can result from the use of dynamic allocation. In this context, several propositions have been done to improve performances of the access scheme for multimedia communications [1, 2]; however, there is currently no solution to significantly reduce initial delay introduced by on-demand capacity assignment type for interactive multimedia applications. Yet, this incompressible delay seriously deteriorates user experience for interactive applications [3].

To fill up a part of this gap, the work presented in this paper is aimed to improve the quality of interactive applications from the end user point of view by enhancing provisioning of on-demand DAMA resources.

The paper proposes a cross-layer approach to provision resource on session start-up taking into account the QoS application requirements. Several experiments making use of the provisioning methods are carried out for different scenarios of interactive session initiation. Their impact on QoS is measured through different application level metrics, i.e. end-to-end (e2e) delay of applicative data.

The rest of the paper is structured as follow. The "QoS Over Satellite" section presents the considered satellite communication framework and the state of art of propositions improving QoS for interactive multimedia applications using on-demand satellite resources. The "Contributions" section shows the core of the paper contribution: The cross layer approach and the various provisioning methods proposed. The "Experiments and Results" section introduces a case study and several experiments that allow showing the improvements of these provisioning methods. Final the "Conclusion" section concludes this paper.

QoS over Satellite

Satellite Communication Context

The main concern in the satellite communication context, and more particularly on satellite return link, is to make an efficient use of the resources, scarce and costly, while taking into account different traffic types (data, voice, video), each having specific QoS performance parameters. Thus, protocols have been designed to optimize the use of these resources and especially to share properly and efficiently the return link resources accessed by multiple distributed Satellite Terminals (ST) which act as satellite boundary nodes.

A combination of static and dynamic techniques has been integrated into the DAMA (Demand Assignment Multiple Access) protocol. This protocol is integrated in the DVB-RCS standard [4], in order to ensure both high utilization of the return link resources and to offer QoS-oriented capacity assignment types.

The return link scheduler of the DAMA protocol supports four main capacity assignment types to reach its objective: **CRA** (Constant Rate Assignment), **RBDC** (Rate Based Dynamic Capacity), **VBDC** (Volume Based Dynamic Capacity), **FCA** (Free Capacity Assignment).

CRA allocation type offers real high QoS guarantees but at a high cost: the delay is reduced to propagation delay and a guaranteed bandwidth is constantly allocated.

However, due to this constant allocation to a ST, even when it is not required, this allocation type leads to a suboptimal usage of resources.

Compared to CRA allocation type, RBDC offers a good trade-off between QoS guarantees and bandwidth efficiency at a lower cost. Indeed, this traffic assignment is based on requests that reflect the average rate of incoming data on the ST. So we will focus more particularly on RBDC assignment type in this paper.

The return access scheme of the satellite is able to provide different types of service. However, QoS differentiation needs architectural solutions at upper layers to be effective. The next section presents an overview of these solutions and the QoS-oriented architecture considered in this study.

QoS in the Satellite Context

In the satellite networking context, the interaction between the IP Layer, where the QoS might be set, and the lower layers where the traffic is finally prioritized is not covered by any standard specification [5]. However, QoS techniques and architectures for satellite networks have been widely studied in the literature. [6, 7] proposes to use Diffserv architecture [8] on both forward and return link. This architecture is well adapted to the return link due to the different classes of service of DVB-RCS capacity allocation. The satellite system, in this study, is assumed to be an access network to the Internet for end-users. Thus, as a boundary node, the ST is the most important component regarding QoS support on the Return Link. It has to implement traffic conditioning/policing functions, in addition to packet classification and per hop forwarding/scheduling according to packet's class of service, as illustrated in the Fig. 6.1.

These recommendations introducing QoS in the ST are the starting point on which our contribution is based. The proposed mapping and admission control based on these recommendations are detailed in the following section.

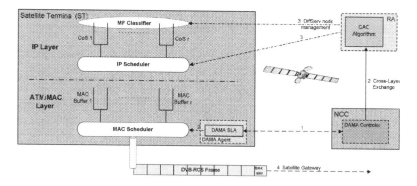

Fig. 6.1 Mapping between MAC and IP queuing in the ST

Considered QoS-Oriented Architecture

We assume in this paper that an e2e QoS Architecture is deployed, integrating IP differentiated services. It includes signaling mechanisms for admission control as well as resource pre reservation, such as proposed in [9, 10].

In order to meet the DiffServ forwarding requirements, the IP classes of service need to be appropriately mapped into MAC QoS classes and then into DAMA capacity categories supported by the Scheduler. Here we consider that EF CoS is mapped to the highest priority MAC buffer DVB-RT using CRA capacity. AF CoS is mapped to the medium priority MAC buffer DVB-VR using RBDC capacity. And BE traffic is mapped to the lowest priority MAC buffer DVB-JT using VBDC and remaining capacity.

The Network Control Centre (NCC), which manages the satellite resources, knows the available resources on the return link. It provides this information to the Resource Allocator (RA) which performs the Admission Control.

A Service Level Agreement (SLA) is passed at logon between ST and the Satellite System. The bandwidth guaranteed for high priority classes (CRA, RBDC) in this SLA, are generally restricted due to their cost. Thus, to avoid the waste of high priority capacity, admission control is based on the remaining satellite resources and limitation per end-user. If bandwidth is available for a specific IP CoS in relation with remaining satellite resources in the corresponding DAMA class, the flow is admitted, if the user is under its contract limitation.

No per-flow admission control is done for BE traffic type, but its global rate is limited to the remaining bandwidth unused by high priority traffic. This ensures full resources utilization and limitations of congestion inside the ST.

This QoS Architecture includes differentiated services and admission control. It enables a flow to use a satellite access type without any interference with concurrent traffic. Thus, a prioritized flow is able to use CRA or RBDC access class, and won't be delayed by Best Effort traffic that would rather use VBDC capacity when available.

However, the pre reservation of resource realized by the QoS Architecture is different from the actual allocation done by the satellite system with internal requests. Indeed, reservation could be pre reserved on the control plane but the connection could finally fail. Thus, immediate allocation of resource in a satellite environment is not feasible because of efficiency and cost considerations.

Contributions in This Context

As previously exposed, CRA assignment type offers real high QoS guarantees but at a high cost. So the proposed QoS guarantees of RBDC access class at a lower cost is of great interest; Thanks to the QoS-oriented architecture previously presented, applicative multimedia flows can take benefit of differentiated access to the satellite system (where AF CoS is mapped toward RBDC).

However, RBDC suffers from initial requests issue. Indeed, initial requests introduce delay, at least 900 ms (data buffering 300 ms + request delay 300 ms + response delay 300 ms), that is not compatible with interactive multimedia application constraints [3]. This delay might be reduced during the communication by requesting more resource. However, this initial delay degrades the overall interactivity of two-way conversational voice and video applications. Indeed, the delay accumulated in the destination receiver's buffers cannot be reduced, except if packets are discarded. In both cases the applicative quality is degraded.

Then, it enlightens the need of anticipation in provisioning resources in order for interactive applications to take benefit of this efficient access class. However, this anticipation must be as close as possible to the real application data transmission, taking into account the main concern of efficiency of a satellite system.

Thus, the contribution of this paper is to propose anticipation for provisioning RBDC resources, based on a cross-layering approach between application and link layer, in order to set up an interactive multimedia session.

Contributions

The OSI (Open System Interconnection) model has been the initial guide for developments in networking. Recent evolutions, notably in the area of wireless communications, have shown that the layered model, in which each layer is implemented independently, may lead to suboptimal performances [11]. In order to avoid such scenarios, one solution is to use cross-layering while designing mechanisms by taking the behavior of other layers into account.

This study provides a downward cross-layering approach, achieved in a control plane, in order to provision RBDC resources when interactive sessions start. The solution should not require modifications of layers interfaces or even to the packet headers. Indeed, this approach should take advantage of information already included within session signaling messages to provide generic QoS and session information (e.g. bitrate, packet size, session id), thus hiding the complexity introduced by dealing with the diversity of multimedia streams.

In this study, we describe a cross-layer approach using SIP (Session Initiation Protocol)/SDP (Session Description Protocol) to signal session start-up and session stream details. However, this cross-layer approach can be used with any session signaling protocol.

This cross-layer approach to provision RBDC is divided in three steps described now:

Application Dependent Mapping (First Step)

In order to do a precise anticipation of resources reservation and request when multimedia session starts, we need to know the amount of resources required by this session. We also need to know the ST Identifier, behind which the session starts, in the satellite network.

However, information integrated in session protocols cannot be directly used to perform this resource provisioning. Then, this proposed first step consists in translating application dependent information (i.e. session protocol syntax) into a normalized representation of this information (e.g. application types, codec names).

During the SIP establishment phase, the SIP User Agents (UAs) exchange and negotiate the type of application, media and codec that is used during data transfer of the session. SIP proxies route requests to UAs because they are initially not aware of their respective location.

Knowing that the gateway (GW) represents the entry point of the satellite network, we propose to install an Outbound SIP Proxy (OSP) within the GW/NCC. This access topology allows the OSP to intercept all SIP establishment messages (SIP INVITE). A control plane is then used to retrieve from these session messages, the application and media types as well as the codec name.

A first translation toward generic application media and codec name is done if needed. In the case of SIP/SDP protocols, audio and video codecs used during session establishment are standardized by the IANA organization [12].

To make resource reservation for multimedia flow, we also need the ST MAC Identifier behind which the session is initiated. In order to obtain this Id, we also use information from establishment session messages. The control plane at the OSP is in charge to extract the IP address and port of UAs. From this transport information, we propose to use the SARP protocol (Satellite Address Resolution Protocol) to retrieve the ST Id giving access to the UA IP Address in the satellite network. The control plane keeps track of the SIP session Id associated to the ST Id.

From Session Description Toward Precise QoS Parameters (Second Step)

The first step provides normalized session information especially on application, media types, codec names and ST Identifier associated to the session.

However, this information are purely qualitative and cannot be used to request resource reservation to the NCC. So we propose to translate this qualitative information into precise quantitative QoS parameters. To achieve this translation, we use standard information on QoS required by generic application types [12] as well as main features of each codec available in standard specifications (like encoding/decoding bitrate). This bitrate corresponds to the throughput of the multimedia flow. The list of correspondence between application, media types and codec names toward associated quantitative QoS parameters (like throughput, maximum e2e delay, and loss rate) is stored in a Media Type Repository (MTR) defined in [10]. We implement an MTR in the architecture, located on the NCC side in the satellite network to reduce access delays to MTR.

Once the control plane has extracted the media type and codec from the establishment session message, it passes this information to the MTR. This later responds with the corresponding QoS parameters.

The obtained QoS parameters (in particular, bitrate and packets size) and the targeted ST Id, allows us (1) to perform admission control, as described in the "Contributions in the Context" section, (2) to provision, later on, the appropriate RBDC resources on satellite return link, at the very moment the data transmission will start. We now focus on the method used to send a RBDC capacity request to the NCC.

Capacity Request for RBDC Resource Provisioning (Third Step)

The TBTP (Terminal Burst Time Plan) represents the resource allocation plan for all STs to access the return link of the satellite system during one super-frame. In the best case scenario, for on-demand capacity (RBDC), the TBTP takes into account new session data 600 ms after it reaches the ST (due to satellite crossings of request and response), implying at least 900 ms of one way delay for the first data packets.

To face this issue, three approaches will be considered to anticipate provisioning of RBDC resources for interactive multimedia sessions. For these three approaches the same process is followed to anticipate RBDC provisioning of resources:

1. Every SIP session initiated behind a ST is tracked by the OSP located on the GW side.
2. If the session is admitted in AF CoS by the RM, part of the QoS-oriented architecture (cf. "Contributions in the Context" section), then on reception of the SIP OK message on GW side, session information is retrieved using cross layering approach [cf. "Application Dependent Mapping (First Step)" section].
3. If the session requires interactivity (known by analyzing session requirements retrieved in the previous step, in particular the required one-way delay):

   – A Capacity Request (based on required bitrate and packet size to evaluate overhead) is directly sent by the control plane to the NCC in the DAMA Server. It uses the same format as usual capacity requests. Two capacity requests are sent for the first two super-frames of the session communication. This avoids any satellite crossings because the OSP and the NCC are on the GW side. For the rest of the communication, the normal capacity request mechanism based on a calculation of incoming throughput on the ST is used. Thus, due to the two first capacity requests, the initial delay should not be experienced by the streams of the session.
   – NCC updates its resource allocation and sends the corresponding TBTP to all STs.

Three solutions are proposed to achieve this third step:

Anticipation of Resource Allocation with No Modification of Access Scheme

The NCC updates its resources distribution to integrate the new Capacity Requests (as it would normally do) but TBTP frequency is not modified (sent every super-frame).

Knowing that a SIP session could start at any moment, two borderline cases have to be considered (cf Fig. 6.2):

- The OSP sends the SIP OK message and the RBDC capacity request just before the generation and emission of the TBTP: this is the best case scenario; indeed, the request is taken into account in the current TBTP computation, and immediately sent on the forward link. Consequently, the ST receives in the same time the authorized SIP OK message and RBDC resources allocation enabling ST to immediately send data of the new session for the next two super-frames on the return link. The satellite e2e delay should be close to the satellite propagation delay (300 ms).
- The OSP sends the SIP OK message and the RBDC capacity request just after the calculation and broadcast of the TBTP: this is the worst case scenario; indeed, the capacity request is not taken into account in the current TBTP that has just been emitted on the forward link. Consequently, the ST receives the authorized SIP OK message but no resource is available to emit data of the new session. Consequently, it has to wait the next TBTP for resources allocation (530 ms later).

The pros and cons of this approach are the following:

pros: No modification of access scheme signaling.
cons: RBDC provisioning anticipation is not optimal.

In order to eliminate this worse case scenario, a solution could be to increase the TBTP emission frequency. Thus, for a super-frame of 530 ms, that corresponds

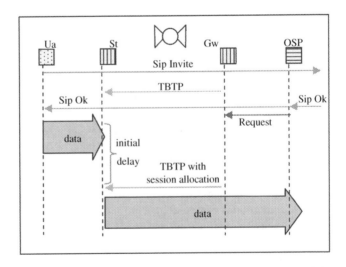

Fig. 6.2 Cross-layering RBDC request

to 10 frames of 53 ms, the TBTP could be proceeding and emitting for each frame. Thus, the maximum delay added to every allocation is of 53 ms. However, it tends to increase signaling traffic on the satellite network.

The pros and cons of this approach are the following:

pros: RBDC provisioning is optimal.
cons: Significant signaling overhead.

Anticipation of Resource Allocation Based on New TBTP Calculation

In order to avoid increasing signaling in the satellite system, we propose to keep TBTP frequency broadcast every super-frame. However, to provision RBDC capacity requested by the OSP, we propose to update the current TBTP to integrate the new session. Thus the NCC broadcasts in the following frame to all STs, the newly calculated TBTP for the current super-frame, including the new distribution of the return link resources (cf Fig. 6.3).

Due to the QoS-oriented architecture considered (cf. "Considered QoS-Oriented Architecture" section), once the STs have received their allocation, they serve MAC queues by priority order: DVB-RT > DVB-VR > DVB-JT queues. Knowing that the RBDC capacity request concerns the DVB-VR class, we propose to reallocate resources dedicated to the service of lower priority that concern DVB-JT class, namely VBDC and FCA assignment types. Indeed, the DVB-JT service is not guaranteed, so the current TBTP can be recalculated, decreasing VBDC allocations to reallocate these resources toward RBDC capacity, requested for the new session. However, when the NCC recalculates the TBTP, it must take care to not reallocate DVB-JT resource already consumed by the current position in the super-frame. When the amount of DVB-JT resources is not sufficient to satisfy the request of the OSP, the remaining resources are computed to be allocated in the next TBTP. This

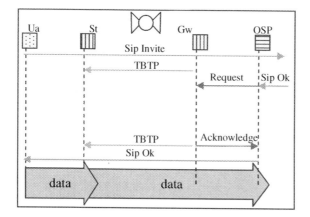

Fig. 6.3 Cross-layering request with TBTP update and emission

method allows the ST concerned by the SIP multimedia session to receive at the same time the authorized SIP OK message and the required RBDC resources allocation, in order to send immediately data of the new session on the return link. However the update of the TBTP for every asynchronous RBDC capacity request coming for the control plane at the OSP requires a significant modification of the DAMA server implementation, and significant computation capacities of NCC.

The pros and cons of this approach could then be summarized as follow:

pros: RBDC provisioning is optimal.
cons: Significant modification of the access scheme.

Anticipation of Resource Allocation Based on SIP Adaptation

This third solution described in Fig. 6.4 permits to avoid the modification of the allocation plan in real time.

When the OSP receives the SIP OK message, we propose to keep it within the OSP until the TBTP is emitted by the NCC; it's equivalent at user level to let the call ringing (one super-frame = 530 ms at most). So the OSP emits an RBDC capacity request to the NCC. Then, when the TBTP is being broadcasted on the forward link, the NCC emits upwards an acknowledgment to the OSP, indicating that the TBTP is about to be sent. Consequently, the OSP releases the authorized SIP OK message; thus, the SIP message and the RBDC resource allocation are transmitted at the same time to the ST. This later is then able to send immediately data of the new session on the return link for the two next super-frames. The pros and cons of this approach could be summarized as follow:

pros: RBDC provisioning is optimal, modification is minimal.
cons: slight modification of access scheme and SIP.

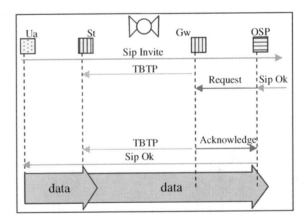

Fig. 6.4 Cross-layering request with Sip adaptation

Experiments and Results

Platform Description

Experiments were conducted on Platine. Platine is a satellite emulation platform proposed in SatSix European project, aiming at functionally validate the DVB-RCS access and network layers definitions. Platine architecture emulates components and implements communication protocols (IP/MAC/DAMA) of a regenerative satellite system.

Fig. 6.5 represents the emulation platform for experiments. The platform is composed by two domains, named "domainsat.org" and "domainter.org". Domainsat.org uses a DVB-RCS satellite access network emulated by the Platine platform.

Behind the ST, which acts as an access router, there is an Ethernet user network. The over-provisioning of the Ethernet network permits us to abstract its characteristics for the experiments. The UA and Proxy of domainsat.org are located in the user network behind the ST. The Sip Proxy, according to QoS requirements of the user, starts the configuration of QoS mechanisms up inside the ST (admission control, classification...).

The terrestrial part of the satellite network is an Ethernet network. It's composed by the GW and the OSP. On the other side, domainter.org uses an Ethernet network. This destination domain is composed by the remote UA and Proxy. This latter is in charge of Sip connection management of this domain. These domains are interconnected by a router, represented by a machine with its routing IP layer activated. Sip Proxies in these experiments are achieved by the **partysip** software and Sip UAs by **minisip** software. Using this platform, we evaluate the impact of our proposal on the e2e applicative delay for a period of time corresponding to the first two superframes of the communication on the return link.

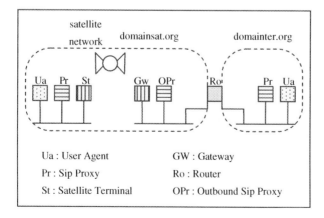

Fig. 6.5 Emulation platform

Test Scenarios Description

In our test scenarios, the UA inside the satellite network initiates a Sip audio communication by calling the remote UA inside the terrestrial network. To measure the performance of the proposed cross-layering RBDC access, we use an audio flow with G711 audio codec at 64 kbit/s. This flow is mapped toward AF 1 DiffServ class and queued in the DVB-VR-RT buffer which used RBDC capacity request at MAC level inside the ST.

Parameters of the emulated satellite system are setup to:

- Forward link Capacity: 1024 kbps
- Frame periodicity: 53 ms
- Super-frame periodicity: 10 frames: 530 ms
- Capacity request period: 1 super-frame: 530 ms
- TBTP emission period: 1 super-frame: 530 ms

For each proposition of the "Cotributions" section is associated an evaluation scenario:

Scenario 0 is the reference test evaluating resources allocation with no anticipation (current access behavior).

Scenarios 1.1 and *1.2* evaluate the first solution with emission of the request respectively before and after the TBTP computation-emission; *scenario 1.3* evaluates the first solution with a TBTP emission period each frame (53 ms).

Scenario 2 evaluates the second proposition (TBTP update and emission) with TBTP emission period at 1 super-frame (530 ms).

Scenario 3 evaluates the third solution evaluation (acknowledgment from NCC).

Results Presentation and Analysis

The Fig. 6.6 represents the satellite e2e delay for the two first super-frames for each scenario.

- In the reference test without resource anticipation (scenario 0), the e2e delay is around 1500 ms which represents the sum of delay for the data buffering (530 ms) in the ST, the delay for the satellite crossings of the request and response (600 ms), and the propagation delay (300 ms). The e2e delay experimented for this scenario is not compatible with the strict time constraints associated to interactive multimedia flow [3].
- The e2e delay of first solution varies between 300 ms when the request is emitted right before the TBTP calculation (scenario 1.1) and 900 ms when the request is emitted just after (scenario 1.2). Then, this approach cannot fully guaranty that the time constraints of interactive flow will be met. However, the packets' e2e delay with TBTP emission frequency set to 1 frame is always around 300 ms

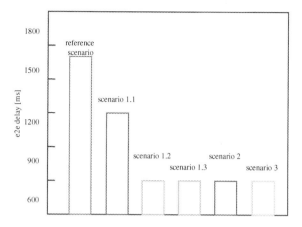

Fig. 6.6 e2e applicative delay for each scenario (mean delay for each scenario)

(scenario 1.3), so the time constraints objective is met in any cases for this scenario, but at the price of a significant signaling overhead.

- For both second and third solutions (scenario 2 and 3) the observed e2e delay is around 300 ms. These two solutions enable to guaranty a delay compatible with time constraints of interactive flows. However, the third solution will be preferred, as the second introduces complex modification of the access scheme mechanisms and might generate significant computation overhead inside the NCC.

Experimental results confirm theoretical behavior and suggest that enhancements are required to guaranty satisfactory quality for interactive application's user over the satellite. The proposed solutions (from scenario 1.3) are satisfactory in terms of provided QoS. However, if we take into account the various overhead, the third solution will be preferred because it introduces slight modifications of the present satellite architecture along with optimal results for the communication.

Conclusions

This paper has introduced and evaluated enhancements for interactive applications' user experience over satellite, using a highly efficient (thus less costly) assignment type of DVB-RCS access scheme (RBDC). This enhancements concern the communication opening where accumulation of delay introduced by this on-demand assignment type may badly degrades applicative quality. The experiments conducted, proved that the proposed anticipation of resource improves communication quality, without loosing efficiency of the satellite system (no capacity is wasted). Among the three proposed enhancements, the third solution will be the preferred one, due to the slight modifications it requires. These enhancements may gain to be coupled with other enhancements proposing to anticipate requests during the communication [1, 2].

Use of dynamic CRA allocation have been foreseen but not studied for three main reasons: 1) not easy to predict release of resources, 2) constant assignment may lead to suboptimal use of resources, 3) implies modification of the global agreement of the satellite system.

In the future, we will study more in details other scenarios (including other enhancements) as well as overhead (signaling and computation) introduced by our contributions.

References

[1] F. Delli Priscoli et al. (2004). Design of a bandwidth-on-demand protocol for satellite networks modeled as timedelay systems. Automatica, Vol 40, Issue 5.

[2] F. Nivor et al. (2006). Optimization of a Dynamic Resource Allocation in DVB-RCS Satellite Networks, Anipla, Roma.

[3] ITU-T (2001). Transmission systems and media, digital systems and networks; end-user multimedia QoS categories, IUT-T G.1010.

[4] ETSI (2003). Digital Video Broadcast (DVB); Interaction channel for satellite distribution systems, ETSI EN 301 790 v1.3.1.

[5] H. Skinnemoen et al. (2005). VoIP over DVB-RCS with QoS and bandwidth on demand, IEEE Wireless Communications Magazine, Vol 12, Issue 5, pp 46–53.

[6] S. Kota et al. (2003). Quality of service for satellite IP networks: a survey, International Journal of Satellite Communications and Networking, Vol 21, Issue 4–5, pp 299–301.

[7] ETSI (2006). Satellite Earth Stations and Systems (SES); Broadband Satellite Multimedia (BSM) Services and Architectures: QoS Functional Architecture, ETSI TS 102 462 V0.4.2.

[8] S. Blake et al. (1998). An Architecture for Differentiated Services, IETF RFC 2475.

[9] IST (2006–2007). Satellite-Based Communications Systems Within IPv6, IST project.

[10] IST (2004–2007). End to End Quality of Service support over Heterogeneous Networks, IST project.

[11] Q. Wang et al. (2003). Cross layer signaling for next-generation wireless systems, Proc. IEEE Wireless Communication and Networks Conference, New Orleans, LA.

[12] Internet Assigned Numbers Authority, unit of Internet Corporation for Assigned Names and Numbers (ICANN).

Chapter 7
Radio Resource Management
for Next Generation DVB-RCS Systems

A. Pietrabissa and C. Baudoin

Abstract The EU-funded IST FP6 Satellite-based communications systems within IPv6 (SATSIX) project aims to study, implement and validate innovative concepts and cost-effective solutions for broadband satellite systems relying on DVB-RCS/S2 standards and services. This paper presents a description of the Radio Resource Management protocols and algorithms developed so far. The major objective is to propose efficient resource management for satellite systems using Fade Mitigation Techniques to face deep fading caused by the attenuation and scintillation effects of atmospheric gas, clouds, rain and melting. The procedures will be simulated as well as tested and evaluated via a satellite emulation testbed developed within the project.

Introduction

The objective of Radio Resource Management (RRM) procedures is to guarantee an efficient use of the valuable satellite capacity and to respect the Quality of Service (QoS) requirements of the in-progress connections.

In general, depending on the QoS requirements, the access to a satellite network is provided by means of two main approaches:

(i) *Static capacity assignment*: At connection set-up the Connection Admission Control (CAC) assigns a certain amount of capacity to the connection for its lifetime. Since this pre-assigned capacity is always available to the connection, low delays can be granted; the drawback is that, since capacity assignments are independent of the actual offered traffic, the satellite links is underutilised. This approach is advantageous for connections characterised by regular traffic and for connections having stringent transfer delay requirements (e.g., real-time traffic such as voice or video).

(ii) *Dynamic capacity assignment*: No capacity is fixedly assigned to the connections. Conversely, the Network Control Centre assigns the capacity to a certain connection on the basis of the capacity requested by the satellite terminals, according to the so-called Bandwidth-on-Demand (BoD) mechanism. Since the

L. Fan et al. (eds.), *IP Networking over Next-Generation Satellite Systems*
© Springer 2008

satellite terminals have to send capacity requests to receive capacity assignments, the latency of this approach is affected by the propagation delay of the satellite links; the advantage is that capacity assignments to the connections can actually track the traffic offered to the satellite system and hence an efficient capacity exploitation can be achieved.

DVB-RCS standard defines a set of different resources allocation schemes at MAC layer that make possible, in principle, to offer different kinds of guarantees to IP traffics [2].

The management of the satellite bandwidth resources has a major impact on the QoS levels a DVB-RCS network is capable to offer. The DVB-RCS standard defines the assignment of satellite resources on the return link, whereas bandwidth assignment on the forward link (i.e., from the Hub towards the RCSTs) is left open.

In fact, the DVB-RCS standard has left many issues open, as for instance the BoD and the admission control strategy. Moreover, the utilisation of Fade Mitigation Techniques (FMT) that allow to counteract these varying attenuations by enabling link adaptation in term of modulation, coding and symbol rate according to the propagation conditions have severe impacts on classical RRM algorithms since the capacity is now variable. This paper describes the RRM procedures and the CAC approach developed for SATSIX.

In particular, after a brief system overview ('Reference System' section) and a description of the Fade Mitigation Techniques ('Fade Mitigation Techniques' section), the following procedures are described:

– Frame constitution algorithms, aimed at assigning the proper DRA scheme to each frame carrier ('Frame Constitution and DAMA Controller Algorithms' section);
– DAMA Controller algorithm, aimed at deciding the capacity shares to be assigned to the RCST on-demand, and the time-slot allocation on the uplink frame ('Frame Constitution and DAMA Controller Algorithms' section);
– Bandwidth-on-Demand algorithm, aimed at determining the capacity request policy ('Bandwidth-on-Demand (BoD) Algorithm' section);
– CAC algorithm, aimed at controlling the blocking and dropping probabilities of the traffic classes supported by SATSIX ('Adative CAC' section).
– Finally, the 'Conclusions' section draws the conclusions.

System Overview

Reference System

The reference satellite communication system (see Fig. 7.1) provides connectivity to corporate, collective or residential users connected to satellite terminals through a unique gateway thanks to a transparent Ka band satellite. DVB-S2 is used on the

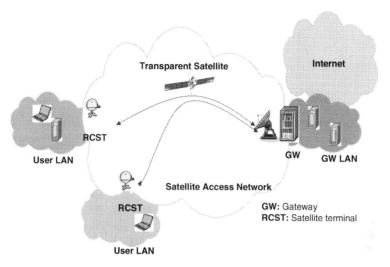

Fig. 7.1 Reference system

forward link (from NCC to RCST) and DVB-RCS on the return link (from RCST to NCC). The network is based on end-to-end IP (Internet Protocol): The medium access is composed of a basic TDM access on the Forward link and a MF-TDMA access on the Return Link.

Fade Mitigation Techniques

Multimedia satellite systems operating at high frequency bands (Ka band and above) suffer a lot from attenuations caused by rain, tropospheric scintillations, Gaussian noise (thermal and system internal interferences). Those atmospheric losses limit the satellite communication link availability, therefore the system performances. In order to avoid fixed margin based on the poorest channel quality (worst case of the coverage), new communication systems with high attenuation levels adapt the system transmission parameters to the signal fading. This allows the channel to be used more efficiently since the power and the rate can be allocated to take advantage of favourable conditions. Those methods that counteract the atmospheric propagation losses and permit link adaptation are referred to as IFMT (Interference and Fade Mitigation Techniques). The basic idea behind adaptive transmission is to maintain the quality of service, and therefore the bit error rate, during a fade event by adapting transmission parameters to the propagation conditions by varying either the transmitted power, the transmission symbol rate, the constellation size, the coding rate/scheme, the space diversity, or any combination of these parameters.

Thus, while maintaining the bit error rate above the required value, these schemes provide a high average spectral efficiency by transmitting at high rates under favourable channels conditions (clear sky) and then reducing the throughput as the channel degrades (rain conditions).

Frame Constitution and DAMA Controller Algorithms

One of the main focus of the SATSIX researches is devoted to the exploitation of the adaptive coding capabilities offered by the DVB-S2 standard [3].

Dynamic Rate Adaptation (DRA) is defined as the mechanism which maintains the return link of each RCST by switching the type of carrier it is allowed to use according to the variable return link conditions experienced by the RCST. The variable return link conditions are due to atmospheric attenuation. A type of carrier is defined by a triplet of symbol, modulation and coding rates.

The advantage of DRA is to use the variation of spectral efficiency and symbol rate to cope with system's power variation instead of requiring a large power dynamic range. It thus helps in reducing RCST cost, since the RCST includes a less powerful amplifier.

Two options are possible to handle the radio resource management: static and dynamic resource configurations.

(1) The former solution relies on a logical separation of the resource according to a given modcod (DRA pool). The RCSTs are dynamically affected to a DRA pool in function of its return link conditions. This kind of approach leads to a non optimal resource affectation, with a statically configured set of fragmented resource that does not correspond to user needs and conditions.
(2) The latter solution is based on a dynamic configuration of resources according to the users' capacity requests and link conditions. The main drawback of this mechanism is that it increases the signalling overhead needed to periodically transmit the frame configuration to the RCSTs.

The frame constitution optimisation problem is strictly linked to the Connection Admission Control (CAC) and the DAMA Controller assignments. In fact, the optimal frame constitution depends on the following inputs:

– DRA scheme usable by each active RCST;
– amount of time-slot allocations reserved to the connections admitted by the CAC;
– amount of time-slot allocations requested by the RCST and subject to the DAMA Controller algorithm.

Given the size of the overall problem, it is not conceivable to have a single optimisation algorithm which, frame-by-frame, assigns the DRA schemes to the carriers in an optimal way. In fact, the number of combinations is huge, considering that in the total available bandwidth of 125 MHz there could be up to 1300 carriers, and that in a single frame of 0.53 sec. there could be up to 17000 time-slots. Moreover, even if it were feasible, this kind of approach would give rise to and excessive amount of traffic overhead, since it would lead to different frame constitutions on a frame-by-frame basis. Thus, the suggested approach is to divide the whole problem in a number of sub-problems and to solve each sub-problem with an heuristic.

First of all, we divide each frame into a number of 'subband' frames, which have the same duration of the whole frame (0.53 s) but a reduced bandwidth with respect to the whole frame. For the sake of brevity, hereafter we will refer to a sub-band frame as a subframe. By defining M subframes, each subframe bandwidth is equal to 125 MHz/M and denoted with $B_{SUBFRAME}$.

Then, we associate each transmitting RCST to a subframe: in this way, the size of the overall problem of assigning the DRA schemes to the carriers is reduced. Even more important is the fact that we will define the number of subframes M in such a way that each subframe bandwidth is lower than 20÷25 MHz. This facilitates the DAMA Controller function, since the frequency hopping constraint is intrinsically verified, and the only remaining constraint in the timeslot assignment is the overlapping constraint.

With this approach, the first problem to solve is the association of each RCST to the most suitable subframe, based on the RCST requirements (DRA scheme, capacity reserved by the CAC and dynamic capacity requirements) and on the load of each subframe. If the requirements of a given RCST change and the subframe it is assigned to is no more capable of guaranteeing its transmission capacity, the NCC tries to change the association of the RCST by checking if there is a more suitable subframe.

Then, the subframe constitution problem is considered, independently for each subframe: considering the generic subframe m, the associations of the subframe carriers with the most appropriate DRA scheme is based on the characteristics of the RCSTs associated to subframe m. A by-product of this procedure is the reserved time-slot allocation: the capacity granted by the CAC to the high priority connections is mapped onto each subframe time-slots.

Finally, on each subframe, the DAMA Controller will decide the assignment of the remaining time-slots over the frame carriers, separately on each subframe, based on the dynamic capacity request of the RCSTs associated to each subframe.

Figure 7.2 shows the modules for the frame constitution procedure and their relation with the CAC and the DAMA Controller modules. The figure shows that the two modules need to know the assignments of the DAMA controller—the so-called Terminal Burst Time Plan (TBTP)—the CAC decisions and the DRA scheme which can be used by each RCST.

Preliminary simulation results show that the proposed dynamic resource configurations outperforms the static configurations. In the simulations, each RCST of a spotbeam is modelled by a Markov chain, characterised by a transition matrix $\mathbf{T}$ whose generic element t_{ij} expresses the probability that DRA scheme j is selected, given that the current DRA scheme is i. From the matrix $\mathbf{T}$, by exploiting the Markov chain properties, we can compute the long-term probability π_i that the DRA scheme i is selected by the RCST.

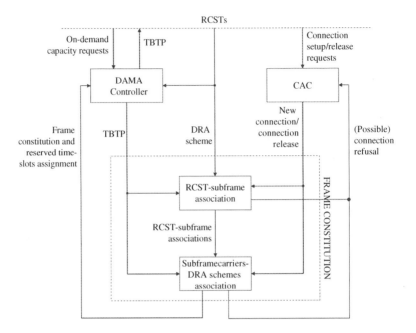

Fig. 7.2 Frame constitution scheme and interactions

The static configuration is set according to the probabilities π_i.

Figure 7.3 shows the simulation results in terms of blocking probabilities and of dropping probabilities[1].

If the simulation is performed by following the same transition matrix $\mathbf{T}$ used to compute the probabilities π_i, as in Fig. 7.3 case **a**), the results of the static configuration are comparable to the results obtained by the dynamic configuration, with an advantage for the dynamic solution which has a smaller dropping probability. On the other hand, if the simulation is performed by following different transition matrixes, the results of the static configuration are considerably worse than the results obtained by the dynamic configuration. For example, in Fig. 7.3 case **b**) the simulation is performed by following a transition matrix $\mathbf{T'}$ modelling the spotbeam in more sunny conditions with respect to $\mathbf{T}$: in this case, both blocking and dropping probabilities are significantly improved if the dynamic approach is used.

---

[1] If the channel conditions are such that the DRA scheme of a given RCST must be lowered, due to the reduction of the available transmission bitrate, it might happen that one or more already active connections must be dropped. Note that connection dropping is more disruptive with respect to connection blocking.

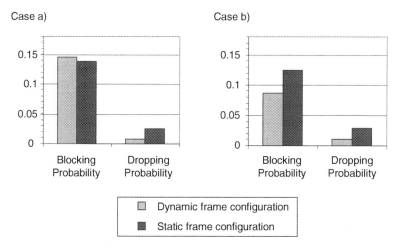

Fig. 7.3 Dynamic vs. static frame configuration

Radio Resource Management (RRM) Algorithms

RRM algorithms deal with the problem of dynamically assigning link capacity to the RCSTs. Two modules are required: the DAMA Agents, located within the RCSTs, are in charge of computing the bandwidth requests via the Bandwidth-on-Demand (BoD) algorithm in such a way that the RCST needs are suitably represented; the DAMA Controller (dealt with in the preceding Section), located within the NCC, is in charge of distributing the link capacity among the RCSTs based on the requests.

Bandwidth-on-Demand (BoD) Algorithm

The work underlying the SATSIX BoD algorithm is based on the results of the SATIP6 project, and represents an enhancement of the BoD algorithm implemented in the SATIP6 emulator. The main approach is the same as the SATIP6 one (see [6]): the system is modelled as a time-delay system, and the bandwidth request is computed based on control-theoretic methodologies. The novelties proposed in SATSIX concern the DVB-RCS adaptive modulation, coding and symbol rate and the enhancing of access delay performances.

With reference to the adaptive coding, from the BoD algorithm viewpoint it has an impact due to the fact that the available capacity varies in time. However, this is not a critical issue, since the BoD protocol is already developed with the capability of dealing with variable capacity: in fact, unpredictable reduction of available capacity might happen also because of concurrent traffic from the other RCSTs, and because of the high priority traffic. The SATSIX algorithm addresses this problem by modelling the available capacity as an unknown disturbance; this allows to deal with the capacity variations due to the adaptive coding as with variations due to concurrent traffic.

The control-theoretic algorithm is developed with the aim of achieving a certain desired behaviour, modelled by a reference queue length; the queue length in the RCST buffer is then driven to this reference queue length by the control algorithm. The queue-based algorithm has the property of congestion recovery: even if during congestions, i.e., reductions of available transmission capacity, the queue length grows, as soon as the congestion terminates the queue length is driven to the reference one. By defining the desired queue length, the RCST might pursue a conservative BoD policy, which is operator-friendly, since no transmission opportunity is lost by the RCST, or a more aggressive policy, which is RCST-friendly, since the queuing delays are lowered at the price of some transmission opportunity is lost by the RCST.

The SATSIX algorithm presents an adaptive multi-Model Reference Control (MRC) approach to adaptively define the reference queue length: based on the incoming traffic characteristics and on the inferred uplink load, the queue length is driven to a reference queue length that will assure a given network utilisation. This target network utilisation efficiency is adaptively varied based on the inferred network load. In fact, if the uplink load is scarce, the BoD policy can be rendered more aggressive—i.e., the target queue length can be lowered—since there is plenty of available transmission capacity; on the other hand, if the uplink is congested, no resourced should be wasted and the BoD policy must be rendered more conservative—i.e., the target queue length must be risen. The critical point is that the RCSTs must be capable of estimating the link status without the help of additional exchange of control messages with the NCC or with the satellite; in fact, the definition of these messages would not be compliant with the DVB-RCS standard, and their transmission would increase the signalling overhead. Thus, the proposed approach is to infer the link status based on measures of the RCST queue lengths themselves. When the RCST transmits a bandwidth requests, it expects to receive it after the transmission and processing time; thus, based on this fact, we can define an expected queue length. If the measured queue lengths are larger then the expected ones, we infer that a congestion is rising and decrease the target network utilisation; if the measured queue lengths are smaller then the expected ones, we infer that the link is lightly loaded and decrease the target network utilisation. The critical issue it such a distributed algorithm is the overall system stability.

Adaptive CAC

Future satellite network infrastructures are expected to support a variety of applications—e.g., data, audio, video, peer-to-peer, and so on—and, therefore, traffic belonging to Traffic Classes (TC), characterised by different requirements [4, 5, 8]. Calls belonging to different traffic classes (TC) compete for bandwidth, and are regulated by an admission control strategy. The objective of the CAC algorithm is to control the blocking and dropping probabilities of the different TCs [4]. To address these objectives, the admission control algorithm proposed in SATSIX is based on a Markov Decision Process (MDP) methodology. By formulating the MDP problem as a Linear Programming (LP) one (as described in [7]), blocking and dropping probabilities can be explicitly taken into account in the problem formulation itself; in other words, it is possible to directly control the blocking and blocking probabilities.

An admission policy consist in associating to each supported TC and to the channel load an admission probability: the MDP approach relies on the fact that the decision to accept or reject a call impacts on whether future calls will be accepted or not [1]. The MDP approach is based on the fact that different admission policies correspond to different probabilities associated to the channel loads (e.g., by applying two different CAC policies, policy A and policy B, the probabilities that the channel load is $x\%$ are generally different); these percentage can be computed by describing the admission control problem as a Markov chain—based on the statistical characteristics of the TCs—and can be used to compute the expected throughput and the expected blocking probabilities of the different TCs. The MDP approach allows one to find the optimal admission policy without the need of explicitly evaluate all the feasible policies.

The MDP approach suitable in satellite networks: it requires a centralised implementation, feasible in the Network Control Centre (NCC); the 'Curse of dimensionality' is limited by the network topology, since the CAC algorithm is executed for each link independently of the other links; the NCC explicitly assigns bandwidth to the Satellite Terminals.

The described standard CAC problem formulation is not totally suited for the SATSIX scenario due to the following problems:

Stationary Hypothesis and Scalability with Respect to the Number of Traffic Classes

The policy obtained by the MDP is optimal under stationary hypothesis. The solution is to estimate on-line the traffic statistical characteristics and to update the admission policy periodically or when the network detects changes in traffic characteristics.

The periodic update of the admission policy gives rise to the scalability problem; In SATSIX an alternative LP formulation of the MDP problem is proposed which is much more scalable with the number of traffic classes (TC) C. In particular, the LP problem size of the standard formulation grows with 2^C, whereas the LP problem size of the proposed formulation grows with $(C + 1)^2$.

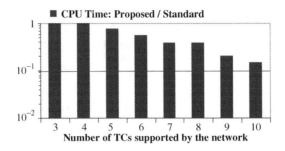

Fig. 7.4 Comparison between CPU times required to compute the admission policy with the standard and the proposed problem formulation

The following figure (Fig. 7.4) shows how the proposed problem formulation improves the algorithm scalability with respect to the standard one as the number of TCs C grows. With 10 classes the difference is about one order of magnitude.

DVB-RCS/S2 Adaptive-Coding

The DVB-RCS/S2 dynamic coding must be taken into account, therefore the channel models must be included within the MDP problem. The adaptive coding capability impacts on the CAC function; in particular, the capacity availability is varying both at the network level and at the RCST level.

If the channel availability is varying, which is the optimal admission policy? If the channel availability is reduced, the network can react in two manners: (1) by lowering the allocated bitrates to adaptive connections (i.e., connections characterised by the fact that they can transmit with different bitrates); (2) when all the adaptive connections are downgraded, it must decide to drop one or more connections (*forced dropping*). Obviously, this last event is rather disruptive and should be avoided. It is clear that, if the operator wants to completely avoid forced connection dropping, he would have to reduce the admission probability of the different TCs—i.e., to increase the blocking probability—and dedicate a large part of the channel capacity to best effort traffic and to the traffic served via the BoD algorithm, which can react fast to the channel variations. The channel model can be used to alleviate this problem.

With the proposed CAC, we define the maximum acceptable forced dropping probabilities (total and/or per TC); based on the statistical traffic characteristics and on the statistical channel model, the CAC then computes an admission policy which respect the constraint on the dropping probabilities. Thus, the operator will be then capable of deciding on the trade-off between channel bitrate utilisation and the tolerable forced dropping percentages. The solution relies on incorporating the channel model in the MDP.

Conclusions

This paper provides an overview of the approaches for the Radio Resource Management and CAC algorithms, with some brief simulation results. There is still on-going work aimed at optimising the algorithm performances and at facilitating the implementation within the SATSIX project.

In particular, the frame constitution heuristics are being comprehensively simulated and evaluated; the Bandwidth-on-Demand algorithm is being rendered adaptive with respect to the network load and to the traffic characteristics; finally, the Connection Admission Control algorithm will be enhanced to cope with the implementation constraints.

Acknowledgements This work is supported by the IST FP6 SATSIX project, funded by European Commission (EC). The financial contribution of the EC towards this project is greatly appreciated.

References

[1] E. Altman (2000). Applications of Markov decision processes in communication networks—A survey. Tech. Rep., INRIA, Available from www.inria.fr/RRRT/RR-3984.html

[2] ETSI EN 301 790 V1.4.1 (2005–09). Digital Video Broadcasting (DVB): Interaction channel for satellite distribution systems.

[3] ETSI TR 102 376 V1.1.1 (2005–02). Digital Video Broadcasting (DVB): User guidelines for the second generation system for Broadcasting, Interactive Services, News Gathering and other broadband satellite applications (DVB-S2).

[4] S. Kalyanasundaram, E. K. P. Chong, N. B. Shroff (2002). Optimal resource allocation in multi-class networks with user-specified utility functions. *Computer Networks*, Vol. 38, pp. 613–630.

[5] J. Liebeherr (1995). Multimedia networks: issues and challenges. *Computer*, Vol. 28, No. 4, pp. 68–69.

[6] A. Pietrabissa, T. Inzerilli, O. Alphand, P. Berthou, T. Gayraud, M. Mazzella, E. Fromentin, F. Lucas (2005). Validation of a QoS architecture for DVB/RCS satellite networks via the SATIP6 demonstration platform. *Computer Networks*, Vol. 49, pp. 797–815.

[7] M. L. Puterman (1994). *Markov Decision Processes*. New Jersey, John Wiley & Sons.

[8] S. Shenker (1995). Fundamental design issues for the future Internet. *IEEE Journal on Selected Areas in Communications*, Vol. 13, No. 7, pp. 1176–1188.

Chapter 8
SATSIX: A Network Architecture for Next-Generation DVB-RCS Systems

L. Fan, C. Baudoin, F. Rodriguez, A. Ramos, J.A. Guerra, B. de la Cuesta,
G. Fairhurst, A. Sathiaseelan, P. Berthou, T. Gayraud, L. Liang, A. Yun,
E. Callejo, I. Melhus, S. Iyengar, H. Cruickshank, and Z. Sun

Abstract Broadband satellite will play an important role to provide universal broadband access for the users. In order to lower the cost, the next-generation satellite systems should support IPv6 and seamlessly integrate with terrestrial networks, including wireless local loops. Based on the studies on the requirements and design constraints, in this paper, a novel network architecture has been proposed as a potential solution to the above problems. The proposed network architecture supports both transparent and regenerative topologies and can seamlessly interworking with WLL. It also integrates QoS, multicast, security, and mobility functions to support a range of transport requirements. How to apply this network architecture to support three different scenarios, namely corporate application scenario, residential application scenario and collective access terminal scenario, are also described.

Introduction

Current broadband satellite services are regarded as a niche market due to the cost of launching a satellite system, and the relatively limited available bandwidth compared to terrestrial counterparts. Cost-effective solutions are essential to improve up-take of broadband satellite and to efficiently accommodate new multimedia applications, integrating satellites into next generation networks. These issues are being addressed in the SATSIX, which will implement innovative concepts and for broadband satellite systems and services.

This paper describes the design of the SATSIX network architecture, with particular reference to support for IPv6 and the integration of hybrid satellite and wireless local loops (WiFi and WiMAX), which together will enable low-cost universal broadband access. Three network architectures have been identified that use this reference network architecture to support three different scenarios, namely corporate application scenario, residential application scenario and collective access terminal scenario. In each case the related protocol stacks are presented.

L. Fan et al. (eds.), *IP Networking over Next-Generation Satellite Systems*
© Springer 2008

Requirements and Constraints

This section focuses on the requirements and design constraints that were faced in the design of the network-layer of the SATSIX system utilising the IP protocol suite.

TCP provides a reliable byte-stream interface to the IP protocol suite. It is used by many popular applications, including web transfer (http, shttp), terminal access (telent, ssh, rlogin), file transfer (ftp, nfs) and email (smtp, pop, imap). TCP may also be used for multimedia services (e.g. streaming over RTP). Although TCP is the most widely used transport protocol and contributes the majority of traffic seen on the Internet backbone, most current Internet multimedia applications use UDP as the transport protocol (e.g. video and audio streaming, gaming). UDP is also used for many signalling applications (e.g. SIP, routing protocols, network management) and for IP multicast (e.g. IPTV).

Multimedia traffic may be associated with an explicit resource reservation, a DiffServ (DS) class, or the Best Effort (BE) service. Each of these approaches has merit. When the entire transmission path is under the control of a network operator, explicit QoS may be configured and used to provide a service guarantee for each configured session. This is the approach favoured in many current IPTV deployments. It is however difficult to manage when the traffic has unknown or highly variable characteristics. An alternative is to use a DiffServ class provisioning capacity to the service as a whole (rather than individual sessions). This can be significantly better than the unrestricted behaviour offered by a Best Effort service, particularly when traffic patterns are well understood. However, in both the BE and DS approaches, several traffic flows can share common capacity, and it is highly desirable to provide a method to fairly distribute the available capacity between the various flows, known as congestion control.

Although for many years TCP and IP offered the only transport protocols within the Internet Protocol Suite, the Internet Engineering Task Force (IETF) has more recently added three new protocols to supplement these specifications: UDP-Lite (a variant of UDP for multimedia optimised for wireless environments), the Stream Control Transport Protocol (SCTP, for transaction-based signalling protocols), and the Datagram Congestion Control Protocol (DCCP) transport protocol as an alternative to UDP for unicast multimedia applications. The features of the various transport protocols are summarised in the Table 8.1 below.

Current UDP-based applications can (and often do) transmit at a constant rate, irrespective of available capacity. Although congestion control is a standard feature of TCP, it is not provided by most current applications that use UDP for multimedia. Growth of such long-lived non-congestion-controlled traffic poses a threat to the overall health of the Internet. Congestion control is particularly desirable for sessions that need to operate over the dynamically changing properties that may result from Internet paths that combine both wired networks and wireless (e.g. satellite broadband). Standards-based Congestion-Control and NAT-traversal are key features of the new DCCP protocol, while supporting the flow-based semantics of TCP, but not the in-order delivery and reliability of TCP. DCCP is designed to operate with any QoS, including the best-effort service, however it can not effectively deliver

Table 8.1 IETF-defined Transport Protocol Features (Note* These features may be provided above the transport-layer, but not in a uniform and standard way)

Transport Protocol / Feature	UDP	TCP	UDP-Lite	SCTP	DCCP
Method	Unicast & Multicast	Unicast	Unicast & Multicast	Unicast	Unicast
Mode	Datagram	Byte-Stream	Datagram	Datagram	Datagram
Setup	None	Connection-setup	None	Connection-setup	Connection-setup
NAT support	None*	Supported	None*	Supported*	Supported
Reliability	None*	Fully-reliable	None*	Partial	Partial
Partial Integrity	None	None	Supported	None	Supported
ECN support	Not defined	Standard	Not defined	Standard	Standard
Congestion-control	None*	Standard	None*	Standard	Standard, Various.
Options	None	Supported	None	Supported	Supported
Typical Applications	Data Multimedia Control	Data Control	Multimedia	Control	Multimedia

real-time multimedia flows over highly variable network paths, since like TCP, it utilises transport layer feedback to determine an appropriate transmit rate, and will result in penalising flows that see abrupt changes in delay and/or capacity.

While the IP transport protocols are designed to operate over any IPv4 or IPv6 network service, there are important network characteristics (network QoS, support for multicast and mobility, security and wireless constraints) that can significantly impact performance when used in a satellite environment. It is therefore important to asses the requirements for each individual application using the network service.

QoS Requirements

Managing bandwidth allocation is a difficult task because different applications, and their associated network traffic, do not usually share the same requirements. This becomes even tougher when we take different users profiles into account. Bandwidth requirements refer to traffic volume/intensity and they will have impact on the configuration of the node mechanisms; they are also important for traffic engineering and/or overall network dimensioning.

The other performance parameters can be directly associated with service categories. Different services exhibit different sensitivities to packet losses, delay and delay variations, from low to medium to high. Some requirements (or performance parameters) are not relevant to some services. The most difficult applications to deal with are those with variable performance requirements. They may lead to either inadequate service or waste of resources.

The degree of symmetry is not considered criteria for class definition, as QoS is in general independently provided on the forward and return links. However, for strongly asymmetric applications (such as broadcasting, VoD, cinema distribution) QoS would make sense only on the forward link. Nevertheless, the degree of asymmetry is considered a factor in network dimensioning on the forward and return links.

To fully understand the different applications requirements and the constraints on bandwidth and delay associated, we will consider the applications behaviour with respect to five characteristics:

- Delay sensibility: Interactive applications are especially sensitive to delay issues, in a scenario like ours, where delay is already high due to the satellite link propagation time ($\cong 270$ ms), delay plays an important role in the link quality. If the end-to-end delay is too high, an interactive communication is difficult or impossible. Applications including gaming, videoconference, or VoIP are the services directly affected by this issue and therefore are the most restrictive applications.
- Jitter: Jitter is usually tied to delay, in the sense that applications requiring small delays will also require small delay variation or jitter. But not only applications requiring small delays (the ones specified earlier) request small values of jitter, also applications like TV Broadcast or VoD need small delay variation to offer really competitive video. A way to compensate for excessive jitter is to increase the size of the jitter buffer which is responsible for reassembling packet streams. When there is an issue with packets, such as arriving out of sequence or too late, the buffer will try and adjust to compensate or fill in with white comfort noise if necessary. Adjusting the buffer can minimise jitter problems, but it can also introduce other issues such as latency, which can cause conversations to be clipped.
- Packet loss sensibility: Packet loss can be caused by many different reasons. Normally, packet loss starts to be a real problem when the percentage of lost packets exceeds a certain threshold, or when packet losses are grouped together in large packet bursts. When a packet is does not arrive the receiver correctly and if TCP is being used, retransmission takes place, making delay and jitter greater and forcing a larger use of bandwidth, but assuring that no packet will be lost. That is the reason that makes us set as a constraint no packet loss for applications like email, web-browsing, ftp, p2p and those using TCP.
- Bandwidth: Bandwidth is by far the most common parameter in users mind thanks to the proliferation of bandwidth consuming applications like p2p, streaming or services like iTunes. This forces us to consider bandwidth as the most important requirement for our network. Some of the applications mentioned above would use as much bandwidth available as possible, and maybe the more the user gets, the more they are going to use, and therefore to demand. It is necessary to take into account that allocating a certain amount of bandwidth (higher than the actual needs of the user) will allow the transmission of higher definition videos, larger files, etc., ... driving us to a higher bandwidth consumption.
- Burstiness: A traffic session that tends to swell to use increasing amounts of bandwidth and produce large surges of packets is said to be 'bursty'. TCP's slowstart algorithm creates or exacerbates the traffic tendency to burst. As TCP addresses

the sudden demand of a bursting connection, congestion and retransmissions occur. Applications such as P2P, FTP, multimedia components of HTTP traffic, and the graphics portion of HTTP traffic (*.gif, *.jpg) are considered bursty since they generate large amounts of download data.

Finally, to sum up, the requirements of each factor involved in the analysis can be summarised in the Table 8.2 below.

IPv6 Network Architecture Requirements

IPv6 solves the Internet scaling problem, provides a flexible transition mechanism for IPv4 which reduces the risk of architectural problems, and it has been designed to meet the needs of new markets such as nomadic personal computing devices, networked entertainment, and device control.

IPv6 provides a platform for new Internet functionality. This includes support for real-time flows, provider selection, host mobility, end-to-end security, auto-configuration, and auto-reconfiguration. And it is designed to run well on high performance networks and at the same time is still efficient for low bandwidth networks (e.g., wireless). Therefore, the use of IPv6 in SATSIX will solve the problem of the shortage of IPv4 addresses, which are needed by all new machines added to the Internet, adding also many improvements to IPv4 in areas such as routing and network auto configuration.

For IPv6 support by the terminals it is required that the terminal network and air interfaces support IPv6, both the terrestrial and satellite networks can use IPv6 packets. The terminal should also include an IPv4/IPv6 gateway. The 'calling' terminal network interface receives the IPv6 packets; they are translated to IPv4 and sent to the air interface. The 'called' terminal receives the IPv4 packets and translates them to IPv6 to send them to the network interface.

- Multicast: Multicast is intended to allow a single device to send datagrams to a group of recipients in the most efficient way and without requiring sending several identical packets to each of the receivers. IPv6 uses a new protocol called Multicast Listener Discovery (MLD), which is embedded in ICMPv6 instead of using a separate protocol. IPv6 multicast supports three service modes, SSM (as in IPv4), ASM (in IPv6, constrained to a single administrative domain) and embedded RP (offering services that would otherwise require MSDP in IPv4).
- Mobility: Mobile Ipv6 is a protocol which allows nodes to remain reachable while moving around in the IPv6 Internet. To reduce the amount of signalling required by MIPv6 a new Hierarchical Mobile IPv6 (HMIPv6) protocol was designed by the Internet Engineering Task Force (IETF). It also improves hand-off speed for mobile connections. The main problem of MIPv6 is its inefficient use of resources in the case of local mobility, since it uses the same mechanisms in both global al local cases. HMIPv6, on the other hand, separates local from global mobility and the later one is managed locally.

Table 8.2 Summary of QoS requirements for various applications

	Bandwidth		Delay	Size	Jitter	Loss (PLR)	Security	Burst Oriented	Protocols
	Return	Forward							
Web-Browsing	16 kbps	64 kbps	<1 sec	10 KB	N.A.	0	Optional (HTTPS)	X	HTTP, HTTPS, TCP & IP
FTP	>14 kbps		N.A.	MB	N.A.	0	Optional	X	FTP, TCP & IP
E-mail	>20 kbps		<5 min	<10 KB	N.A.	0	Required	X	POP3, SMTP, IMAP, ESMTP, TCP & IP
Messenger	As available		N.A.	250 B	N.A.	0	Optional	X	MSNP8, MSNP9, MSNP10, TCP & IP
Internet Services									
E-Government	32–384 kbps		<1 min			0 (data)			Web, Messenger & Streaming
E-Commerce	32–384 kbps		<2 sec	N.A.	<1 mseg	1% (video) 3% (audio)	Required (VPN)	X	
E-Health	2 Mbps		<2 sec						
VoIP	4–82 kbps		<0.4 sec (Echo Control)	N.A.	<1 mseg	3%	Optional (VPN)	X	H.323, SIP, RTP, UDP & IP
Videoconference	50–460 kbps		<0.4 sec (Echo Control)	N.A.	<1 mseg	1% (video) 3% (audio)	Optional (VPN)	X	H.323, SIP, RTP, UDP & IP
P2P	256 kbps	64 kbps	15–60 sec	N.A.	N.A.	0	Required	X	MFTP, Fast Track, TCP & IP
Streaming									
Video	1–2 Mbps (CBR)		<2 sec	N.A.	<1 mseg	1%	Required		RTP, UDP & DCCP
Audio	64–128 kbps		<2 sec	N.A.	<1 mseg	3%	Required		RTP, UDP & DCCP
Alert Messages	10–20 kbps		<1 sec	250 B	N.A.	0	Required	X	Multicast Protocols

Interactive Gaming	64–500 kbps	< 1 sec	N.A.	< 1 mseg	0 (data) 1% (video) 3% (audio)	Optional	X	H.323, SIP, RTP, UDP, RTSP, DCCP….
Remote Control	500 kbps	< 300 sec	N.A.	N.A.	0	Required		TCP
Shared Applications	200 kbps	N.A.	N.A.	N.A.	N.A.	Optional		UDP
Video Surveillance	20 kbps	2–3 sec	N.A.	N.A.	N.A.	Optional		UDP
Web Conference	34–64 kbps 128 kbps (video) 20 kbps/user (audio)	< 300 msec	N.A.	N.A.	N.A.	Optional		UDP/TCP

- Security: IP Security (IPSec) is a mandatory feature in terrestrial IPv6 networks. IPSec offers strong and complete security services. In theory, it can be used in point-to-multipoint scenarios: in such a case, IPSec Security Associations (SA) have to be shared by the source and the concerned receivers. However, the Internet Key Exchange (IKE) protocol, which is always used with IPSec for key exchange and SA establishments, is a point-to-point oriented protocol. It does not allow establishing shared Security Associations between multiple network equipment. SAs have to be configured statically and manually in the source and in each receiver. This is not scalable for large groups.

Overall Network Architecture Definitions

The SATSIX system presents an integrated architecture that has been applied to both a transparent star and regenerative mesh topology using a DVB-RCS satellite link. The link layer is provided by the DVB-RCS standard [1], is derived from the highly successful Digital Video Broadcast (DVB) standard for Satellite [2]. In DVB-RCS, the Forward Link Subsystem (FLS) is provided by DVB-S, and the Return Link Subsystem (RLS) uses Multi-Frequency TDMA. The Time-Division Multiplexed (TDM) FLS is also used to transmit control tables that configure and control the operation of each RCS Terminal (RCST). SATSIX also considers systems that use the DVB-S2 [3], a second-generation standard, (designed to replace DVB-S, and recently incorporated into the DVB-RCS specification) as an alternate technology on the FLS, and can employ Dynamic Rate Adaptation for the RLS link. These new waveforms offer higher efficiency for transmission and also introduce more flexibility when designing the network service.

DVB-S utilises the ISO MPEG-2 Transport Stream [4]. It defines the physical and link layers that together provide a fixed rate simplex transmission of data by fragmenting data into fixed-sized frames (known as TS Packets) [5]. The most widely deployed method for transmission of IP data over DVB-S uses the Multiprotocol Encapsulation (MPE) [6]. An alternative, the Unidirectional Link Encapsulation, ULE [7], provides a more simple and flexible interface to the IP layer. Both are supported over the SATSIX air interface.

The DVB-S2 standard defines a set of advanced coding and modulation waveforms that offer a significant improvement in efficiency over that provided by DVB-S. S2 specifies two adaptive methods: the pre-provisioned Variable Coding/Modulation (VCM) method and a more sophisticated dynamic method called Adaptive Coding/Modulation (ACM). To take full advantage of the adaptive VCM and ACM methods requires using new methods based on the Continuous Generic Stream. A new Generic Stream Encapsulation (GSE) [8] allows a transmitter to directly transport IP packets without using TS Packets. The interface to IP resembles that offered by ULE. The most significant benefit is the flexibility that it will allow for operators to change the waveform on a frame-by-frame basis. This method is particularly appropriate to the two-way service provided in DVB-RCS systems. In these systems, the reduction in operational cost offered by flexible use of ACM is seen as crucial to successful competition with other satellite and terrestrial Internet services. The remainder of this section describes the architectural components required to design an RCS system that is able to offer QoS and other IP services over the S2 physical layer.

Network Architecture Overview

The SatSix network architecture is derived from the ETSI Satellite-Independent Service Access Point (SI-SAP) reference model for IP-based satellite systems, and introduces a satellite-independent QoS architecture [9]. In this section, the overall reference network architecture is described, and then applied to two specific DVB-RCS topologies, namely transparent star [10] and regenerative mesh [11].

Figure 8.1 shows the physical elements of the network reference architecture. These are described below:

- RCST: Return channel satellite terminals act as an interface between the system and external users providing bi-directional services through the satellite network. It can operated in is the interface between DVB-RCS and external users/ networks, such as WiFi and WiMAX.
- Satellite: It provides the backhauling link between the RCST and the hub or other RCSTs. This can use either a transparent satellite or one that has OBP capability.

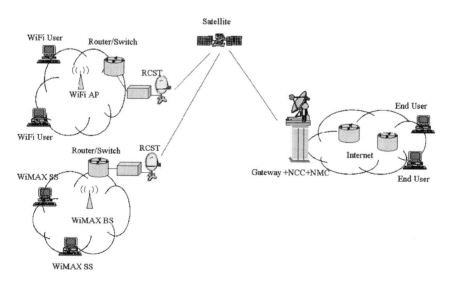

Fig. 8.1 Reference network architecture

- NCC: Network Control Center provides session control, routing and resource access to the RCSTs and manages the OBP configuration [11] and DVB control tables.
- NMC: This controls the management of all the system elements. The AmerHis NMC is split in two subsystems:
- The Element Manager (EM): It is responsible for the management of the redundant NCC (including NCC-RCST) and of the GWs.
- The Network and Service Manager (NSM): It is responsible for the management of the VSNs, Service Providers, RCSTs, and telecom services and NCC-initiated connections.
- Gateway (GW): This provides the interconnection with terrestrial networks (ISDN/POTS, Internet, and Intranet). The GW is comprises the following subsystems:

  - NMS: Interfaces between ISP/operator and Hub to manage user traffic and service provision. The NMS also performs control and supervision of HUB equipment, the network, and the RCSTs.
  - FLS: Encapsulates IP packets in MPEG-2 frames and transmits these frames on a TDM carrier, base-band modulated. In the RF equipment, the modulated bitstream is upconverted and sent to the antenna, which transmits it up to the satellite at frequencies in the Ku band.
  - RLS: The Gateway antenna receives the downlink signal at Ku band and downconverts it to the 950–1,450 MHz IF band. Thereafter, RLS filters the different return channels that compose a superframe, demodulates the signal, decodes the TDMA bursts form the terminal and extracts and forwards the frames containing the IP packets.
  - Reference and Synchronisation Subsystem (REFS): Receives time signals from GPS and provides synchronisation and timing for the different Gateway subsystems.

- Access router/Switch: This provides the access point with terrestrial networks.
- WiFi Access point: A wireless access point (AP) is a hardware device that acts as a communication hub for users of a wireless device to connect to a wired LAN. APs extend the physical range of service a wireless user has access to. An AP is identified by the broadcast SSID (Service Set Identifier) and must provide wireless security.
- WiFi user: Terminal or end user who accesses the network through WiFi connection.
- WiMAX BS: A WiMAX base station connects the terrestrial networks to the WiMAX subscriber station.
- WIMAX SS: A WiMAX subscriber station provides service through a wireless connection.

Figure 8.2 shows the network architecture for the transparent star topology. The main features are: the satellite does not have OBP and RCSTs can only communicate with the gateway/NCC via the satellite. WLLs are connected to the DVB-RCS via RCSTs.

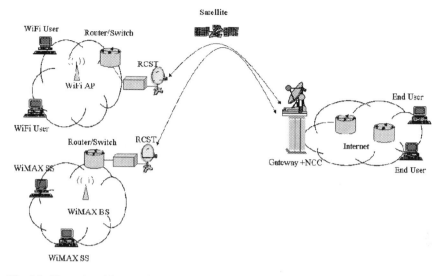

Fig. 8.2 Network architecture for the transparent star topology

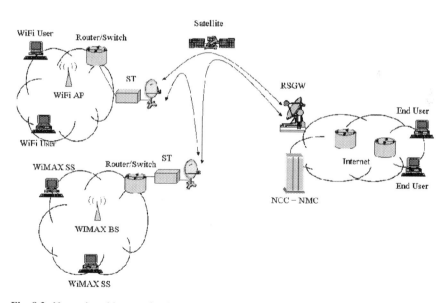

Fig. 8.3 Network architecture for the regenerative mesh topology

Figure 8.3 shows the network architecture for the regenerative mesh topology. It main features are: the satellite has OBP, and a RCST not only can communicate with the gateway/NCC via the satellite, but also can communicate with other RCSTs.

Interworking with WLL

Interworking with WiFi

The complexity of a Wi-Fi network depends on the user scenario. For personal use, it can be limited to a Basic Service Set (BSS), i.e. a collection of Wi-Fi end terminals associated to a single Wi-Fi access point (AP). For companies or service providers, it is likely to be an Extended Service Set (ESS), i.e. a collection of access points (with their associated terminal stations) interconnected by means of a distribution system. This latter can be based on a wired technology, typically Ethernet, or on a wireless technology (Wi-Fi or WiMAX). In the former case, the RCTS could simply integrate a Wi-Fi interface configured as an access point or could be connected to an access point by means of an Ethernet link. In the second case, the RCTS is connected to the distributed system by means of a switch/router as shown in Fig. 8.4. Note that, as with some commercial solutions, the above-cited switch/router can be a wireless switch controller which centrally manages all access points. By so doing, QoS provision, security, mobility is managed by the wireless switch controller and not fully deported in the access points.

Wi-Fi stations (access points and end terminals) should support some QoS functionalities at the Medium Access Control (MAC) Sublayer in order to support SATSIX classes of services. This is the case of many recent Wi-Fi products that are already available on the market and that are labelled with Wi-fi MultiMedia (WMM) (Fig. 8.4).

Interworking with WiMax

The WiMAX base station has two interfaces and is connected to a RCST. It is noted that the WiMAX users (i.e. Subscriber Station) can only communicate with the

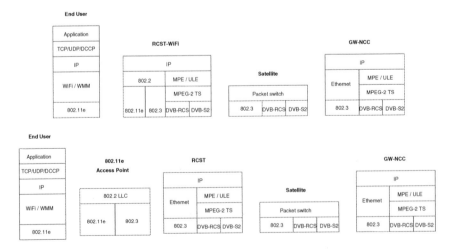

Fig. 8.4 Protocol stack of interworking between DVB-RCS and WiFi or WMM

WiMAX base station, and the WiMAX base station is connected to the DVB-RCS system via a RCST. The backbone of the DVB-RCS system is the IPv6-based Internet. In this way, there is no modification on the DVB-RCS system, and the main modification is on the MAC scheduler of the WiMAX base station. Since the design of WiMAX MAC scheduler is not standardised the IEEE 802.16 and is up to the operators, this approach adds flexibility.

The enhanced functionalities of the MAC scheduler in the WiMAX base station is as follows:

- Map the WiMAX QoS category onto the DVB-RCS SATSIX traffic classes.
- Provide end-to-end IPv6 communications over both DVB-RCS and WiMAX air interface. It includes how to perform segmentation and reassembly of IP packets at the WiMAX base station, which is connected to the satellite terminal, and how to perform address mapping/translation. The end-to-end protocol stacks are also described.
- Contact the DAMA Manager in NCC for admission control.

The negotiation of the SLA is realised through the SIP signalling protocol. By exploiting the Session Description Protocol, the SIP client located in the end user declares the session parameters (user address, type of media, bandwidth) to the stateful proxy SIP present in the satellite terminal. At this point the satellite terminal, which represents the edge router of SATSIX domain, makes classification and marking of the incoming flows according to each associated Per Hop Behaviour. Once the user has sent with the SDP application information about its traffic to the Proxy SIP in the RCST, the DAMA agent can read these parameters directly in the stateful proxy, and on the basis of the traffic is going to enter network, it can forward the requests of resources reservation before the flow start to fill the queue, and monitor continuously the state of the buffers in order to dynamically change the bandwidth requests. Mobility management is another interesting topic in the interworking between DVB-RCS and WiMAX (Fig. 8.5).

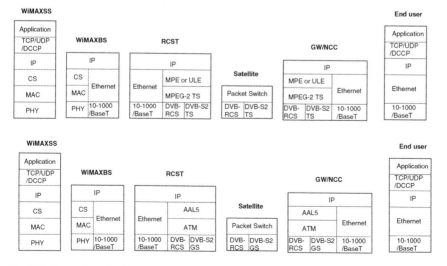

Fig. 8.5 Protocol stack of interworking between DVB-RCS and WiMAX

Functional Architecture Definition

Based on the overall network reference architectures, Figs. 8.6 and 8.7 shows the overall reference functional architectures for transparent star [10] and regenerative mesh [11] topologies, respectively. These functional architectures have integrated QoS, multicast, security, mobility and transport functions.

The main elements of the functional architectures are:

- This functional architecture not only support end-to-end QoS, but also support dynamic QoS according to applications and users needs. The satellite segment can interwork with Internet QoS DiffServ in order to provide end to end QoS at network level. The terminal model can perform this interworking in terms of signalling and QoS parameters mapping. The involved entities are SIP proxy, QoS agent, QoS server, PEP, IP compression, IP QoS, MAC QoS, RRM, C2P agent and C2P server.

- This functional architecture can provide up-to-date multicast management for both IPv4 and IPv6. The RCSTs should act as an MLDv2 multicast router proxy to forward the MLDv2 messages between listeners and the remote multicast router in the NCC.

- This functional architecture support security at application level such as TLS, SSL, and DTLS, key management protocols such SATIPSec and GSAKMP, key distribution systems like LKH and layer 2 security enhancing security level provided by L3 solutions such as SatIPSec. The Group controller/key Server will be co-located within the NCC and all the key management group members will be

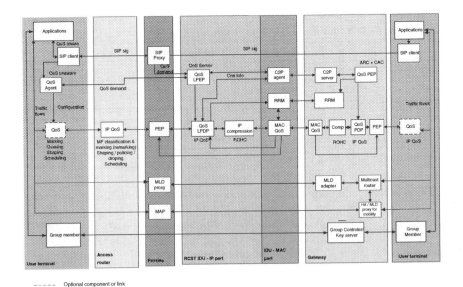

_ _ _ _ _ Optional component or link

Fig. 8.6 SATSIX overall reference functional architecture for the transparent star topology

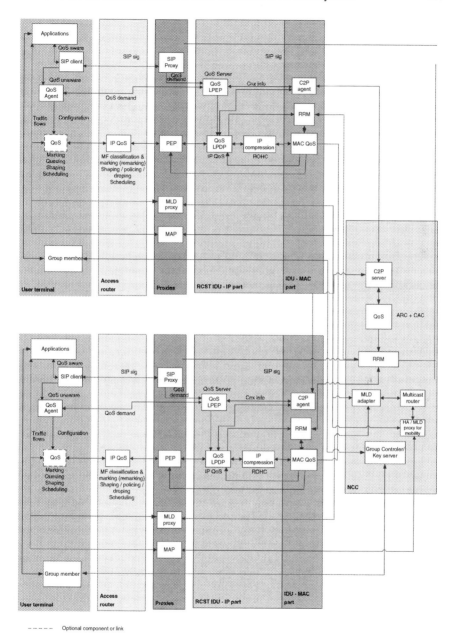

Fig. 8.7 SATSIX reference functional architecture for the regenerative mesh topology

co-located within user terminals. All key management messages will flow between the Group controller/Key server/NCC and all the secure data will be between the user terminals in mesh and user terminals and Gateway in a star configuration.

- This functional architecture has enhanced standard IPv6 mobility in a satellite system, with the use of Mobile IPv6. Mobility Anchor Point (MAP) is located in RCSTs and Home Agent (HA) is co-located in gateway. This design can reduce the signalling message during intra-domain movement and handover.
- This functional architecture supports PEP enhancement.

Applying the Network Architecture

In this section, the proposed network and functional architectures are applied to three different scenarios, namely corporate application scenario, residential application scenario and collective access terminal scenario.

Corporate Application Scenario

This type of scenario and architecture can concern a company having distant offices, warehouses, manufactures, subsidiaries or individual workers located in rural or hardly reachable areas by wired backbone.

It composed of different subnetworks belonging to the Headquarters and to 'distant offices'. The interconnections among the subnetworks is based on VPN end-to-end. It also includes tunnelling between the distant offices located beyond the satellite access on the local loop and the Headquarters located on the terrestrial backbone.

The scenario covers the following main categories:

- Intranet: provides internal site-to-site or LAN-to-LAN connectivity over public network.
- Internet: provides access to the internet.
- Extranet: allows secured connections with business partners, suppliers and customers of e-commerce. It is an extension of Intranet VPNs completed with firewalls & protections.

In the framework of each category, different traffic types are allowed:

- Internet protocols: TCP (ftp, http, ...), DCCP (VoIPO, VoD, ...), UDP (IPTV, ...)
- Control protocol (SIP, H323)
- Applications protocols: Video conferencing, VoIP
- Applications-signalling based on frameworks like SAP or web services

The distant offices do not have direct access to the Internet. Their only entry point is the Headquarters.

The distant offices which are not located in the same spot can however 'talk' to each other without passing through the Headquarters, provided they are using their Intranet facilities only.

The network architecture shall be composed by the following elements

- Distant offices: user equipment with wireless capabilities (Wifi or WiMAX), WiFi AP or WiMAX BS, Router/switch, and Firewall with VPN capabilities (if not provided by the router)
- Headquarters: Gateway, Router/switch, Firewall with VPN capabilities, WiMAX BS (optionally also WiFi AP), Content/application servers (es SAP), DHCP server, Proxy server, and Firewall for the access to the Internet (Fig. 8.8).

The protocol architecture takes into account the different types of applications and control protocols at user level; considering the presence of a wireless connection the simplest architecture foresees the interconnection of a Access point in the WiFi case, or a base station, in the WiMAX case, with routing capabilities (so with the presence of the IP layer) in order to implement the interworking strategy, basically classifying the traffic at IP level and creating a mapping between the wifi/WiMAX classes and the SATSIX one (Fig. 8.9).

Residential Application Scenario

Nowadays the principal residential application of satellite network is the provision of triple play services. The principle of the 'Triple Play' concept is to provide the three

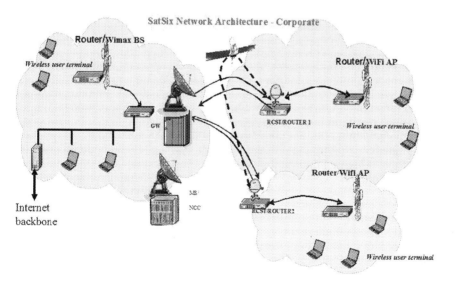

Fig. 8.8 Corporate application network architecture

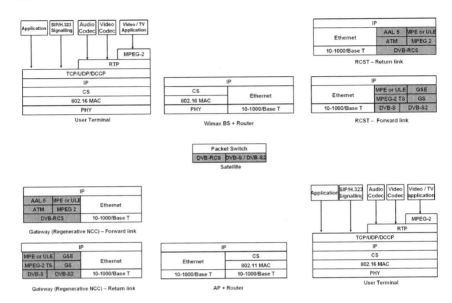

Fig. 8.9 Corporate application protocol stack (Regenerative payload)

basic access services (Television, Internet access and Telephony) through one single satellite communication network. The services considered in this scenario are:

- Digital TV
- Video broadcasting
- Interactive television
- Web Browsing
- VoIP Telephony & Videoconference
- P2P applications

Residential application scenario uses both transparent and regenerative satellite systems.

In order to provide Triple Play, IP connectivity must be used on top of a layer-2 adaptation layer based on DVB Standards. The uplink channel specification has been defined by ETSI in the DVB—Return Channel via Satellite (DVB-RCS) standard, and the downlink channel, also defined by ETSI, is the DVB by Satellite (DVB-S2) standard (Fig. 8.10).

Television broadcasting, video and audio broadcasting are ideally suited to the broadcast nature of satellite networks and become more cost effective the larger the end user population is, assuming that the same pieces of content need to be delivered to all (or large groups of) end users. Others triple play services as web browsing, VoIP and P2P applications involves a mixture of terrestrial and end users.

The satellite system architecture is compound by:

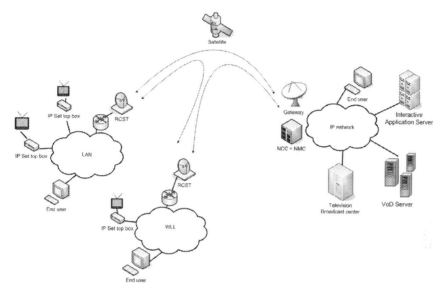

Fig. 8.10 Residential application network architecture

- On-Board Processor (regenerative system): The OBP combines DVB-RCS and DVB-S standards into a single multi-spot satellite system allowing cross-connectivity between different uplink and downlinks thanks to the signal processing.
- Network Management Center: in charge of the management of all the system elements.
- Network Control Center: controls the network, provides session control, routing and resource access.
- Satellite Terminal: is the interface between the system and external users.
- Gateway: provides interconnection with terrestrial networks.

In order to provide complete Triple play services, some equipment is needed:

- Home Gateway: It interconnects and integrates all kind of products in the home.
- IP Set-Top Box: A set-top box is a device that enables a television set to become a user interface to the IP world.
- Video on Demand Server.
- Streaming Server.
- Video service Provider.
- Television Broadcast Center.

The proposed protocol stack for the residential application scenario is shown in next figure. The proposed protocol stack differs from different services depending upon service need. The presence of wireless technology has been considered in the protocol stack (Fig. 8.11).

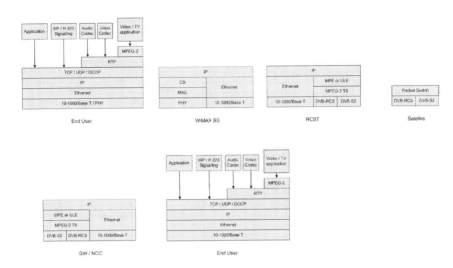

Fig. 8.11 Residential application protocol stack

Collective Access Terminal Scenario

The collective scenario aims to provide a high level of connectivity to citizens and small companies (or distant offices of bigger companies) belonging to rural communities that are not ADSL eligible. This solution is based on hybrid network combining a satellite segment and terrestrial wired as well as wireless access such as WiFi, WiMAX or Ethernet.

It is considered that the final users do not have to take care of the technology used for transmission of the traffic. The connection to satellite on the WLL side is single for the whole community that shares the satellite segment.

So the collective scenario is represented by all these public places in which any kind of people can access to the service without any specific requirements but only with a subscription that allows access to the broadband wireless network and applications.

Examples of this scenario are: Hotel, public places, mountain hut, islands harbours, emergency situations (war or natural disasters).

Three types of interconnections may be provided on the local loop and may be used simultaneously:

- Wired access: the Return Channel Satellite Terminal (RCST) interfaces with Ethernet wiring providing access in a building.
- WiFi: the RCST interfaces with WiFi Access Point, which spreads and collects the traffic over the local area. The PCs are equipped with WiFi cards that allow them to move within or at a short distance from the premises.
- WiMAX: the RCST interfaces with a WiMAX Access Point, which spreads or collects the traffic over the local area. This may concern users close to the AP

that are equipped with WiMAX cards, as well as users far from the AP, depending on the legally allowed transmission power. WiMAX may also be used a backhaul for WiFi, where a 802.11 network is connected to the 802.16 Subscriber Station (SS). These solutions can be put in place for quite big local areas, as dock for instance.

In this type of scenario, the link between the RCST and the Satellite Access Provider's Gateway is seen as a LAN. The local loop can be seen as a sub-LAN.

All the traffic passing over the satellite must pass through the Satellite Access Provider's Gateway. This means that no direct Internet flow is possible between two RCSTs connected to a local loop.

The baseline network is composed mainly by the following elements:

- User equipment with wireless capabilities (Wifi or WiMAX)
- Hot spots composed by Wifi access point (switch) with WPA capabilities as minimum or WiMAX BS
- Satellite access subnetwork: Router (to the RCST) with embedded firewall, and RCST
- Service provider: Satellite HUB/Gateway, and Router (to external internet network) with embedded Firewall (Figs. 8.12, and 8.13).

The protocol architecture takes into account the different types of applications and control protocols at user level; as in the case of the Corporate scenario, considering the presence of a wireless connection, the simplest architecture foresees the interconnection of a Access point in the WiFi case, or a base station, in the WiMAX case, with routing capabilities (so with the presence of the IP layer) in order to

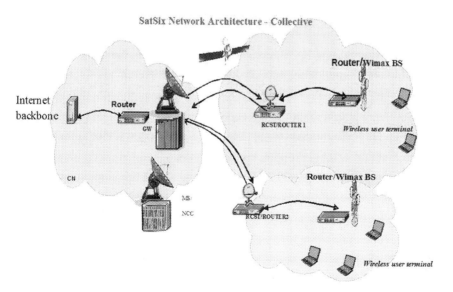

Fig. 8.12 Collective access scenario network architecture

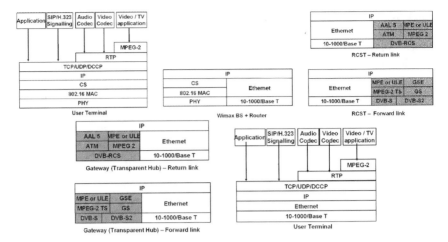

Fig. 8.13 Collective access scenario protocol stack (Transparent payload)

implement the interworking strategy, basically classifying the traffic at IP level and creating a mapping between the wifi/WiMAX classes and the SATSIX one.

Conclusions

The SATSIX system presents an integrated architecture that has been applied to both a transparent star and regenerative mesh topology using a DVB-RCS satellite link. This paper focuses on the requirements and design constraints that were faced in the design of the network-layer of the SATSIX system. The reference functional architecture integrates QoS, multicast, security, and mobility functions to support a range of transport requirements. Issues impacting integration of wireless local loop and network-layer QoS have been highlighted and solutions have been briefly outlined. Three network architectures have been identified that use this reference network architecture to support three different scenarios, namely corporate application scenario, residential application scenario and collective access terminal scenario. In each case the related protocol stacks are presented.

Acknowledgements This work is supported by the IST FP6 SATSIX project, funded by European Commission (EC). The financial contribution of the EC towards this project is greatly appreciated.

References

[1] EN 301 790, "DVB Interaction channel for satellite distribution systems", ETSI.
[2] EN 301 421, "DVB Modulation and Coding for DBS satellite systems at 11/12 GHz", ETSI.

[3] EN 302 207, "Second generation framing structure, channel coding and modulation systems for Broadcasting, Interactive Services, News Gathering and Other Broadband Satellite Applications".

[4] ETSI.ISO/IEC 13818–1, "Generic Coding of Moving Pictures and Associated Audio Information Systems".

[5] M. J. Montpetit, G. Fairhurst, H. Clausen, B. Collini-Nocker, and H. Linder, "A Framework for Transmission of IP Datagrams over MPEG-2 Networks", RFC 4259, 2005.

[6] ETSI/EN 301 192, "DVB Specifications for Data Broadcasting", European Telecommunications Standards Institute, 1999.

[7] G. Fairhurst and B. Collini-Nocker, "Unidirectional Lightweight Encapsulation (ULE) for Transmission of IP Datagrams over an MPEG-2 Transport Stream (TS)", IETF RFC 4326, 2005.

[8] "Digital Video Broadcasting (DVB); Generic Stream Encapsulation (GSE) Protocol For Interactive and Professional Applications", ETSI Work in Progress, 2007.

[9] TS 102 462, "Satellite Earth Stations and Systems (SES); Broadband Satellite Multimedia (BSM) Services and Architectures: QoS Functional Architecture", ETSI.

[10] TS 102 352, "Broadband Satellite Multimedia (BSM); Transparent Satellite Star—A (TSS-A); DVB-S/DVB-RCS for transparent satellites", ETSI, 2005.

[11] "Broadband Satellite Multimedia (BSM); Regenerative Satellite Mesh B (RSM-B); Profile of DVB-RCS and DVB-S Standards", ETSI, 2006.

Chapter 9
Interworking Strategy Between DVB-RCS and WiMAX

F. Rodriguez, I. Melhus, L. Fan, A. Pietrabissa, C. Baudoin, and Z. Sun

Abstract Aiming to provide low-cost universal broadband access, this paper describes the strategy that supports the integration of hybrid satellite and WiMAX, based on the network architecture proposed in the IST Satsix project. In this context, WiMAX networks can be used as wireless local loops integrated with a satellite network, lowering the cost of the broadband internet connections and providing a universal access to rural and marine areas. The paper presents the problems concerning the interworking between DVB-RCS and WiMAX technologies and the solutions adopted in the SatSix research project, partially funded by European Commission within the Information Society Technologies (IST) 6th Framework Programme.

Introduction

Broadband Internet access has becoming increasing ordinary in the urban areas, but there are still some rural or marine areas that are less accessible to the terrestrial broadband access. Satellite-based communication systems, by interworking with Wireless Local Loop (WLL), can provide the best and quickest solution to this problem. This approach not only can provide universal broadband access since satellite systems can connect any WLL to the backbone Internet no matter where the WLL is, but also can lower the cost of satellite systems.

The SATSIX project focuses on the use of satellite systems to offer attractive solutions for the access networks in three key scenarios, targeting the corporate, the collective and residential users. New application scenarios have been defined for both corporate and consumer markets (including 3-play and TV over DVB-S2).

The corresponding user requirements and services definition define the overall SATSIX network, in particular at the network layer, WLL with a satellite access using WiFi or WiMax will enhance network services including mobility, multicast, security, and QoS.

This paper will focus on the interworking between two most promising satellite and terrestrial broadband systems, namely Digital Video Broadcasting – Return Channel via Satellite (DVB-RCS) [5, 6] system and Worldwide Interoperability for

L. Fan et al. (eds.), *IP Networking over Next-Generation Satellite Systems*
© Springer 2008

Microwave Access (WiMAX). To integrate with the next generation Internet, end-to-end IPv6 communications is supported in both of these two systems.

Firstly, the reference network architecture of interworking between DVB-RCS and WiMAX is presented. It is noted that the WiMAX users (i.e. Subscriber Station) can only communicate with the WiMAX base station, and the WiMAX base station is connected to the DVB-RCS system via a Return Channel Satellite Terminal (RCST). The backbone of the DVB-RCS system is the IPv6-based Internet.

Secondly, the main requirements of the smooth interworking between DVB-RCS and WiMAX is analysed in this paper.

The Satsix Network Architecture

The SATSIX network architecture may be applied to two network scenarios: the transparent star and the regenerative mesh/star. The main difference between these two is the support for a regenerative mesh topology, using satellite has the on-board processing (OBP), allowing a Return Channel Satellite Terminal (RCST) to not only communicate with the gateway/NCC (Network Control Centre) via the satellite, but also to directly communicate with other RCSTs (in a single satellite hop).

Both WiFi and WiMAX terrestrial access networks may be connected to the RCSTs to provide shared access to the traffic forwarded on the satellite link (Fig. 9.1).

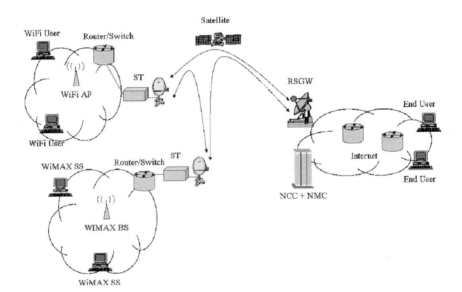

Fig. 9.1 Network architecture for the regenerative mesh topology

The RCSTs for a next generation communications satellite system must be low cost and high-performance satellite terminals, and support IPv6. All terminals will use adaptive DVB-RCS air interface for transmission and DVB-S/S2 for reception (DVB-S has to be maintained for backwards compatibility). At present the terminals transmission is based on non adaptive DVB-RCS and the reception on DVB-S.

Problem Analysis

Both QoS and mobility management will be a challenge for the interworking between DVB-RCS and WiMAX; in the following some details about Wimax QoS management and mobility management will be shown.

Problem on QoS

In the Satsix scenario, the Base Station (BS) can be connected through a LAN interface to the SATSIX terminal; the access of the subscriber station (SS) is made at MAC level through the set up of connections.

A connection set up can be accepted at MAC level without any parameter exchange with the SATSIX node and no mean to negotiate the QoS parameters compatible with the more bandwidth critical node except from the use of the session initiation protocol (SIP).

The user equipment shall negotiate the QoS parameters with the WiMAX during the SIP session exchange as described in the next paragraph, and while the WiMAX connection is being established (best effort), the SIP data go to the SATSIX SIP proxy where it is intercepted and processed by the connection admission control (CAC) at the SATSIX user terminal.

Summarizing, if the SATSIX CAC has success, the WiMAX SS can establish a connection, negotiating the needed QoS parameters with the BS and it can start to send IP packets.

Each packet will be classified according to the traffic classes supported by WiMAX and the mapping between IPv6 classes and WiMAX MAC classes shall take place. At BS level MAC PDU are assembled to IP Packets and forwarded to the SATSIX terminal where there will take place classification and segmentation. At DAMA (Demand Assignment Multiple Access) level the scheduling will take place taking into account the SIP parameters for each satellite connection and the classification of packets according to the mapping between the Ipv6 traffic classes and the DVB classes.

So in principle it is necessary to define a mapping between the WiMAX and DVB classes in order to maintain coherence in the managing of each IP flow and so to maintain edge to edge QoS.

That is to say, the mapping of the WiMAX QoS category [1] (i.e. Unsolicited Grant Service (UGS), Real-Time Polling Services (rtPS), Extended Real-Time Polling Services (ErtPS), Non-Real-Time Polling Services (nrtPS) and Best-Effort Service (BE)) onto the DVB-RCS capacity requests (i.e. Continuous Rate Assignment (CRA), Rate Based Dynamic Capacity (RBDC), Volume Based Dynamic Capacity (VBDC), Absolute Volume Based Dynamic Capacity (AVBDC) and Free Capacity Assignment (FCA)) [2].

The scheduling algorithm must take into account the response time of DAMA algorithm (due to satellite transfer time) in order to avoid queues saturation due to too high reactiveness of the WiMAX. Problems can arise, if after negotiation with NCC the WiMAX connection could not be available. So the reservation mechanism could provoke a waste of bandwidth at DVB interface. Even saturation at the WiMAX BS could have impacts on the behaviour of the reservation strategy at DVB level.

The above analysis is applied to static definition of service flows.

A possible strategy may be allowing a dynamic definition and change of service flows, foreseen in WiMAX specifications, but in this case problems could arise at DAMA level, because the possible overriding of the provisioned QoS parameters negotiated after SIP negotiation. That is to say, after the SIP set up a set of QoS parameters has been accepted at SATSIX level, and correspondingly at WiMAX service provision. The SS can modify its behaviour and renegotiate the QoS parameters of the active connection ID (CID), but this renegotiation cannot be intercepted by the DAMA manager.

So ideally there should be an information exchange between the WiMAX 'Authorization module' and the DAMA managing function for a dynamical redefinition of QoS parameters before allowing any change in the service flow profile (Fig. 9.2).

Problem on Mobility Management

The current standard 802.16–2004 for fixed installations, promote bandwidths of 70 Mbps. The latest mobility amendment, 802.16e adds support for nomadic roaming at vehicular speeds. Data rates are envisioned at 2–3 Mbps/user for portable and

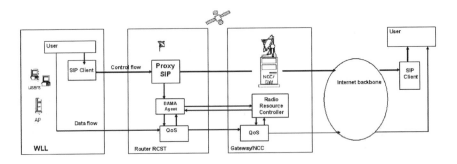

Fig. 9.2 Baseline high level architecture of the SATSIX system

1–2 Mbps/user when mobile. WiMAX defines four levels of user mobility: fixed, nomadic, portability and mobility. There are two working groups (i.e. WiMAX Networking Group [3] and IETF 16 ng working group) working on the mobility management in WIMAX.

IEEE 802.16 is a connection oriented access technology for the last mile without bi-directional native multicast support. It has only downlink multicast support, there is no mechanisms defined for mobile stations to be able to send multicast packets that can be mapped to downlink multicast connection. This may be a problem for some IPv6 protocols, e.g. IPv6 ND (Neighbour Discovery). This is additionally complicated by the definition of the commercial network models found in WiMAX, where the WiMAX transport connection is defined to extend the 802.16 MAC transport connection all the way to an access router. This is performed by establishing a tunnel between the base station and the access router. This leads to multiple ways of deploying IP over 802.16 based networks. Mobility management for the interworking between DVB-RCS and WiMAX has to be considered in the following models:

Figure 9.3 shows a WiMAX and satellite integrated network were the WiMAX basestation is anchored with an access router (AR) located at the satellite terminal (the satellite is not included in the WiMAX network). The WiMAX Subnet consists of a single BS/AR and multiple MSSs (Mobile Subscriber Station), and where a single address prefix is allocated to each attached MSS. In this scenario the IP Network Service is performed by native IPv6 transport over the satellite link.

A second alternative is shown in Fig. 9.4 where the access router is located at the satellite gateway separated from the WiMAX BS (The WiMAX transport mechanisms are extended to the satellite network—satellite is included in the WiMAX network). The WiMAX Subnet consists of a single AR with multiple BSs and MSSs and where a shared address prefix is used by all attached MSSs. In this scenario the IP network service is performed by an IP-tunnel over the IPv6 Convergence Sublayer (802.16e/WiMAX solution) over the satellite link. The IPv6 link between the MSS and the AR is considered as a point-to-point link.

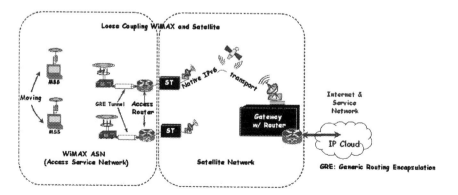

Fig. 9.3 WiMAX with access router located at satellite terminal (connected to BS)

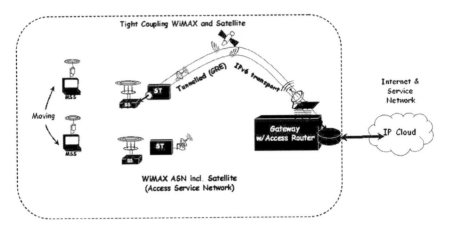

Fig. 9.4 WiMAX with access router located at satellite gateway (separated from BS)

The access router in Figs. 9.3 and 9.4 might be a 'WiMAX access gateway router' or a 'standard router'. In the first alternative PHY and partly MAC is implemented in the WiMAX BS, while the router is responsible for handover-Ctrl, etc. In the second alternative all WiMax networking functionalities are implemented in the WiMAX BS.

So two alternative scenarios has to be studied for WiMAX mobility management [4]:

- Client Mobile IP (CMIP), where the Mobile Client resides in the MSS. For mobility within the WiMAX network using CMIP, the point of attachment CoA (Care-of Address) changes when subnet changes.
- Proxy Mobile IP (PMIP), where the Mobile Client resides in the access router. The Proxy MIP solution does not involve a change in the point of attachment address when the MSS moves. There is no need for the terminal to implement a client MIP stack.

WIMAX Specification Related to QoS

WiMAX Principal Procedures During Service Activation

Connection Setup—Service Flow Setup

WiMAX adopts a two phases activation model in which the resources assigned to a particular admitted service are not actually committed until the service flow is activated.

Each admitted or active service flow is mapped to MAC connection through a unique CID.

The service flows are preprovisioned and the BS initiates the setup of service flows during user terminal initialisation.

In addition the user station and BS can dynamically establish service flows, as more deeply analysed in the following.

The establishment of a service flow connection is performed by a three way handshaking protocol; in a similar way the user station can renegotiate flow parameters (dynamic service change).

Each user sends a connection establishment request to BS; the request is then analysed in Call admission control and if accepted attribute of QOS and also two CID for each direction of the connection are registered in the Service flow data base.

In order to perform QOS process packets are classified according to their mentioned CID in MAC entrance point by classifiers.

Service Flow QoS Parameters (Static or Dynamic) Definition and Change

A configuration and registration function for preconfiguring SS based QoS service flows and traffic parameters is characterized by:

A signalling function for dynamically establishing QoS enabled service flows and traffic parameters.

- Utilization of MAC scheduling and QoS frame parameters for uplink (UL) service flows
- Utilization of QoS traffic parameters for downlink (DL) service flows

The principal mechanism for providing QoS is to associate packets traversing the MAC interface into service flows as identified by the CID.

A service flow is unidirectional flow of packets that is provided a particular QoS

The SS and BS provide this QoS according to the QOS parameter set defined for the service flow.

The primary purpose of QOS features defined here is to define transmission ordering and scheduling on the air interface.

Service flows in both the UL and DL direction may exist without actually being activated to carry traffic.

All service flows have 32 bit flow id (SFID). Admitted and active flows have 16 bit CID.

Service Flow Characterization

A service flow is characterized by a set of QoS parameters such as latency, jitter, and throughput assurance.

- AdmittedQoSParamSet = A set of QOS parameters for which the BS is reserving resources. The principal resource to be reserved is bandwidth.
- ActiveQoSParamSet = A set of QOS parameters defining the services actually being provided by the service flow. Only an active service may forward packets.

- Authorization Module = A logical function within the BS that approves or denies every change to QOS parameters and classifiers associated with a service flow. As such it defines an envelope that limits the possible values of the AdmittedQoSParamSet and ActiveQoSParamSet.

A service flow can be so:

- Provisioned (its AdmittedQoSParamSet and ActvieQoSParamSet are both Null.)
- Admitted (in this case the service has resources reserved by the BS for its AdmittedQoSPramSet but these parameters are not active.)
- Active (in this case the flow has resources committed by the BS for its ActiveParamSet (that is not Null).)

As service flow may be associated with from 0 to N PDU, but a PDU is associated exactly to one service flow.

The flow is identified by 32 bits of the service flow ID (SFID). A convergence sublayer (CS) process submits user data to the MAC SAP for transmission to the AC interface.

The information delivered to the MAC SAP includes the CID identifying the connection across which the information is delivered, so the service flow is mapped on that CID (MAC connection)

The service flow may be defined explicitly by traffic parameters or by a reference to a Service class, which is characterized by a service class name and QoS parameters. Such parameters can be also modified by the particular service flow.

An **Authorization module** will approve every change to the service flow QoS parameters.

The authorization can **be static** that is to say admission and activation requests for these provisioned service flows shall be allowed as long as the Admitted QoS parameters set and the Active QoS parameters set is a subset of the Admitted QOS Param Set. Requests to change the Provisioned QOS parameters set will be refused.

All the possible services are defined in the initial configuration of each SS.

Service flows can be also **dynamically** created changed or deleted.

This is accomplished through a series of MAC management messages:

- DSC: Dynamic service change
- DSD: Dynamic service delete
- DSA: Dynamic Service Activate

SATSIX Interworking Strategy on QoS

As seen before WiMAX is based on the set up of connections at MAC level, and depending on the particular CS used, one can realize particular mapping of the traffic flows of the upper layers into the MAC traffic classes provided by the WiMAX specification.

So a scheduling strategy for the BS polling task can be defined to be adapted to the scheduling strategy used in the Return Channel Satellite Terminal (RCST) for serving the input queues, after the mapping between the traffic classes used in WiMAX and those used in the RCST.

Every time a new connection is set up the user application sends SIP parameters that can be intercepted by the RCST and if the WiMAX grant process has success a MAC connection is established at WiMAX level, starting then to send packets without a real relation between the two sub networks.

Moreover, if the dynamic bandwidth allocation is allowed at WiMAX node then the application should renegotiate the QoS parameters through SIP but stopping the services it is supporting.

In fact when the dynamic definition and change of service flows take place, a possible overriding of the provisioned QoS parameters negotiated can exist.

The SS can modify its behaviour and renegotiate the QoS parameters of the active CID, but this renegotiation cannot be intercepted by the satellite subnetwork.

So ideally there could be an information exchange between the Authorization Manager and the satellite bandwidth allocation management process for a dynamical redefinition of QoS parameters before allowing any change in the service flow profile.

In the following the possible WiMAX–satellite interworking scenario is described:

Two groups of users can interface to the WiMAX network.

- Users managing connections inside the local loop network
- User managing connections outside the local loop network and possible going to the satellite subnetwok.

In the first case the generated traffic is networked by the local nodes; in the second case the traffic must be networked also by the satellite nodes.

When a new application starts a SIP session must be set up; the SIP INVITE message must contains the following header fields: To, From, CSeq, Call-ID, Max-Forwards and Via.

The session description protocol (SDP) messages include informations about QoS requirements for that SIP session, that are used by the RCST to negotiate the bandwidth allocation with the NCC.

The connection characteristics (i.e. the QoS parameters) in the SDP message are intercepted by the SIP proxy (in the RCST) and forwarded to the QoS agent. Then the RRM module is triggered in order to manage the dynamic bandwidth allocation in the satellite subnetwork. This can be performed both at IP and MAC level, by MAC QoS module. In the latter the class of services are mapped onto the DVB-RCS capacity requests (i.e. Continuous Rate Assignment (CRA), Rate Based Dynamic Capacity (RBDC), Volume Based Dynamic Capacity (VBDC), Absolute Volume Based Dynamic Capacity (AVBDC) and Free Capacity Assignment (FCA)). The capacity requests are sent to the NCC via SYNC burst, according to DVB-RCS standard, and the capacity, if available (a connection admission control algorithm is required) is allotted by NCC in the TBTP (Terminal Burst Time Plan) after a round trip delay.

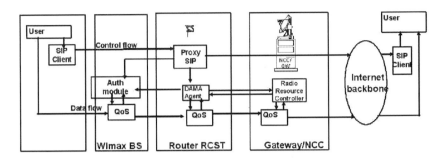

Fig. 9.5 Functional architecture for the WiMAX interworking

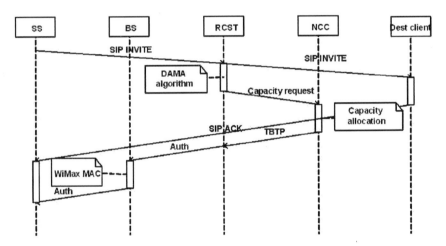

Fig. 9.6 WiMAX–satellite subnetwork interaction sequence diagram

Recalling the initial scenario description in the first case the WiMAX node behaviour is the same as in a terrestrial network; the Authorization Manager on the basis of the available resources can accept or not the establishment of new service flows.

For the second case the architecture foresees a SIP interaction also with the BS node. In this case the QoS parameters must be compatible between the wireless terrestrial network and the satellite network; so that the WiMAX management of the incoming connection requests must wait for the protocol interaction between RCST and NCC (Fig. 9.5).

Bandwidth allocation response is then returned to the BS and if enough band and QoS guarantee is stated, the Authorization Manager can accept the connection requests and if the grant has success it can establish the MAC connection at WiMAX level, allowing then the application to start sending packets (Fig. 9.6).

By this way if the dynamic bandwidth allocation is allowed at WiMAX node then the application now can renegotiate the QoS parameters through SIP.

SATSIX Interworking Strategy on Mobility Management

In these sections three scenarios for WiMAX network mobility using the satellite for backhauling will be described.

Mobile IP Client Resides at Mobile Subscriber Station

Scenario A1

The satellite link is not incorporated as part of the WiMAX access network, the CMIPv6 resides at MSS, access router at WiMAX BS/satellite terminal, and tunnelling is performed via Home Agent, see Fig. 9.7. It should be noted that the Access Router is connected to the Base-station via a GRE tunnel.

The MIPv6 capable MSS needs information about the Home Agent or home link and/or its Home Address (HoA) in order to initiate MIPv6 signalling towards the Home Agent. The MIPv6 client in the MS has to be bootstrapped with this information. The MSS acquires the information required for establishing a MIPv6 session via DHCPv6 (Dynamic Host Configuration Protocol). Prior to the MSS initiating DHCPv6 it has to authenticate itself to the network.

After obtaining the HoA (global scope IPv6 address) via the DHCP response, the CMIPv6 client shall send a BU (Binding Update) to the Home Agent to register it's binding to the CoA. If the MSS received the HoA in the DHCP reply message, the MSS shall set the HoA field in the BU to the received HoA. If the MSS did not receive the HoA in the DHCP reply message, but received the HL prefix info, the MSS can perform stateless address autoconfiguration.

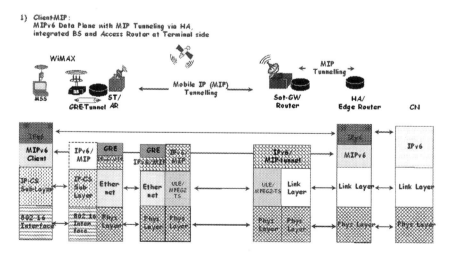

Fig. 9.7 Scenario A1: Mobility in WiMAX–satellite network

An IP-in-IP tunnel is established between Home Agent-Sat-GW/Router – BS/
RCST/Access Router (MIPv6 tunnelling over the satellite link).

An ongoing session by an MSS that is using MIPv6 may incur an inter Access
Router Handover. This may happen due to the MSS incurring handover to a
WiMAX BS that has connectivity to a new Access Router. The operational details
of such handovers are:

- The MSS/CMIPv6 client shall reset it's binding with CoA as soon as the MSS
  receives a new Router Advertisement from a new Access router containing a
  prefix other than the one received in the router advertisement which was used
  for address autoconfiguration.
- In case of stateful IPv6 address configuration scenario for CoA with DHCPv6,
  the MSS will not be able to send and receive any data unless it reconfigures the
  IPv6 stack with a new CoA via DHCPv6.
- Upon configuring a new CoA, the MS shall perform MIPv6 BU/BA (Binding
  Update/Binding Acknowledge) procedures.

Scenario A2

The satellite link is not incorporated as part of the WiMAX access network,
CMIPv6 resides at MSS, access router at WiMAX BS/satellite terminal, and route
optimization has been performed (no tunnelling via Home Agent), see Fig. 9.8.
Details of the connection between Access Router and Basestation are not shown.

Due to the MIPv6 route optimization there is normal IPv6 packet transport
between the CN (Correspondent Node) and the MSS and vice versa, without any
MIPv6 tunnel. This means less overhead over the satellite link.

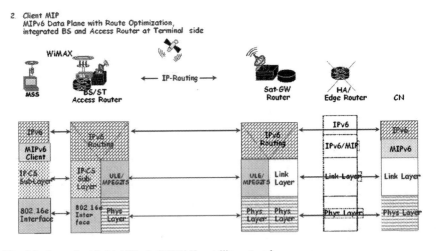

Fig. 9.8 Scenario A2: Mobility in WiMAX–satellite network

Mobile IP Proxy Client Resides at Access Router

Scenario B1

The satellite link is incorporated as part of WiMAX network, no Mobile IP client at MSS, the PMIPv6 resides at access router/satellite gateway and there is tunnelling via Home Agent, see Fig. 9.9.

The WiMAX network architecture currently supports Proxy MIPv4 (PMIPv4) for enabling mobility for hosts that may not have a MIPv4 client. There exists a requirement to support mobility for IPv6 hosts without having any explicit expectations of the host IPv6 capability. Proxy MIP6 (PMIPv6) is a solution that is aligned with the architectural direction of WiMAX, given the specification of Proxy MIP4 for IPv4 mobility and hence a similar approach for IPv6 mobility. The approach has not been specified yet as the work since the work with PMIPv6 definition still is in progress.

PMIPv6 is one of the network-based mobility management protocols which can avoid tunnelling overhead over the air/satellite-link, as well as mobile host's involvement in mobility management [PMIPv6]. The PMIPv6 protocol is an extension to MIPv6, which intend to provide seamless mobility within a locality, and where the work is still under progress. There are several key benefits to using PMIPv6 as the network-based mobility management protocol:

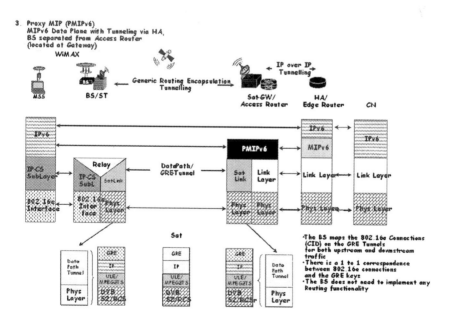

Fig. 9.9 Scenario B1: Mobility in WiMAX–satellite network

- The same MIPv6 Home Agent can be used be used at the same time with little or no modifications for clients that support MIPv6 and for clients that don't have host mobility software (does not require to have the MIPv6 protocol stack) and would like the network to handle mobility for them.
- Re-use of supporting work that has been done for MIPv6. This includes the ability to establish security associations, RADIUS and Diameter extensions, bootstrapping, dual stack support, reliability, fail over and HA switching.

PMIPv6 introduces a new entity, Proxy Mobile Agent (PMA), which is a functional element on the access router acting as a relay node between the Home Agent and the mobile station. The PMA performs the mobility signalling on behalf of the mobile station by establishing a bidirectional tunnel between the Home Agent and the proxy mobile agent. The details of the tunnel are given below:

- Tunnel source address is the Home Agent's address
- Tunnel destination address is the proxy mobile agent's address
- Tunnel encapsulation mode is IPv6/IPv6

The Home Agent typically can use the routing table for routing packet to the mobile station through the established tunnel. The PMA emulates the mobile station's home interface on the access interface, making the mobile node that it is connected to its home link. It will also update the Home Agent about the current location of the mobile station.

The Home Agent (just as in MIPv6) is the anchor point for the mobile station's home prefix and thus will receive all packets sent to the mobile station's home address. The Home Agent will route all the packets over the tunnel to the mobile proxy agent and in turn will route it on the access link to the mobile station. For packet sent from the mobile station, the proxy mobile agent will act as a default router and will route all the received packets over the tunnel to the Home Agent which in turn will route it to the destination.

In Fig. 9.9, a PMIPv6 is placed at the Sat-Gateway/Access Router. The connection between the Home Agent and the Access Router is set up by an IP-in-IP tunnel, while there is a GRE (Generic Routing Encapsulation) tunnel over the satellite link between the Access Router and the WiMAX BS per MSS basis. The BS maps the IEEE Connections (CID) on the GRE tunnel for both downstream and upstream traffic. There is a 1 to 1 correspondence between the IEEE 802.16 Connections and the GRE Keys. The BS does not need to implement any IP routing functionality.

Conclusions

The focus of this paper was on the interworking between DVB-RCS and WiMAX. Firstly, the reference network architecture has been presented, with the relations between each node, in particular the WiMAX users (i.e. Subscriber Station) and the WiMAX base station, and the WiMAX base station and the DVB-RCS system. Secondly, the main requirements of the smooth interworking between DVB-RCS and WiMAX are analysed. Ideally there could be an information exchange between the Authorization Manager and the satellite bandwidth allocation management process for a dynamical redefinition of QoS parameters before allowing any change in the service flow profile through the use of a SIP interaction also with the BS node. Following this, a interworking strategy on QoS, considering the communication of different types of traffic sources covered by a Wimax BS and connected to satellite DVB-RCS networks, is proposed. Finally, different aspects of IPv6 WiMAX mobility, aspects related both to the mobility within a standalone WiMAX network and mobility when WiMax network is loosely or tightly coupled to a satellite network.

Acknowledgements This work is supported by the IST FP6 SATSIX project, funded by European Commission (EC). The financial contribution of the EC towards this project is greatly appreciated. The authors also thank the partners from IST-SatSix project for their support to the architecture definition and to carry out of this work.

References

[1] H.S. Alavi Dept. of ECE University of Tehran, "A quality of service architecture for IEEE 802.16 Standards", Asia-Pacific Conference on Communicaton, Perth, Australia, 3–5 October 2005.
[2] F. Ohrtman "Wimax handbook, building 802.16 Wireless Networks", May 24, 2005.
[3] "WiMAX End-to-End Network System Architecture", Stage 3: Detailed Protocols and Procedures, August 8, 2006 Release 1V&V DRAFT.
[4] M. Riegel, "Mobile WiMAX Network Architecture".
[5] ETSI EN 301 790: Digital Video Broadcasting (DVB); Interaction channel for satellite distribution systems.
[6] ETSI EN 101 790: Digital Video Broadcasting (DVB); Interaction channel for Satellite Distribution Systems.

Chapter 10
The Satellite Communications Network of Excellence 'SatNEx' Removing Barriers, Integrating Research, Spreading Excellence

Anton Donner

Abstract SatNEx aims to rectify the fragmentation in satellite communications research by bringing together Europe's leading academic institutions and research organisations in a cohesive and durable way. The resultant pan-European network provides a collective grouping of expertise and state-of-the-art laboratory facilities that would otherwise remain dispersed throughout Europe.

Introduction

Today, Europe is a space power—the space industry, satellite operators and space organisations are major players in their respective fields, which are global in nature. Over the years, European R&D in satellite communications has encompassed a large number of activities spanning many programmes and organisations, however, such initiatives have been restricted in their strategic value, with only limited collaboration and coordination and further exasperated by a lack of the critical mass required to make an impact on the world stage.

As a first approach to integrate European research in satellite communications, the Satellite Communications Network of Excellence 'SatNEx' was established.

Removing Barriers, Integrating Research and Spreading Excellence are the primary objectives of this project which is supported under EU Framework Programme 6 with the overall goal of rectifying the fragmentation in satellite communications research by bringing together Europe's leading academic institutions and research organisations in a durable way. The creation of the Network aims to establish critical mass and allow access to a range of expertise currently distributed across Europe.

The origins of the SatNEx project can be traced back to the early 1990s, with the establishment of several COST actions. These satellite specific actions brought together a number of the partner institutions that now comprise the SatNEx consortium. Integration between satellite and terrestrial networks, mobile issues, fixed networks were covered by several separate COST Actions and consolidated finally by COST 272 (Packet-oriented Service Delivery via Satellite). Propagation impairment mitigation techniques (PIMT) for Ku-band, Ka-band and above were addressed by another set of COST Actions, which resulted in COST 280 (Channel Modelling and Propagation Impairment Mitigation for Millimetre Wave Radio Systems). This action aimed to improve the design and planning of present and future mm-wave broadband telecommunications systems and services (including broadcast and multimedia). COST 272, 280, and all related preceding actions can be seen as the parents of SatNEx.

The first phase of the project started in January 2004 and ended in March 2006; the second phase started subsequent to the first phase and will end in spring 2009. Preparations to continue the project without EU support are ongoing.

All satellite communications activities in SatNEx are placed in the context of related ESA efforts, mainly the ARTES R&D programmes, and are carried out in coordination with the activities in the thematic priority on 'aeronautics and space'. ESA/ESTEC plays as a partner an active role in SatNEx and many project partners are also participating in ESA projects concerning fixed and mobile satellite communications.

Implementation

This section describes the joint programme of activities (JPA) and shows how the JPA brings about the expected degree of integration. The overall JPA is described and broken down to three levels of activities, activity streams, workpackages, and Joint Activities (JAs), a flexible and powerful element designed during the first phase of SatNEx to reflect its peculiarities.

Figure 10.1 shows the structure of the Joint Programme of Activities, broken down to the second level of specified elements. The overall JPA is described and broken down into activity streams (WPs x000) and work packages (WPs xy00).

The Integrating Activities (WP 1000) support the Jointly Executed Research (WP 2000) by organising the exchange of personnel (WP 1100), performing integrated management of knowledge and intellectual property (WP 1200), providing a communication and collaboration platform (WP 1300) and providing multi-purpose web services (WP 1400). In the opposite direction, WP 2000 supports WP 1000 by proposing suitable cases for personnel exchange, and providing new knowledge and expertise. Moreover, WP 1000 supports the Spreading of Excellence (WP 3000) by offering the communication and collaboration platform.

Similarly, there is a strong connection between the Jointly Executed Research (WP 2000) and the Spreading of Excellence: WP 2000 produces new knowledge and intellectual property which gives input for standardisation and regulation (WP 3300). In the opposite direction, WP 3300 provides information from standardisation and regulation bodies to WP 2000.

Also, there are strong relationships between the work packages in the WP 2000 activity: the visions and system studies (WP 2100) influence the research into networking (WP 2200) and air interface (WP 2300), as well as on the research trials (WP 2400). WP 2400 itself spans horizontally over WP 2200 and WP 2300.

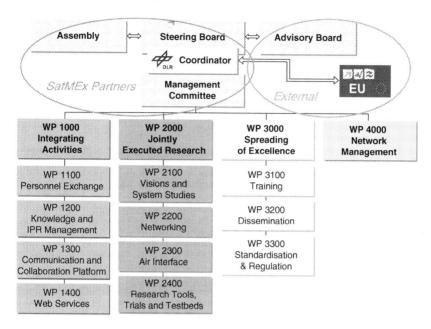

Fig. 10.1 SatNEx JPA, organisation and management

Finally, WP 3000 primarily aims at the Spreading of Excellence to Europe beyond SatNEx. This activity provides training opportunities, disseminates information and transfers knowledge, influences standardisation and regulation, and enhances public awareness of the benefits of satellite communications.

The philosophy underlying the SatNEx approach revolves around the selection of focused actions in order to capitalise on the expertise present within the Network and ensure that the integration is effective and durable. These focused actions, named 'Joint Activities (JAs)' are carried out jointly by the partners and include research, integration and dissemination activities. The research programme, in particular, focuses on addressing knowledge gaps that may be present within the network and on extending the existing knowledge base, which was established at project start-up by the various partners. The emphasis is on the challenge to existing concepts and ideas, in the never-ending search for improvements and breakthroughs, which can only be achieved by leaving the field open to exploratory research activities. Concerning the implementation of JAs, it is important to note that they can be flexibly set up and closed during the project phase, self-initiated (bottom-up) or top-down under control of the Steering Board (SB). This flexible work organisation structure on the third level, replacing earlier and classical fixed sub-work-packages, has shown to be well adapted to the specific needs and objectives of the project. At time of writing the following JAs are active in the network:

- JA-2110 Research Strategy and Visions
- JA-2120 Satellites in 3G and Beyond

- JA-2130 Broadband Access Networks
- JA-2210 Satellite Networking in Challenging Environments (EGGS)
- JA-2220 Integrated Multi-Platform Systems (IMPS)
- JA-2230 Cross-Layer Protocol Design
- JA-2240 Network Security
- JA-2250 IP-Networking Optimization
- JA-2310 Channel Modelling and Propagation Impairments Simulation
- JA-2320 Physical Layer
- JA-2330 Radio Resource Allocation and Adaptation
- JA-2410 Access, Network and Transport Layer Trials
- JA-2420 Application Layer Trials

In this respect, mobility is an important aspect of SatNEx's work, with academic staff and researchers being encouraged to move between institutions to allow access to specialised research equipment and to facilitate knowledge transfer and research integration.

SatNEx benefits from a fruitful relationship with industry. Representatives from the European space industry, satellite operators and regulation/standardisation organisations involved in satellite communications are members of the Advisory Board, which emphasises the importance SatNEx places on contact with industry and standardisation bodies.

The remainder of this article highlights selected work packages without going into technical details.

The SatNEx Communication Platform

The purpose of the SatNEx platform (WP 1300) is to demonstrate and deploy satellite-based virtual meetings and lectures/seminars services over Europe. This activity also promotes 'by example' the added value of the daily use of SatCom technologies. The platform uses Internet-based applications for communications and presentation of the meetings. An IP-based approach is indispensable to support a large number of collaborative applications, although the initial use of the platform is mainly for audio/video transmission. A solution simply based on replication of unicast multimedia traffic would not make efficient use of the satellite capacity. Hence, multicast capability was a major requirement to leverage the benefits of native satellite broadcast capability. To ease rapid deployment, the use of off-the-shelf components was necessary. Reliable support by the vendors also remains a key requirement.

There are two key scenarios in which the platform may be used: (a) a remote lecture scenario and (b) a distributed interactive meeting.

In the first scenario, a partner from the SatNEx community organises (and broadcasts from its own site) a 'Master lecture' on a specific topic. During the session, he/she needs to address all interested remote SatNEx partners and would appreciate real-time feed back to collect 'instant' messages or questions (e.g., jabber messages with questions to be answered at the end of a lecture). The interactivity is low and therefore does not necessarily require bi-directional satellite connectivity for all the attendees, many may instead utilise their existing terrestrial Internet access as a return channel.

In the second scenario, a group of SatNEx partners wish to organise a distributed conference. A few partners, the 'foreground attendees', are 'active' while the others, the 'background attendees', are only interested in 'passive' observation, i.e. watching and listening to the session from the foreground attendees. This scenario covers a number of cases with different levels of interactivity where the speaker frequently changes.

When there is a frequent change of speaker, the dynamics of interactive teleconferences over GEO-satellite are strongly impacted by the appreciable satellite propagation delay combined with the video processing delay. The interaction is then considerably improved by coordination, for example the ability to ask for, and grant, access to the 'floor'.

Participation in both scenarios is supported using the same SatNEx platform equipment. For an attendee that only requires participation as a background observer, a simpler setup may be sufficient. For example, sessions can be announced as SDP (Session Description Protocol) records using email or from a web site. The SDP records can be directly opened on a PC with standard multicast-enabled application software, such as Quicktime Player™. This approach is supported on a variety of different operating systems.

The SatNEx platform architecture combines an Internet-based terrestrial segment for collecting input streams and a satellite segment that aggregates the traffic into a feed and broadcasts the streams using DVB-S. This achieves the best of both worlds. On the one hand, a terrestrial-only solution would raise the issue of multicast connectivity and the necessity to connect every partner to the multicast-enabled Internet backbone. On the other hand, a bi-directional satellite solution is not affordable because of the need to equip all partners.

The MCU (Multipoint Control Unit) is the meeting point that translates unicast, terrestrial traffic into a single multicast stream that is suitable for satellite broadcasting. In addition, the MCU acts as a videoconference 'babel fish', transcoding the different contribution media formats into a single standard supported by all receiver stations. For content transmission, from an active partner to the central MCU, a H.264 codec is preferable. H.264 has been optimised for transmission over networks without guaranteed bandwidth reservation (most current Internet paths).

For multicast data transmission from the MCU over the satellite, SatNEx rents permanent satellite capacity with guaranteed bandwidth. Currently, H.263 encoding is used to ensure successful decoding by all partners including PC systems that may have limited CPU resources. However, hardware requirements for H.264 decoding are constantly decreasing, and future transmissions plan to also use a H.264 codec over the satellite, decreasing bandwidth requirements and the associated cost.

To summarise, the journey of the communication flow is as follows: (a) content is pushed via the terrestrial Internet from the partner sites (i.e. live lectures or content from tape and data streams generated by Polycom VSX 7000s videoconference devices deployed at all partner sites) to the MCU (Codian), (b) this is multicast and uplinked to the satellite, (c) received through an IP/DVB-S router (IPricot) and then (d) multicast onto the destination LAN for viewing (e.g., Quicktime Player or VLC). To make interaction during lectures possible among the platform users, a Jabber chat server was installed and clients at each site.

Training, Dissemination and Standardisation

Training represents an important part of SatNEx's remit and is supported through a number of initiatives including the hosting of internship projects, the establishment of an annual SatNEx Summer School, the dissemination of papers of a tutorial nature, and live and recorded lectures transmitted via the satellite platform. In the past years the Summer School took place in Pisa (2005), Salzburg (2006) and Pisa (2007) again. From 2007 a certificate of attendance valid for 1 ECTS (European credit transfer system) point will be provided to those attendees who take each daily test. Details of the training opportunities offered by SatNEx, together with other ongoing activities, can be found at the SatNEx website http://www.satnex.org.

Conferences and workshops provide a mechanism for researchers to develop their professional skills, e.g. public speaking and scientific writing, as well as providing a means for discussing with their peers aspects of their research. One of the aims of SatNEx is to promote through sponsorship and/or organisation, satellite communications conferences and workshops to allow on-going research activities to be presented to a wider audience. The SatNEx flagship conference is the International Workshop on Satellite and Space Communications (IWSSC). While the call for contributions is open to any organisation, a high share of presentations from within SatNEx is usually achieved. The event took place in Siena (2005), Madrid (2006) and Salzburg (2007).

The mission for WP 3300 'Standardisation' is to promote participation in the work of the standardisation and regulatory groups appropriate to satellite communications. The aim is to inform partners of the work of the various organisations, enabling them to focus their research activity on key requirements of these groups. The WP also seeks to encourage participation at standards meetings, and enable contributions to standards and regulatory documents.

WP 3300 promotes good practice in participation in standardisation and regulation. It helps to direct the Network's research, by bringing into focus the anticipated needs and recent developments in standardisation and regulation. The work package also influences the standards process, by targeting improving results to key working groups participating in the development of satellite standards. The Network promotes this participation in standardisation and regulation by supporting attendance at meetings and co-ordination of activities, and through improved communication between researchers and standardisation and regulation working groups. The Network co-ordinates these activities with those of other initiatives, such as SatLabs and the Integral Satcom Initiative (ISI), and other Framework projects appropriate to the satellite community.

Summary

The SatNEx project has brought together partner organisations (see Table 10.1) from across the European Union with the aim of de-fragmentation in the area of satellite communications. This is achieved through the performance of a joint programme of activities, which include integrating activities, jointly executed research and spreading of excellence.

Table 10.1 SatNEx partner organisations and responsibilities in the Network

Deutsches Zentrum für Luft- und Raumfahrt e.V. (DLR), Germany	Project co-ordinator, SB member, WP 1000, WP 4000, JA-2130
Aristotle University of Thessaloniki, Greece	—
Budapest University of Technology and Economics, Hungary	—
University of Bradford, Mobile and Satellite Communications Research Centre, UK	SB member, WP 3000, WP 3100, WP 2200
Centre National d'Etudes Spatiales (CNES), France	SB member, WP 1300, JA-2420
Consorzio Nazionale Interuniversitario per le Telecomunicazioni (CNIT), Italy	SB member, WP 1400, WP 2400, JA-2230, JA-2250, JA-2330, JA-2410, IWSSC'05
European Space Agency (ESA)	SB member
Fraunhofer Gesellschaft zur Förderung der Angewandten Forschung e.V., Germany	—
Groupe des Ecoles des Télécommunications (GET/ENST), France	JA-2210
National Observatory of Athens, Greece	WP 3200, JA-2120
Istituto di Scienze e Tecnologia dell 'Informazione "Alessandro Faedo" (ISTI), Italy	SB member, Summer School'05 &'07
Jozef Stefan Institute, Slovenia	JA-2220
Office National d'Etudes et de Recherches Aérospatiales (ONERA), France	JA-2310
Paris Lodron Universität Salzburg, Austria	Summer School'06, IWSSC'07
TéSA Association, France	SB member, WP 2300
Technische Universität Graz, Austria	—
Universitat Autònoma de Barcelona, Spain	—
Universidad Carlos III de Madrid, Spain	IWSSC'06
The University of Surrey, UK	SB member, WP 1200, WP 2100, JA-2110, JA-2240
University of Aberdeen, UK	WP 3300
Università di Bologna, Italy	SB member, WP 2000, JA-2320
Universita Degli Studi Di Roma "Tor Vergata", Italy	—
Universidad De Vigo, Spain	—
Institute of Communications and Computer Systems (ICCS), Greece	—

The author wants to take the opportunity to thank the consortium for the continuous high effort and motivation they bring into the Network. Without these ambitious partners the Network would have never been as successful as it is now.

Acknowledgements The EC funds SatNEx under the FP6 IST Programme.

Chapter 11
The Use of Novel Satellite Broadcast Technologies for the Provision of Integrated Services

Evangelos Pallis, George Mastorakis, Athina Burdena, Ahmed Mehaoua, and Yassine Hadjadj Aoul

Abstract Digital satellite broadcasting, primarily targeted to unidirectional services, soon expanded to the interactive domain, utilising uplink technologies such as DVB-RCS. In this context, due to their wide and uniform coverage, satellites can provide an ideal medium for the provision of triple play services (voice, video and data) to remote areas not covered by terrestrial infrastructures. This paper discusses and demonstrates the delivery of triple play services over a fully functional platform utilising the second-generation satellite broadcasting technology (DVB-S2) for the downlink combined with DVB-RCS for the uplink.

Introduction

The widespread adoption of satellite links as carriers of integrated digital services, combined with the development and launching of novel infrastructures, equipped with broadband transponders and state-of-the-art signal processing capabilities, have accented the role of the satellite technology, mostly based on DVB-S, as an efficient end-to-end communication medium, targeted not only to network and TV operators, but also to end consumers [1]. The use of satellite infrastructures is necessary in cases where terrestrial solutions are proved inadequate, for instance in isolated, rural and less developed areas or in when a large coverage area is required.

Since the introduction of DVB-S, many technological advances have been achieved in the areas of satellite broadcasting technologies. The interaction protocol (DVB-RCS) [2] provides a standardized and efficient solution for the satellite return channel, thus allowing for the delivery of interactive services. Also, the recent second version of the DVB-S specification (DVB-S2) [3], which seems to be the most promising solution in satellite communications, features many improvements over its predecessor, including efficient support for data services, near-optimal coding schemes and Adaptive Coding and Modulation (ACM).

The combination of DVB-S2 and DVB-RCS constitutes a very efficient interactive access platform for IP-based satellite applications [4]. In this context, a very promising approach is this of the migration of triple play services (i.e. TV/video, data and voice telephony) to the satellite sector. Triple play is today offered mostly via terrestrial wired access infrastructures, like xDSL. There exists, however, a considerable percentage of customers living in rural/isolated areas which are not covered by terrestrial infrastructures. The same holds for customers on the move, such passengers as in trains, airplanes or ships. For all these cases, the satellite access solution is very promising for the delivery of ubiquitous broadband integrated services.

In this context, this paper presents the design and development of an interactive satellite access platform based on DVB-S2/RCS and equipped with the appropriate modules for the provision of video, voice and data streams to remote nodes. Service distribution from the satellite node to nearby customers is performed via a WiMAX infrastructure. A real on-air validation is carried out, via the HellasSat II satellite.

This paper carries on as follows: Section II presents the principles and benefits of DVB-S2 networks and discusses the nature of satellite IP-based triple play services. Section III describes the architecture of the implemented network and refers to the integration/validation process. Finally, Section IV concludes the paper.

DVB-S2 and Triple Play

Following the worldwide dominance of the DVB-S specification, ETSI proceeded in adopting its successor, which was designed after extensive research. DVB-S2, featuring optimal turbo coding along with per-stream adaptive coding and modulation, enables for significant improvements in satellite communications. DVB-S2 promises to improve transmission performance and add flexibility to existing broadcasting networks. DVB-S2 is intended to replace DVB-S in the forward link of next generation broadband satellite systems.

With the aim of promoting terminal cost reduction, DVB-S2 has been conceived as a single standard for addressing different application areas and market segments. The new standard owes its flexibility in the adoption of advanced coding and modulation schemes. DVB-S2 features a FEC system based on LDPC (Low-Density Parity Check) codes concatenated with BCH codes, which enables Quasi-Error-Free operation at about 0,7 dB to 1 dB from the Shannon limit, depending on the transmission mode (AWGN channel, modulation constrained Shannon limit). Furthermore, DVB-S2 includes higher order modulation schemes (QPSK, 8PSK, 16APSK, 32APSK). The combination of the above coding and modulation schemes, may offer a 30% increase in spectral efficiency over DVB-S, for all the above-mentioned application areas.

Finally, support of individual quality-of-service targets has been recommended for interactive applications. For the latter case, no recommendation is included in the DVB-S and DVB-S2 standards, as far as the return path is concerned. Therefore, interactivity can be established either via terrestrial connection or via satellite. DVB offers a variety of return link specifications, such as for example via satellite (DVB-RCS), via PSTN/ISDN (DVB-RCP), via GSM (DVB-RCG), and via cable (DVB-RCC). This paper utilises a DVB-RCS return channel to implement a fully interactive satellite-based broadband access infrastructure.

Since the infrastructure presented in this paper is IP-enabled, it can offer a "bouquet" of multiplexed IP heterogeneous streams, to constitute a triple-play environment:

Digital Television over H.264/MPEG-4. MPEG-2 has been for more than a decade the standard for digital satellite television. However, recent advances in coding techniques have come up with technologies which are remarkably more effective.

The MPEG-4 standard (ISO/IEC 14496) defines an audiovisual coding standard to address the emerging needs of the interactive and broadcasting service models as well as the needs of the mixed service models resulting from their convergence. Taking MPEG-4 a step beyond, MPEG and ITU created the even more efficient H.264 or MPEG-4 Part 10 Advanced Video Coding (AVC) codec.

Exploiting these advances, digital satellite TV can be provided over the H.264/RTP/UDP/IP multicast protocol stack, resulting in a 50% bandwidth saving over legacy MPEG-2.

VoIP Telephony. Despite all advances in broadband multimedia communication, the plain, PSTN-like voice conversation is still essential to everyone. A triple play satellite platform offers the capability to extend the coverage of the earth telephone network to remote areas, even where the mobile PLMN service is unavailable. Instead of using proprietary audio conferencing technologies of questionable effectiveness, open and widely adopted protocols (H.323/SIP) are preferable.

Data (Internet) access. Internet access has always been the main service offered by DVB-based interactive satellite networks and is an essential part of a triple play "package". A dedicated firewall/proxy can be utilized for interconnecting the satellite provider platform to the Internet and to other wired or wireless IP local networks. User-to-user data exchange is also possible, although with high delay, due to the double-hop connection. In all cases, standard TCP acceleration techniques can be employed with positive impact to the performance of the connection.

Last but not least, since in a satellite downlink all streams from heterogeneous services are multiplexed in the same bouquet, a resource management mechanism is essential to provide the per-stream required QoS. Multicast DTV streams must be delivered with as low losses as possible, whereas in the voice traffic, delay and jitter must be kept to the minimum. On the contrary, Internet (TCP) traffic can be assigned to the lowest QoS requirements.

An Infrastructure for Efficient Satellite Integrated Services

Based on the aforementioned issues, a satellite integrated services access solution has designed and realized in the frame of IST IMOSAN project [5], which constitutes a complement and an extension to triple play terrestrial platforms in the cases where the latter are proved inadequate. The developed system, whose overall architecture is depicted in Fig. 11.1, provides a flexible and viable broadband communications path for individual users and small wired or wireless local networks which are geographically isolated or, in general, are in a condition which prevents them from connecting to terrestrial network infrastructures.

The platform is based on a DVB-S2 communication chain (IP-to-DVB Encapsulator, Multiplexer, Modulator) and a remote (commercial) DVB-S/RCS Hub which collects the uplink data. A VPN (Virtual Private Network) tunnel from the remote Hub feeds the data into the provider platform. There, a Proxy/3play Router feeds the triple play streams (destined to the end users) to the Encapsulator for processing and transmission, and routes appropriately the IP datagrams which arrive via DVB-RCS from the users' sites. The Encapsulator/Multiplexer operates in compliance to [6] and treats each traffic stream individually and can apply different queuing priorities to each service. This differentiation is performed in a static manner. In case that dynamic bandwidth management is required, a mechanism like the one described in [7] can be employed.

Video streams are served by a real-time H.264 encoder fed by a live source, and a VoIP Gateway utilising H.323/SIP acts as an interface to the public PSTN network. Internet connections are firewalled and served via a Web proxy.

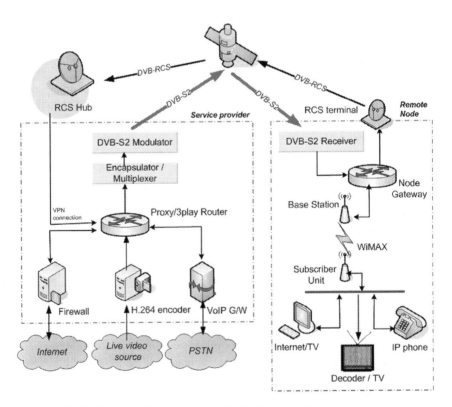

Fig. 11.1 The developed satellite infrastructure for triple play services

At the remote node, since integrated DVB-S2/RCS terminals are not fully deployed, reception and transmission is undertaken by two separate modules – a DVB-S2 receiver and a separate, standard DVB-RCS terminal. A Node Gateway undertakes the routing and the policing of the traffic within the node. Each node can serve one or more users, and distribution of services can be done by any LAN or MAN technology. In our platform, a WiMAX infrastructure is used to connect the end users to the satellite node, as shown in Fig. 11.1. At the user site (featuring a WiMAX Subscriber Unit), integrated services are presented using separate devices: a VoIP phone, an H.264 decoder/TV screen and a standard PC with a Web browser for the voice, video and service respectively.

The platform was tested in a real scenario during a 2-hour transmission via the HellasSat II satellite. A single customer, located near the remote satellite node, received triple play services via its WiMAX SU. For the DVB-S2 downlink, 18 MHz of satellite bandwidth were used, utilising 8PSK modulation and code rate of 3/5. For the uplink, the HellasSat commercial RCS infrastructure was exploited. Web access was satisfactory despite the increased response time, and voice connection was validated via a standard call to an external PSTN number. A live TV feed was served via the H.264 encoder, over multicast IP, and presented at the user's decoder.

Figure 11.2 depicts a screenshot from the GUI of the DVB-S2 Multiplexer, where the three different service streams are combined into a single Multiplex. The upper graph shows the video traffic (CBR H.264 multicast at 2.5 Mbps), the middle

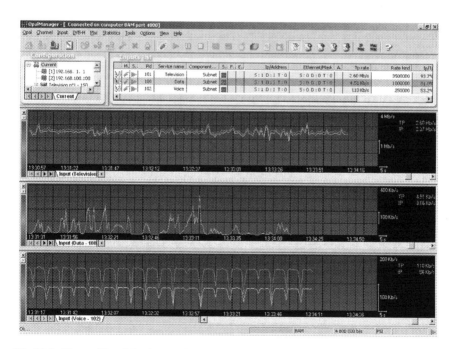

Fig. 11.2 The satellite triple play multiplex, as monitored in the DVB-S2 Multiplexer; per-service traffic view (video, data, voice)

corresponds to Web access, and the lower shows the voice-related activity (a single voice call at 64 kbps). In each graph, the lower line corresponds to the actual IP bit rate whereas the upper shows the bit rate measured after encapsulation (at MPEG-2 TS level). The difference between the two is most observable in the voice case, where the use of small-sized IP packets results in a relatively high overhead inserted by the encapsulation procedure.

Conclusions

The provision of integrated services over satellite provides a promising and viable solution in cases where fixed terrestrial access is not applicable (e.g. in remote/rural areas or in mobile use). The exploitation of the newly standardised DVB-S2 technology -which has been designed taking interactive services into consideration- adds efficiency and flexibility. This paper proposed and implemented an infrastructure for the provision of triple play services (voice, video and data) over a DVB-S2/RCS network, in an all-IP environment, featuring Web access, interfacing with the public PSTN network for voice calls and encoding/re-transmission of live TV feeds using H.264. The operation of the platform was validated in a real on-air trial over the Hellas Sat II satellite. The infrastructure proposed provides a quite efficient solution and, when optimised and combined into a single commercial platform/solution, is bound to add a new dimension to satellite-based communications.

Acknowledgements The design and implementation of the DVB-S2/RCS platform for triple play services provision is being carried out in the frame of the EU-funded IST IMOSAN research project (Integrated Multi-layer Optimisation in Broadband DVB-S.2 Satellite Networks). IMOSAN will also study issues concerning per-service multi-layer optimisation, exploiting the ACM (Adaptive Coding and Modulation) features of DVB-S2

References

[1] H. Clausen, H. Linder, B. Collini-Nocker, "Internet over Direct Broadcast Satellites", *IEEE Communications Magazine;* June 1999; pp. 146–151.

[2] ETSI EN 301 790. *Digital Video Broadcasting (DVB); Interaction Channel for Satellite Distribution Systems.* ETSI European Standard, 2000; v. 1.3.1; March 2003.

[3] ETSI EN 302 307. *Digital Video Broadcasting (DVB); Second Generation Framing Structure, Channel Coding and Modulation Systems for Broadcasting, Interactive Services, News Gathering and Other Broadband Satellite Applications.* ETSI European Standard; v.1.1.; March 2005.

[4] A. Morello, V. Mignone, "DVB-S2: The Second Generation Standard for Satellite Broad-Band Services" *Proceedings of the IEEE;* Jan. 2006; Vol. 94, 1; pp. 210–227.

[5] The IMOSAN Project Homepage. *IMOSAN—Integrated Multi-layer Optimization in Broadband DVB-S.2 Satellite Networks.* Available: *http://www.ist-imosan.gr.*

[6] ETSI EN 301 192. *Digital Video Broadcasting (DVB); DVB Specification for Data Broadcasting.* ETSI European Standard; v.1.4.1; Nov. 2004.

[7] G. Gardikis, A. Kourtis, P. Constantinou, "Dynamic Bandwidth Allocation in DVB-T Networks Providing IP Services", in *IEEE Trans. on Broadcasting;* Sep. 2003, pp. 314–318.

Chapter 12
Service Integration in SatLife Regenerative Network

I. Moreno, A. Yun, E. Callejo, F. Vallejo, J.A. Torrijos, R. Giménez,
C. Miguel, Dr. H.S. Cruikshank, S. Iyengar, J.A. Guerra, and M. Catalán

Abstract AMAZONAS satellite, launched in the summer of 2004, includes AMERHIS payload, being the first regenerative system using DVB-RCS/DVB-S standards to inter-connect four coverage areas located in Europe and America, in only one single satellite hop.

In the frame of SATLIFE, the different regenerative satellite terminals have been enhanced, including standard technologies to optimize performance through the satellite link. Management Station upgrade is in line with these RCST improvements, supporting new video terminals and multicast routing apart from broadcast.

Using the new developments in the prototypes, different multimedia services and applications have been integrated over the regenerative satellite network.

Index terms: Digital TV, DVB-RCS, IP, Multicast, Multimedia, Network, OBP, Regenerative, Satellite, Video Broadcasting.

Introduction

Within the 6th Framework Program, Satlife constitutes a successful try of validating DVB-RCS [1] regenerative systems for a great variety of interactive services. The trials will end on June 2007.

Satlife is an EC consortium composed of several companies and universities: Hispasat, Telefonica I+D, Alcatel Alenia Space, Indra, NERA, EMS, Shiron, Thales, Universidad Politecnica de Madrid and University of Surrey.

Regenerative concept is based on OBP behaviour. Following a MF-TDMA access scheme, uplink DVB-RCS signals are switched on board and modulated into DVB-S [2] transport streams. The satellite includes four transponders for AmerHis [8–10] payload, corresponding to the same number of coverage areas in Europe, North America, Brazil and South America.

The Regenerative System includes the following elements in the ground segment:

L. Fan et al. (eds.), *IP Networking over Next-Generation Satellite Systems* 157
© Springer 2008

- Management Station. MS is in charge of session control, on-demand or permanent connection control, dynamic capacity assignment and terminals management. It is also possible configuring the OBP from the MS.
- Regenerative Terminals. Located at subscriber premises, RCSTs act as IP unicast/multicast routers.
- Regenerative Gateways. RSGWs, managed by SPs, provide seamless integration with terrestrial networks and voice/video services.
- SP-RCSTs. Provides native MPEG-2 broadcasting services, when connected to Video SP, Digital TV, or SW Download servers.

The Project objectives were divided into a four-step procedure:

1. New services and applications definition.
2. System specifications to comply with such service performance.
3. Integration of the developed subsystems in a Laboratory System Prototype.
4. Real Trials in OBP System using the integrated prototypes.

Subsystems Developments and Laboratory Integrations

As shown in Fig. 12.1 and 12.2, Satlife prototypes developments are based on AmerHis ground segment network elements. Some of them are related to terminal capabilities improvements, others involve integration of new equipment in the regenerative satellite network.

The strategy followed for these developments and integrations was:

1. In-factory testing was performed by the manufactures,
2. Prototypes and applications integration in the System Laboratory.
3. Then, they were moved to the real scenario to complete in-orbit trials.

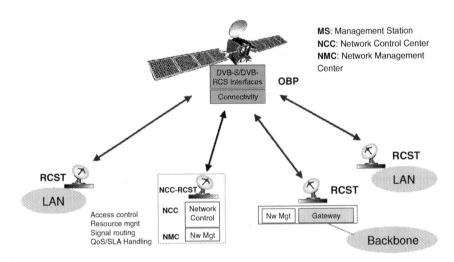

Fig. 12.1 Regenerative System Architecture

In this section, evolutions of the elements constituting the satellite ground segment are presented.

NCC

In the MS, improvements involve mainly the NCC, and are focused on multicast connectivity [4] and multi SP-RCST support. These developments are reflected in Fig. 12.3, and provide some new functionality:

- Multicast routes support. Possibility to define traffic routes with one or more destination TDMs, usable for IP multicast traffic. Bandwidth savings are achieved comparing to having only broadcast routes.
- Dynamic multicast PID range definition. By configuring the NCC with such ranges, it eases multicast flows administration at MPEG level, allowing multicast reception in the RSGW.
- VSP tables re-assembling. The System contemplates the presence of several VSPs simultaneously. To keep compatibility with standard IRDs, it is necessary that certain tables, like BAT, CAT, SDT, EIT, and RST, are built by the NCC using contributions of different SP-RCSTs

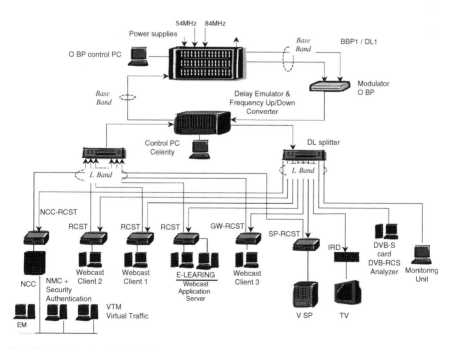

Fig. 12.2 Laboratory System Prototype

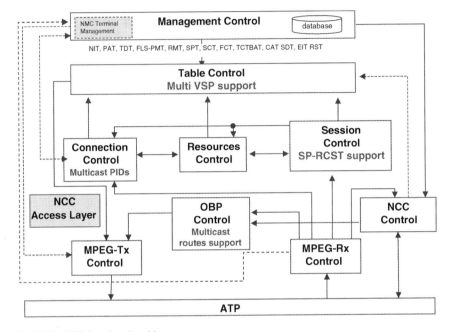

Fig. 12.3 NCC functional architecture

- SP-RCST support. This kind of terminal, used for native MPEG-2 video services, has particularities impacting NCC. Guaranteed bandwidth resources, together with Return Path Interaction descriptor at terminal logon, and PID reservation for multicast routes and DVB-SI tables, must be provided.

RCST

The RCST has been upgraded to improve performance and functionality. Some functions have been added:

- PEP functionality, to accelerate TCP traffic in mesh scenarios.
- Multicast has been enhanced in order to support mesh and star topologies simultaneously.
- QoS features of the terminal have been boosted by increasing the number of flow types, and include the possibility of blocking certain outgoing flows (reverse firewall feature).
- NAT. Network Address Translation provides additional security and saving of global IP addresses for clients located behind the terminals.
- New MIB objects to support security authentication handshake.

At the same time, compatibility with the rest of the System terminals from different manufacturers was required.

RSGW

As shown in Fig. 12.4 in regenerative systems, RSGW station has a similar function to the one of the hub in the transparent one, giving access to terrestrial networks (Internet, ISDN) and providing other services like dynamic multicast or Intranet access. Perhaps the main difference lies in the low-cost concept and the possibility of multiple installations in one or more system IP networks. Moreover, it does not support routing functions, and manages subscribers only from the service point of view.

In Satlife, the RSGW has been evolved with the following functionality:

- SIP integration. SIP and H.323 clients are supported simultaneously, even behind the same subscriber RCST. VoIP and videoconference services include ISDN incoming/outgoing calls, pure Internet calls, and conversations between Satlife satellite users. Accounting is performed by the RSGW, but the generated files will be later grabbed by the MS.
- H.323 enhancements. NATed endpoints support gives the possibility of establishing calls directed to the ISDN, Internet or the satellite network, while being registered at the RSGW Gatekeeper. Another interesting new feature is the multi ITSP assignment, having access to external ITSPs located in the Internet or in other RSGWs.
- Multicast reception. Multicast scenarios with sources at the satellite side and receivers in the Internet Multicast Bone were tested, in addition to the star multicast service.

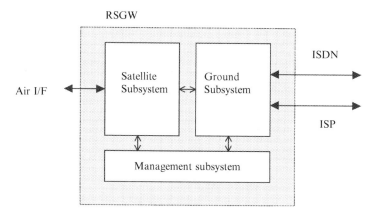

Fig. 12.4 RSGW architecture

SP

The Service Provider equipment multiplexes MPEG-2 and H.264 video contributions over MPEG-2 TS, and send them on an RTP stream to the SP-RCST through the Ethernet interface. It connects to SW Download and an Interactive Applications servers through an ASI interface.

An H.264 monitoring unit was also integrated in the laboratory to analyze video quality.

SP-RCST

This terminal has been developed to fully support DVB-RCS. being introduced to integrate native video services in the regenerative platform.

During logon, SP-RCST obtain some parameters from NCC in the TIMu message. Just before traffic transmission, a handshake is started from the SP, and following data is provided from the SP-RCST:

- Number of assigned TBTP bursts, and number of MPEG packets per burst.
- SYNC burst period and content
- Usable PID range for traffic as specified in the Return Path Interaction descriptor.
- DVB-RCS tables configuration: table presence and period.
- Management information related to the SP-RCST sent by the NCC.

Several tests were carried out:

- Management. It has been verified that SP-RCSTs can be provisioned in the System, and operated, using the current management platform.
- VSP tables tests. Multi-VSP management tests to validate DVB tables re-assembling by NCC and verifying compatibility with standard video receivers (IRDs).
- **Digital TV,** consisting on broadcast of digital TV channels and interactive applications over MPEG-2 TS from a single RCST directly to user premises STB or H.264 analyzer.
- **SW Download** test, multicasting multimedia data over MPEG-2 TS from a single RCST directly to satellite receive-only computers.

Signalling and Traffic Simulators

As shown in Fig. 12.5 and 12.6, focused only in Laboratory testing,Virtual Traffic Module (VTM) was designed to perform simulations regarding RCST- MS interface by using 'virtual' satellite terminals. This terminals were provisioned in the MS database, just like any other real RCSTs. VTM connects to the NCC Ethernet interface for signalling.

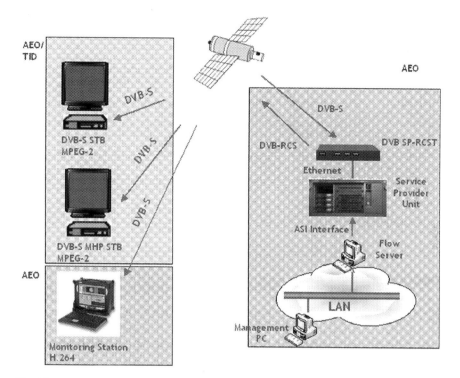

Fig. 12.5 Digital TV test bed

Several messages were reproduced:

- Session logon/logoff and synchronization maintenance, allows to simulate the maximum load of logged subscribers.
- C2P connectivity [3] [6–7]. The virtual terminals establish different-profile connections in order to stress the NCC.
- Capacity requests. Virtual terminals dynamically asked for RBDC and CRA capacity.

From these tests several conclusions were reached, NCC robustness has been checked, and there have been identified possible evolutions in the QoS schema. Since the AmerHis regenerative system has started commercial exploitation short ago, results are of crucial importance for NCC laboratory debugging.

Some of the recommendations derived from simulation results are:

- Minimizing the *number of required channel IDs* will increase link utilization.
- *Diffserv classification* according to traffic classes (EF, AF, BE), being the non-elastic (EF) flows the ones with higher priority.
- *Real time services require no buffering* to minimize jitter. DropHead buffer management is recommended.
- Within EF flows, it is advisable using *different queues for VoIP and Video*, due to the length of Video packets. VoIP priority must be higher than Video.

- A BoD scheme is required for AF flows, providing efficiency in the link sharing capacity, fairness and stability/quick convergence.
- For EF (non-elastic) traffic, the BoD recommended solution is based on session level signalling (for example, SIP/COPS/RSVP+Admission Control). Moreover, *the NCC should be notified of the required resources for each RCST*, differentiating between guaranteed resources for non-elastic traffic and desired resources for elastic traffic. Currently, NCC uses CAC according to RCST QoS service, and differentiates between RBDC and CRA dynamic capacity requests.
- *Priorities and weighted fair queuing (WFQ) algorithms* are recommended to assure no starvation of Best-effort flows. This is satisfied by the token-bucket algorithm in the RCST.
- *FCA policy* proportional to the requested bandwidth is proposed to improve performance of AF/BE traffic in non congested situations.

Results: Subsystems and Services integration, Trials

After Satlife prototypes integration in the System laboratory, they were installed in the real System to perform the trials, with participation and support from all the partners, from three coverage areas located in Europe, Brazil and South America.

Integration of end user applications and services has been done as a way of testing end-to-end solutions as E-learning, multi-videoconference, streaming, Internet access, digital TV, etc. The integration of final user services for AmerHis [5] and transparent RCS is of essential importance since, added to the satellite ubiquitous advantage, these applications will be hugely used by both professionals and home customers.

Targeted users, objective of the performed tests are:

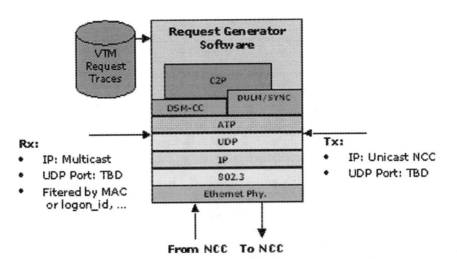

Fig. 12.6 VTM protocol architecture

- Every citizen that requires the service.
- Small and medium companies, providing the necessary communications.
- Schools, for an Internet access without problems.
- Hospitals, for telemedicine applications and connections with other hospitals.
- Public offices and town halls, for an effective and correct administrative work.
- Rural zones and small villages.

Nomadic Terminal

As shown in Fig. 12.7 this transportable terminal, integrated with a GPS unit, is carried in a MPV vehicle with a self-deployable antenna. A complete set of tests were performed:

- The nomadic RCST was completely integrated in the real System, being able to perform logon from different geographic locations, using automatic antenna (1.2 m) orientation. The operator only needed to manually enter in the GPS unit, the downlink frequency and the satellite position in degrees.
- Multiple regression tests were performed, showing total compatibility with the rest of System terminals. These tests included: DVB-RCS tables decoding, unicast and multicast connectivity, FTP and Web browsing.

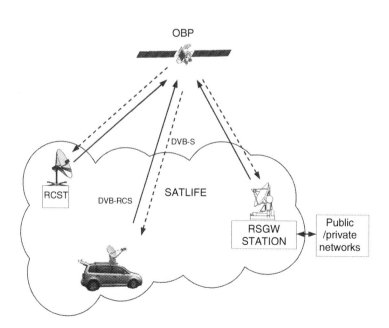

Fig. 12.7 Nomadic terminal testing

Video–on-Demand (VOD)

As shown in Fig. 12.8, VOD service has been tested keeping in mind other kind of applications, as for example remote learning or lessons (for tele-education), remote demonstrations, or any other kind of act that requires a visual media remotely available.

This comment has to be kept in mind when considering the bandwidth requirements, of at least **64 Kbps** of **SDR**. The test ran demonstrated a consumed dynamic bandwidth (RBDC) oscillating around **1 Mbps** while playing a content coded at this bitrate.

As far for the playback quality, technology integration and easiness of use, we have to state that everything is more than correct and perfectly implemented, so the final user will have a complete and working VOD STB fully functional in a few steps.

In case an inter-beam scenario had to be evaluated, it should be considered the option of installing two or more remote servers, one in each beam, or assess the possibility of accessing to only one VOD server but with only intra-beam satellite link, so the rest of the access should be made by other means.

It is also to be noted the perfect synchronization of the STB with the multicast channel, so once hooked on it, the needed parameters are recovered, and then the platform starts to operate showing the available contents to the user.

The playback is also correct for both MPEG2 and MPEG4 contents, being the required bandwidth lower for the second, or a higher quality using the same bandwidth, as it was expected.

We could also appreciate the use of the standard protocols. The captured traced packets shows the use of the HTML protocol for the information page display, and the RTSP for the video/audio control communication between the STB and the server so it its possible the use of any standardized STB.

Summarizing, we can say that this test positively tested the multicast support, the VOD session control and the dynamic and efficient resources management.

Video-on-Demand provides unicast MPEG-2 TS over IP on user request to standard IP STBs, with RTSP support; and Near-Video-on-Demand works based on multicast, optimizing satellite bandwidth. Standard user RCSTs are used for sending and receiving traffic.

Middleware

As shown in Fig. 12.9 it is generally accepted that not everyone will want to have a PC to use multimedia services. Making sure that services, especially online public services, are available over different terminals, such as TV sets or mobile phones, is crucial to ensure inclusion of all citizens. The Satlife project will contribute to achieve these objectives using the regenerative DVB-RCS system, that will allow users to access multimedia services from different devices (PC, TV, etc).

This battery of tests was in charge of probing the development of multimedia services using this middleware, such as, TV broadcasting, VOD and Internet access.

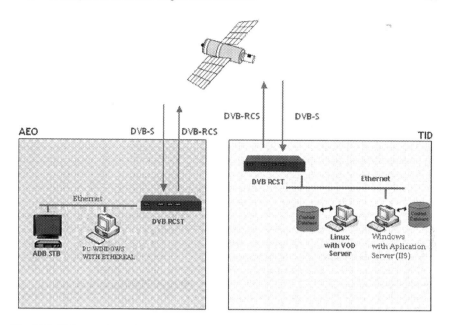

Fig. 12.8 Video-on-demand test

It was tested here again the correct multicast support, VOD authentication, and dynamic and efficient resources management. Other aspects that could be appreciated were the standardized protocols used for the data: HTTP for the user interface, and web browsing and RSTP for the VOD contents.

The web browsing positive test also showed the correct placing and displaying of the images, tables, forms…, apart from the correct execution of the JavaScript code, the correct use of the cascade style sheets and the proper work of the Java applets. The HTML 1.1 support was also verified.

We can say that, being able the STB, as it was, of acknowledging and performing the remote control keystrokes, this is a perfect solution for web browsing for those users who will not want to have a PC, while maintaining at the same time its rights to access to the Information Society.

ADSL and Satellite Internetworking

Another target of Satlife project is the seamless integration with terrestrial networks. An exponent of this integration is the interoperability between satellite and ADSL networks.

Successful integration of both technologies allows to provide broadband access to rural environments or small communities, which can not afford a single user terminal. However the cost of an ADSL concentrator plus the RCST equipment can be easily assumed by all the community.

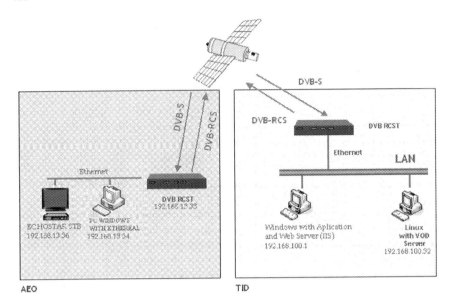

Fig. 12.9 Middleware test platform

Satlife has integrated a solution based on a RCS terminal and a mini-DSLAM to efficiently attend rural regions, as a way to deploy satellite solutions for Internet mainly, like connectivity and other services of interest.

This test proved that, until then, correct integration with the already existing equipment is also possible, and the seamless and transparent network integration was achieved.

It was tested the proper functionality of the following situations: PPPoE single user access, when the customer is using a USB ADSL modem, PPPoE multi-user scenario, using an ADSL modem-router and both situations of single/multi-user scenario making use of bridging and routing protocols. These four scenarios are nowadays the most used in the ADSL configurations and perfect integration with the RCST satellite technologies was verified.

The following diagram shows the typical ADSL and Satellite Internetworking architecture proposal.

Home Gateway

As shown in Fig. 12.10 and 12.11 in this set of test it was exploited many of the functionality that a Home Gateway device could offer for the remote surveillance and management of a home, using the satellite regenerative platform as access.

The used device, a Pylix™ Siemens gateway proved its support for all the home tested devices under the LonWorks protocol. These domestic devices comprised in our demonstrator the lights, a shutter, a dishwasher machine and an oven.

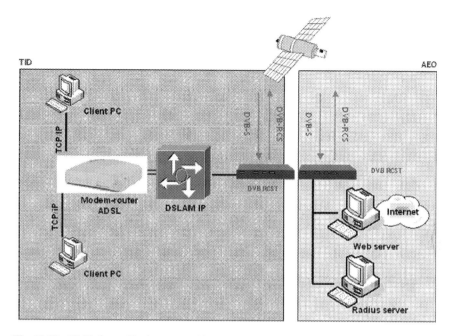

Fig. 12.10 ADSL & satellite internetworking

It was also tested a surveillance equipment conformed by an IP camera with the appropriate software, connected to the Pylix™ gateway via its private network through RS-232.

We have to say that the required bandwidth for all these tests is not very high, being enough with 100 Kbps for the video surveillance which is the application with a higher bandwidth demand.

We could correctly appreciate the interoperation and seamless integration of all the involved networks: the Pylix™ private network to which was attached the IP camera, the LonWorks network also connected to the Pylix™, the home private network managed by the RCST and the satellite link and remote satellite platform.

It was also tested the platform easy of use. All the operations were performed via an HTML interface displayed to the remote user. It also demonstrated the possibility of using the IP camera not only for real time surveillance, but also for establishing alarms and actions to be taken in case one of them is triggered.

Streaming

As shown in Fig. 12.12, we could consider this set of test as one of the most resource-demanding. The required bandwidth is of at least **100 Kbps** of **SDR** for the smallest video capabilities and of about **1 Mbps** if more quality is to be considered.

It is also to be noted that a high PDR and a very low SDR does not suffice for this service, as for the video reproduction in a correct fashion requires a high sus-

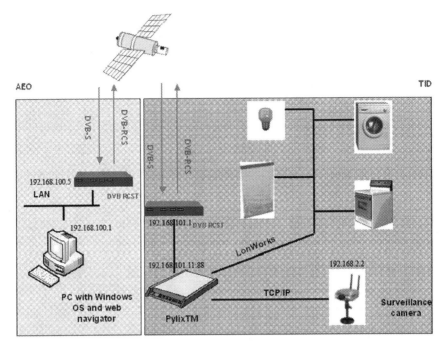

Fig. 12.11 Home Gateway test setup

tained bandwidth during all the content playback, even taking into account the buffers implemented by the players.

The conclusions obtained from this section are the following: the service is functional for low-medium size-quality, but when the quality is crucial a higher SDR is needed, in other case, the results will not always meet expectations.

We have also to note that the use of multiple bitrate contents makes possible serving with only one content to a varying range of clients with different download speeds each one of them, achieving the best quality in each case.

The possible applications of this service are real time retransmission of remote events or for example video surveillance of sensible or dangerous areas. Taking into account that the bandwidth requirements for this last option are not very high (of about 100 Kbps, as demonstrated during the test), this option reveals as a perfect one for some specific situations.

The test also proved the seamless integration of all the technologies involved: various media servers, the satellite link and the corresponding media players, apart from the IP camera.

Multiconference and SIP

As shown in Fig. 12.13 and 12.14 these tests consist on MCU multicast scenarios integration in the satellite network. Up to three MCU are used, two of them located

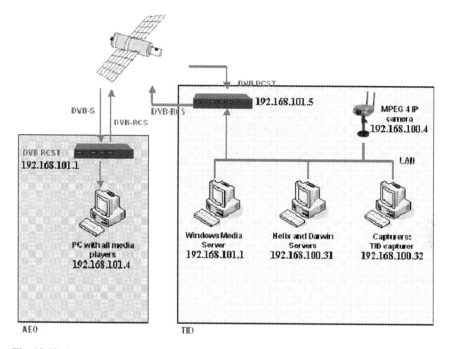

Fig. 12.12 Streaming test platform

in the Satlife satellite network, and the third in the Internet, accessing to the system through an RSGW. SIP clients behind user RCSTs and from the Internet participate in the multiconferences.

The MCUs register in two SIP proxies placed in the satellite network and in the Internet. UDP flows among them are multicast, taking advantage of the enhanced RCST multicast features and the satellite on-board replication. This implies less than **1 Mbps** of capacity needed for the three-MCU call. Only one satellite hop is needed employing multicast traffic, high interactivity being possible with **300–400 ms** of delay.

Other multiconference scenarios are possible, without MCUs, using SIP/H.323 clients supporting multicast. This implies several RCSTs transmitting and receiving to/from the same multicast group, avoiding echo issues, meaning savings in MCU licensing. They would be applicable for conferences with less participants (<6).

Webcast for Corporate Services

These trials provided Web-based Corporate oriented services, including a wide range of functionality for synchronous and asynchronous modes with low or high interactivity. They were tested in a meshed scenario. The applications supported are:

• Live events. The audio and video was multicasted from the teacher RCST and received in all the terminals in a synchronized way, using a stream rate of **300 Kbps**. PowerPoint slides were also synchronously presented.

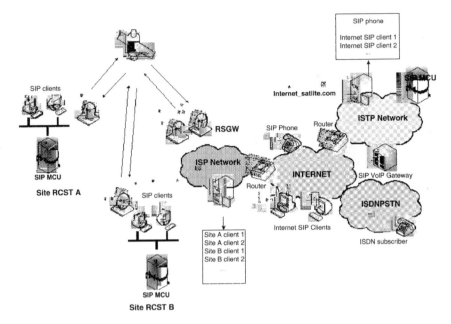

Fig. 12.13 MCU multiconference scenarios

- E-Learning. The students could interact in the event chat, being the chat server placed at the Application server, and in a multiconference collaboration environment, with the MCU at the Application server. The delay was slightly higher than 1 sec due to the double satellite hop, but suitable for the interaction. The unicast flow from each terminal to the server is of **100 Kbps**, and from the server to the attendants **is N**

***100 Kbps**.

- Media Repository
- Webcasting Channels

Internet Access

As shown in Fig. 12.15, this section has positively tested the scenario of users accessing the Internet with private IP addresses.

The RSGW Service Provider selected an ISP, which was in charge of providing the required Internet access link and supplying a range of public IP addresses. NAT functionality is provided by the RSGW itself.

Being this service a basic and absolutely necessarily one, we could say that the successful results guarantee first hand a high probability of success of the whole project.

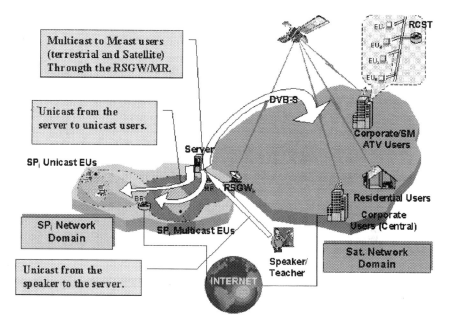

Fig. 12.14 E-Learning live events/courses multimedia flows

E-mail, FTP, HTTP and Instant Messaging are the most used applications nowadays. If we also add Audio/Video Streaming, all these account for more than 80% of all the potentially applications used by a residential or even professional user.

As the tests have proved, all the services are high quality functional and operative, being the only one really dependant on the speed granted the audio/video streaming, but a **SDR** of about **100 Kbps** would suffice for this last one.

For the rest of the services, a minimal SDR is enough while maintaining a reasonable PDR, so it is available when needed.

The tests have proved that it is both possible to receive or send mails, to browse the web, receive or send large archives via FTP, make use of the Instant Messaging facility or even watch a video served from elsewhere.

We should also note that having been proved for both the European and American beam, the availability is granted anywhere, but for the Audio/Video streaming service, a minimum SDR needs to be defined at the NCC for the European/American uplink routes.

Intranet Access

Simulating a corporate environment, different applications were tested in the Satlife network, using mesh connectivity:

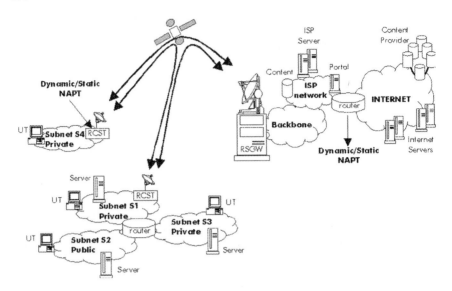

Fig. 12.15 Internet access scenario

- Remote desktop management. Using Windows XP remote desktop and Tight VNC client/server applications. This is one of the most delay-sensitive services and therefore the performance was better with SDR > 0 (guaranteed capacity).
- FTP. File transfer TCP performance was checked, comparing the effect of PEP in the RCST. Results were very clear: with PEP function disabled, the transmitting terminal did not reach the PDR of the C2P connection, or took a long time to achieve the maximum capacity. When PEP was enabled, peak bitrate (2 Mbps) was reached in less than 5 secs, and capacity spent for ACKs was negligible.
- Videoconference. Calls were established with low bandwidth consumption, separating signalling and data packets according to DSCP IP header value. RCSTs opened high-priority jitter sensitive connections, in a separate channel, for a defined range marked with Windows Packet QoS Scheduler. Results were very satisfactory for G.723/H.263 codecs, with total bitrates under **100 Kbps**.
- Web browsing performance was verified, with IS HTTP server and IE client. Navigation was smooth, with an only initial delay in the first accessed web page.

The test demonstrated that both intrabeam and interbeam operation was possible and functional, as it was expected from the Amazonas versatility.

Security Authentication Application

As shown in Fig. 12.16, this utility can be considered as a first step in the implementation of the solution proposed in [1]. Through the management channel between the MS and the RCST, set at terminal logon, this mechanism uses SNMP protocol and Quick Key Exchange (QKE) to authenticate any Satlife terminal provisioned in the System.

The implementation affects MS and RCST:

- In the RCST, a new OID has been included to store the generated nonce, a random generator module and a function to create the hash value to be sent to the MS, were included in the firmware. The static value to create the hash was a secret cookie place in the RCST file system.
- In the MS, an application was designed to generate the nonce value to be sent to the RCST, and verify the hashed message authentication code sent by the RCST in the QKE response. The same static values than the ones in the RCSTs are stored in the MS. This application is able to access the System database, to obtain a valid listing of the RCSTs/GW-RCSTs, together with their IP satellite addresses.

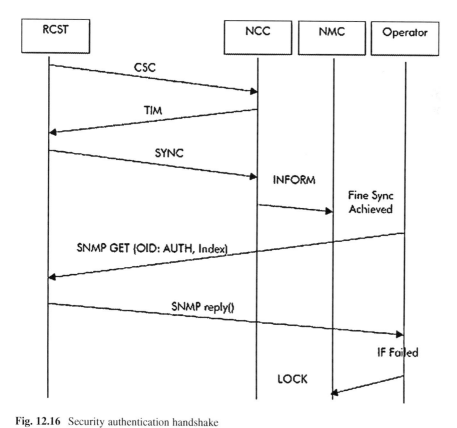

Fig. 12.16 Security authentication handshake

The method demonstrated to provide additional security to the existing MAC authentication procedure, and could be used in future implementations of C2P protocol, for authentication and encryption.

Conclusion

New satellite generations are including on-board processor payloads to make more efficient the provision of multimedia services. SATLIFE project has designed and built advanced applications and services to operate with AMERHIS, the DVB-RCS on-board processor of AMAZONAS satellite.

Conventional services only possible today in star configuration have been launched on the base of SATLIFE, leaving a direct way to future development. As a best approach for real exploitation, the trials used real satellite capacity, with end users in different coverage areas, to demonstrate service efficiency and low cost of deployment, being the main project objective the achievement of a 'broadband for all'.

Development and adaptation of DVB-RCS satellite terminals to other technologies to improve performance and security, while keeping interoperability, is the key for internetworking with terrestrial networks, to transparently offer multi-service access to home and corporate users.

Acknowledgements This work is supported by the IST FP6 SATLIFE project, funded by European Commission (EC). The financial contribution of the EC towards this project is greatly appreciated.

References

[1] EN 301 790, DVB Interaction Channel for Satellite Distribution Systems, ver 1.2.2, ETSI.
[2] EN 301 192, DVB Specification for Data Broadcasting, ver 1.2.1, ETSI.
[3] TR 101 790, DVB Interaction Channel for Satellite Distribution Systems, Guidelines for Use of EN 301 790, ver 1.2.1, ETSI.
[4] A. Yun, I. Moreno, and F. J. Ruiz Piñar, "IP Multicast over New Generation Satellite Networks. A Case Study: AmerHis", IWSSC 2006.
[5] I. Jiménez and I. Moreno, "AmerHis: Exploitation and Service Definition in an Interactive Regenerative Satellite Network For Mesh Communications", 12th Ka Band Conference, 2006.
[6] A. Yun, E. Callejo, S. Combes, and D. Mignolo, "Harmonization of Connection Control Protocol (C2P) for DVB-RCS Systems", AIAA, 2007.
[7] P. Conforto, S. Pellegrino, A. Yun, D. Mignolo, and S. Combes et al., "Harmonised Network Management in DVB-RCS System with Connection Control", AIAA, 2007.
[8] F. Vallejo and A. Yun "AmerHis: Triple Play Over an OBP-Based DVB-RCS Satellite Platform", 23rd AIAA ICSSC, 2005.
[9] A. Yun and F. Vallejo, "DVB-RCS Meets Direct Mesh Connectivity", Satlabs DVB-RCS Symposium, Sep. 2005.
[10] M. Wittig and J. Maria Casas, "A Communications Switchboard in the Sky: AmerHis", Esa Bulletin, No. 115. Aug. 2003.

Chapter 13
A Novel QoS Architecture for Next Generation Broadband Satellite Systems

A. Ramos, B. de la Cuesta, B. Carro, J. Aguiar, D. Pérez, C. Baudoin, P. Berthou, S. Abdellatif, and T. Gayraud

Abstract This paper presents the architecture developed in Satsix project in order to improve the current QoS level in satellite/terrestrial networks. The dynamic QoS support is one of the major achievement of SATSIX in the QoS area, with the extension of the QoS agent/server architecture proposed in the previous SATIP6 project. This QoS architecture is based on DiffServ architecture, and it has been proposed aiming to tackle different QoS level sessions.

Introduction

Several past and on-going EU and ESA initiatives aim at developing QoS (Quality of Service) architectures for integrated satellite/terrestrial networks, however SATSIX completes that earlier work thanks to a dynamic QoS support that allows improving the QoS level achieved.

The paper is organised as follows: firstly in Section B a class definition has been proposed meeting application requirements based on several other traffic engineering standards such as BSM. After presenting the QoS requirements in Section C, Section D explains the overall QoS architecture based on the DiffServ architecture which has been proposed aiming to tackle different QoS level session. The session control and connection control mechanisms are also considered, and interworking issues with either WiFi and WiMAX networks are presented as well as several techniques/solutions that have been proposed to ensure a integrated QoS support on both segments. Finally, in Section E the Gateway and terminal QoS architectures are detailed where enhanced IP scheduling techniques are identified, and a new HDLB (Hierarchical Dual Token Bucket) scheduler is defined allowing a hierarchical repartition of the guaranteed/booked throughputs along with an optimized scheduling.

L. Fan et al. (eds.), *IP Networking over Next-Generation Satellite Systems*
© Springer 2008

Traffic Classes

Having very different characteristics from services deserves different grade of service; several standardization bodies have tried to define service categories (also called QoS classes, to be intended at application layer).

The definition of traffic classes for SATSIX network relies on the service requirements defined in terms of performance parameters and on the specifics of level 3 and level 2 mechanisms. The definition of different traffic classes has been based on the existing traffic model BSM [1, 9] and SatLabs [2] ETSI [10], in Table 13.1.

Table 13.1 Proposal for the SATSIX traffic classes

	Traffic Class	Service Categories	Services
0	Pre-emption	Reserved capacity(CRA)	EF
1	Real-Time – High interaction	Reserved or requested capacity(CRA, RBDC). strict conformance	EF Vo IP.
2	Real-Time - Interactive	Reserved or requested capacity(CRA, RBDC). loose conformance	EF Real time video, Interactive gaming.
3	Transaction Data - Highly Interactive, high bandwidth, medium delay and jittertolerant	Guaranteed + best effort dynamic capacity (G-VBDC, VBDC, FCA)	AF1 Signallling, Interactive TV, TV digital.
4	Transaction Data – Interactive, medium band width, medium delay, low jitter	Guaranteed + best effort dynamic capacity (G-VBDC, VBDC, FCA)	AF2 Instant messaging, Content distribution, broadcasting, streaming media.
5	Low Loss Only(Short Transactions, Bulk Data, Video Streaming), low bandwidth, low/medium loss, medium delay	best effort dynamic capacity (VBDC, FCA)	AF3 Bulk data, File Sharing, Web browsing, E-mail, Peer to peer.
6	Best Effort	Best effort dynamic capacity (VBDC, FCA)	BE

QoS Requirement

IP QoS Requirements

At network level SATSIX system provides different qualities of service in the RCST (Return Channel Satellite Terminal). "QoS assured" or "QoS enabled" application sessions will be enabled allowing QoS guarantees and wireless and satellite bandwidth optimisation.

- A "QoS-Enabled" session allows the endpoints to complete the session establishment either with or without the desired resources. Such session will use dedicated resources if available, and use a best-effort connection as an alternative if resources cannot be dedicated. In cases where resources are not available, the originating and/or terminating user end-point might check with the user to obtain guidance on whether the session should complete or not.
- A "QoS-Assured" session will complete only if all the required resources are available and assigned to the session. A provider may choose to block a call when adequate resources for the call are not available. For voice service, public policy demands that the phone system provide adequate quality at least in certain cases: e.g., for emergency communications during times of disasters. Call blocking enables a provider to meet such requirements. This mode should be preferred for real-time services especially for a full telephony service and will be assumed in the developed scenarios.

This distinction is done thanks to a cascaded architecture of Bandwidth Brokers which are elements in the Diffserv architecture that process the on-demand admission requests for IP resources and they have Connection Admission Control functions. These brokers can interface with the applications in different ways in order to know the QoS capability to offer. First, the QoS agent and QoS servers developed in SATIP6 [3] allow to intercept application flows and assign a given QoS class to each of them according to user or administrator configuration. Second, a centralised Bandwidth Broker in the access network (e.g. located in a satellite Gateway or NCC, Network Control Center) can directly interface with some application session server.

MAC QoS Requirements

MAC layer QoS support consists in capacity request calculation/generation and scheduling/dispatching of packets from the supported queues to fill-up the assignments in TBTP (Terminal Burst Time Plan). The MAC/DAMA (Demand Assigned Multiple Access) algorithms concern the satellite procedures for bandwidth assignment. The objective is to guarantee an efficient use of the valuable satellite capacity and to respect the QoS requirements of the in progress connections.

The Connection Control Layer handles the bearer setup for each media component, including admission control and resource reservation. However, bearers are considered end-to-end here, or at least edge-to-edge inside a network domain. The connection control requirements must apply over all the different DVB-RCS scenarios defined within the frame of SATSIX project.

QoS Architecture

Functional Architecture

The QoS architecture used in SATSIX has two approaches:

- "MAC-oriented", where the communication between the RCST and the gateway is based on C2P (Connection Control Protocol), what adds complexity and delay in the communication, but in advantage it can address both star and mesh topologies.
- "IP-oriented", where specific signaling is used between the RCST and the gateway. It is possible only in star systems Fig. 13.1.

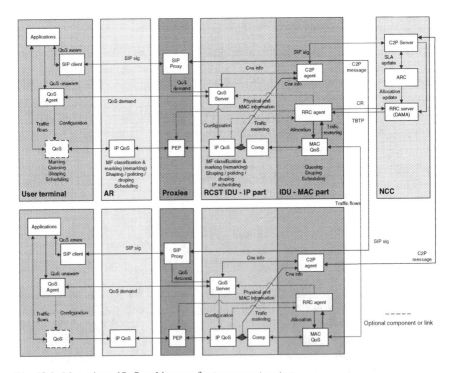

Fig. 13.1 Mac orientedQoS architecture for transparent system

The satellite segment shall provide QoS in terms of Packet Loss, Delay and Jitter and shall interwork with Internet QoS DiffServ in order to provide end to end QoS at network level. The terminal model shall perform this interworking in terms of signalling and QoS parameters mapping.

This functional architecture not only supports end-to-end QoS, but also dynamic QoS according to applications and users needs. The satellite segment can interwork with Internet QoS DiffServ in order to provide end to end QoS at network level. The terminal model can perform this interworking in terms of signalling and QoS parameters mapping.

The satellite network model implements the following Traffic Management procedures:

- (Dynamic) Traffic Resources Management
- Traffic scheduling
- Traffic shaping/policing

Entities and Components

The elements in the QoS SATSIX architecture are the following:

- SIP (Session Initiation Protocol) proxy. It will inform the QoS server about signalling information (new session establishment, SIP parameters, etc.). In the IP-oriented it will also inform the Access Router controller (ARC) about SIP signalling.
- QoS Agent. It will allow the user to select a QoS level for each application or for a given traffic flow and it will send this wanted QoS level to the QoS Server.
- QoS Server. It is in charge of collecting the QoS information from the SIP proxy and the QoS Agent on traffic flows and to configure accordingly the IP and MAC layers.
- PEP (Performance Enhancing Proxy). It is in charge of the TCP and HTTP acceleration using physical layer information in order to limit the delay.
- IP Compression. It carries out a header compression in order to decrease the overhead introduced by the successive stacks and thus decrease the load and the delay. It shall proceed to a negotiation to configure the needed parameters.
- IP QoS, It has the functions of classification, marking; policing/shaping/dropping and the scheduling at IP layer:

  - The classification selects which flows are to receive a particular service.
  - The marking function sets the DSCP (Differentiated Service Code Point) field according SLA (Service Level Agreement) policies and the results of the classifier functions. The DSCP value will determine further PHB (Per Hop Behaviour) actions to be applied on this packet.
  - The policing/shaping/dropping function ensures that a traffic stream conforms exactly with the parameters given by a particular traffic profile. This

may cause some packets to be delayed, or even dropped if the particular packet stream largely exceeds its traffic profile. When a conventional router has to forward a packet, the packet enters into a FIFO (First In First Out) queue in order to wait transmission of preceding ones. In a DiffServ architecture the FIFO is replaced by several queues (one for each PHB class) with their own specific service policy.

- A scheduler will serve the different queues and distribute available transmission capacity among them according their corresponding service class policies.

- MAC QoS. Its functions at MAC layer are:

  - Queuing the incoming packets in the corresponding MAC FIFO.
  - Dropping according to the MAC FIFO configuration.
  - Scheduling the packets stored en the queue according to the allocation provided by the RRC (Radio Resource Control) agent.

- RRC agent is responsible for generating capacity requests according to IP and MAC information, and for the extraction of the uplink packets function of the allocation carried in the TBTP. Requesting capacity becomes more troublesome in satellite scenarios, due to their large bandwidth-delay product and to the scarcity of bandwidth resources. Therefore it is important to consider that the delay between a user's request and the network's response causes difficulties in the design of a stable protocol and the protocol should guarantee that the requested bandwidth is every time fully utilised, in order not to waste link capacity.

- RRC server will receive the capacity requests of each terminal, and the SLA modification from the C2P server or the ARC (according to MAC-oriented or IP-oriented approach respectively) and it will calculate the allocation plan sending back the TBTP to the terminals.

- Access Resources Controller (ARC) that will perform the connection acceptance control (CAC) and will inform the RRC server about connection modifications and the required level of QoS.

2 other components are present in the MAC-oriented architecture:

- C2P agent, which handles the MAC connection at the terminal side.
- C2P server, which handles the MAC connection at the NCC side.

Session Control

Very few applications that are implemented today are aware of the QoS provided by the underlying network. As the applications are rarely able to define their own requirements, a user-oriented solution allowing any application to take benefits from network services has been defined in the SatIP6 previous project. The solution proposed is called a "QoS Agent".

Implemented on the user terminal, the "QoS Agent" waits for application streams, maps them statically or dynamically to the chosen QoS level according to the user's choice and, through interaction with the QoS Server module that configures the RCST in order to take the request into account.

The entity in charge of QoS reservation is a standard SIP proxy with enhanced functionalities. The main functionality of a SIP proxy is to route the SIP messages from a SIP entity to another SIP entity and so on to the final destination. If we want the QoS reservation to remain transparent to the application, the enhanced SIP proxy must intercept session descriptions (defined using SDP, Session Description Protocol, syntax) from SIP messages exchanged between caller and callee, translate the parameters of each media in the session in QoS characteristics and manages QoS reservations with the QoS server on behalf of the application.

The functionality added to the standard SIP proxy is:

- A SDP analyzer to enable the proxy to understand the session descriptions defined in SDP format included in the SIP messages.
- A media table updated at SIP session establishment. Each media is defined by a 4-uple (IP source address, IP destination address, source port, destination port), the kind of media (audio, video...) and the Real-Time Protocol (RTP) profile. The media negotiated by caller and callee are gathered using the Call-ID, which is a unique identifier of the related SIP session.
- A SDP/DiffServ mapping from the kind of media (and the codec) to the DiffServ service (and the corresponding peak rate), essential to the QoS module.
- A QoS module that takes into account the resource reservation corresponding to one media in a SIP session. It manages QoS reservations with the QoS Server.

Connection Control

The philosophy of network connection control functionality is formed by the following elements:

- PCF (Policy Control Function): The QoS and traffic requirements are carried in the SDP (Session Description Protocol). The PCF function translates, for each media component, the media description into a set of IP QoS parameters.
- AC (Admission Control): function is requested by the PCF function to verify the actual availability of requested resources in the core transport network.
- AG (Access Gate): is located in the transport data path and performs per micro-flow processing (policing, monitoring statistics, etc). The AG performs all the interworking between the Access Network specific procedures and the Core Network.
- ARC(Access Resource Control): if there is no a protocol to carry QoS parameter between the Access Gate and end user, so it is necessary a similar admission control block for the access network. In the case where no ARC block is

implemented, the admission control in the satellite access network will be based on static configuration of QoS parameters.

Interworking between satellite and WLL

QoS Interworking with WiFi

The main contribution of 802.11e is the introduction of a new coordination function called Hybrid Coordination Function (HCF). HCF integrates two new access techniques: a contention based channel access technique named EDCA (Enhanced Distributed Channel Access) and a controlled channel access technique named HCCA (HCF Controlled Channel Access). EDCA adds to DCF medium access differentiation by distinguishing four classes of service (named Access Categories): VOice (VO), VIdéo (VI), Best Effort (BE) and BacKground (BK). Each targets a different class of applications with different QoS requirements.

EDCA access categories were designed to correspond to 802.1D traffic classes [4] which were incorporated to 802.1D to enable layer 2 switches and bridges to support time-critical traffic, such as voice and video, effectively (the materiel defining 802.1D traffic classes is sometimes referred 802.1p which was a supplement to the earlier version of 802.1D). 802.1D defines 7 traffic classes, each of which can benefit of simple segregation form the others. In descending importance, these classes are: network control, voice, video, controlled load, excellent effort, best effort and background. Based on these traffic classes, 802.11e provides guidance on the mapping of these traffic classes into EDCA access categories [6] Table 13.2.

When comparing SATSIX traffic classes to 802.1D traffic classes, it is easy to draw a parallel between SATSIX classes of traffic 0,1,2,6 and respectively network control, voice, video and best effort 802.1D traffic classes. Based on table 3–7, these classes of traffic can be mapped respectively into EDCA access categories VO, VO, VI, and BE.

Concerning traffic class 3 which supports highly interactive data transactions, we propose to map it into access category VO since such services have a delay requirement in the same order of voice applications and generate a small amount of traffic.

Concerning traffic class 4 which corresponds to interactive (but delay tolerant) data transactions, we propose to map it into access category VI since such services have a delay requirement in the same order of those devoted to traffic class 1.

Table 13.2 Recommended 802.1D traffic class to EDCA access category mapping

Network control	Voice	Video	Controlled load	Excellent effort	Best effort	background
VO	VO	VI	VI	BE	BE	BK

Finally, traffic class 5 is mapped to access category BE. Note that the medium access service provided to access category BK could have been sufficient for some of the services associated to SATSIX traffic class 1. This is particularly the case of bulk data transfer, file sharing and peer-to-peer but not the case of web browsing which requires at least best effort service. This is the reason of this mapping Table 13.3.

Concerning the EDCA parameters, we propose to follow the default values proposed by the IEEE 802.11e amendment and also adopted by the Wi-Fi alliance.

QoS Interworking with WiMax

WiMAX is based on the set up of connections at MAC level, and depending on the particular CS (Convergence Layer) used, one can realize particular mapping of the traffic flows of the upper layers into the MAC traffic classes provided by the WiMAX specification.

So a scheduling strategy for the BS (Base Station) polling task can be defined to be adapted to the scheduling strategy used in the RCST terminal for serving the input queues, after the mapping between the traffic classes used in WiMAX and those used in the RCST.

Every time a new connection is set up the user application sends SIP parameters that can be intercepted by the RCST and if the WiMAX grant process has success a MAC connection is established at WiMAX level, starting then to send packets without a real relation between the two sub networks.

Moreover, if the dynamic bandwidth allocation is allowed at WiMAX node then the application should renegotiate the QoS parameters through SIP but stopping the services it is supporting. In fact when the dynamic definition and change of service flows take place, a possible overriding of the provisioned QoS parameters negotiated can exist.

The SS (Subscriber Station) can modify its behaviour and renegotiate the QoS parameters of the active CID (Connection identifier), but this renegotiation cannot be intercepted by the satellite subnetwork.

So ideally there could be an information exchange between the Authorization Manager and the satellite bandwidth allocation management process for a dynamical redefinition of QoS parameters before allowing any change in the service flow profile Fig. 13.2.

Table 13.3 Proposed SATSIX traffic class to EDCA access category mapping

&$$$;	0	1	2	3	4	5	6
&$$$;	VO	VO	VI	VO	VI	BE	BE

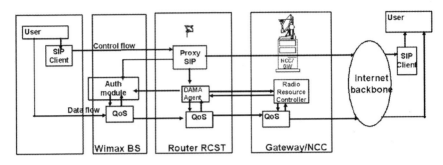

Fig. 13.2 Functional architecture for the WiMAX interworking

QoS Support in Terminal and Gateway

QoS Support in Terminal

At IP layer, Satellite Terminal in SATSIX acts as DiffServ node, and from a logical point of view it has two functionalities: packet classification and traffic conditioning. At MAC layer, the RCST includes classical MAC queues controlled by the RRC agent in charge of the resource allocation. The MAC layer relies on the C2P which is responsible of the access connection establishment, modification and release Fig. 13.3.

Thanks to information provided by the SIP proxy or the QoS agent, the QoS server inform the C2P agent to create the corresponding access connection or modify their characteristics accordingly. This leads to modification of the CRA or RBDC (Rate Based Dynamic Capacity) parameters at the NCC side. The QoS server also adds filtering rules to manage these flows and modify if needed the IP traffic class parameters. In short, as a consequence of an incoming SIP multimedia flows, access connection parameters are modified, leading to a modification of the associated CRA and RBDC parameters handled by the RRM as well as the IP scheduler configuration.

In order to avoid the impact on the establishment delay in the star topology case, connections between the RCST and the GW are established during the logon phase and their parameters are dynamically modified according to SIP proxy or QoS server information using C2P. This solution will allow to have the same QoS architecture independently from the target system, to avoid the impact on the establishment delay in the transparent case and the need for a SIP proxy in the GW.

SATSIX IP QoS relies on the notion of Classful Queuing Discipline (CQD). In order to implement CQD, a new scheduling algorithm called Hierarchical Dual Leaky Bucket (HDLB) based on HTB [5] has been defined to bring better support for burst traffic, with the introduction of [6] scheduling inside the hierarchical

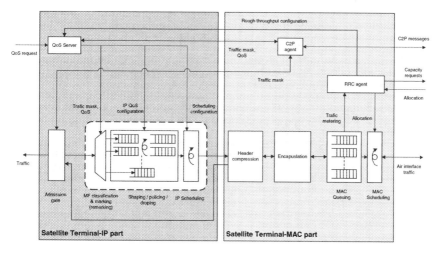

Fig. 13.3 Terminal QoS architecture

classes. Obviously, this new scheme benefits from the hierarchical approach which is admittedly well suited for fairly share capacity between different classes taking into account guaranteed bandwidth and capacity borrowing. The last advantage is related to the high level of configuration, that allows to handle different traffic classes definition such as SatLabs [3] or BSM [2] and to modify dynamically the guarantee offered for each classes as well as the output throughput of the scheduler. Hierarchical Token Bucket (HTB) ensures that the amount of service provided to each class is at least the minimum that it requests and the amount assigned to it. When a class requests less than the amount assigned, the remaining (excess) bandwidth is distributed to other classes which request service. The introduction of DLB (Dual Leaky Buckets) inside the HTB infrastructure will improve the scheduling performances with a better management of bursts. Thus, DLB accepts at the minimum, the compliant traffic which has been agreed for a service class. Nevertheless, even non-compliant traffic can be admitted partially or totally depending on network load conditions. As a matter of fact, the goal of the traffic control module is to admit as much as non-compliant traffic pertaining to a service class as possible, without infringing the negotiated QoS requirement of other service classes.

The overall scheduling is optimized by improving the coordination between the IP and MAC scheduling stages. Indeed, lack of coordination leads to scheduling decision that are neither relevant nor optimal. The link between both layers relies on a back pressure feedback based on the MAC FIFOs size. This information indirectly aggregates MAC and physical information such as capacity requests, allocations, queue sizes, and used DRA scheme (modulation, code and symbol rate) on the uplink air interface since queues size depends on the resource allocation and FMT loops. The feedback is then used to adapt the output throughput of the I P scheduler.

QoS Support in Gateway

The QoS architecture in the gateway is an evolution of the one implemented in AmerHis gateways [7, 8] where only two classes of traffic where handled: real-time and non-real-time.

The traffic classification policy identifies the subset of network traffic which may receive a differentiated service by being conditioned and/or mapped to one or more behaviour aggregates within the DS domain. The policing rules should be consistent with the SLA defined for given service class. A specific queuing behaviour is applied (PHB) based on the results of the classification. Once traffics to the terminals have been conditioned, they enter in its corresponding PHB queue. These queues store packets corresponding to all terminals.

An efficient scheduling algorithm is considered in SATSIX gateway architecture in order to address an efficient up-to-date satellite communication systems. For this purpose the three traffic categories (Expedited Forwarding, Assured Forwarding and Best Effort) are served by a scheduler using a simple priority queuing (PQ) discipline and a Weight Fair Queuing (WFQ) in the way showed in next figure Fig. 13.4.

With PQ, packets belonging to one priority class of traffic are sent before all lower priority traffic to ensure timely delivery of those packets. Therefore, packets from BE and AF queues will be served only when EF queue is empty. WFQ assigns a weight to each flow, which determines the transmit order for queued packets. Given this handling, WFQ ensures satisfactory response time to critical applications, such as interactive, transaction-based applications, that are intolerant of performance degradation, since it is able to detect higher priority packets marked with precedence by the IP Forwarder and can schedule them faster, providing superior response time for this traffic. Thus, as the precedence increases, WFQ allocates more bandwidth to the conversation during periods of congestion.

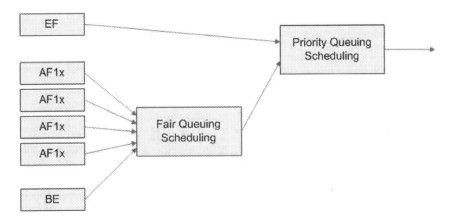

Fig. 13.4 Scheduling in SATSIX gateway architecture

Conclusion

SATSIX completes the earlier work developed in previous projects thanks to a dynamic QoS support that allows improving the QoS level achieved. The QoS architecture developed in SATSIX is based on the DiffServ architecture, and aims to tackle different QoS level sessions (QoS-enabled and QoS-assured) by using a cascaded architecture of Bandwidth Brokers. This dynamic QoS support is one of the major achievement of SATSIX in the QoS area, with the extension of the QoS agent/server architecture proposed in the previous SATIP6 project by an additional SIP proxy the goal of which is to dynamically feed other IP/MAC QoS components with the relevant information. Another key issue is a new HDLB scheduler inside the terminal architecture that allows a hierarchical repartition of the guaranteed/booked throughputs along with an optimized scheduling.

Acknowledgements This work is supported by the IST FP6 SATSIX project, funded by European Commission (EC). The financial contribution of the EC towards this project is greatly appreciated.

References

[1] ETSI TS 102 295, "Satellite Earth Stations and Systems (SES); Broadband Satellite Multimedia; Services and Architectures: BSM Traffic Classes".
[2] "SatLabs Harmonised Short-Term QoS Specification", ref.: sl_322, Version 1.1.1.
[3] IST SATIP6 Project (Contract IST-2001–34344), Home Page on http://satip6.tilab.com
[4] IEEE 802.11e/D13.0, Part 11, "Wireless LAN Medium Access Control (MAC) and Physical Layer (PHY) specifications: Medium Access Control (MAC) Enhancements for Quality of Service (QoS)," draft supp. to IEEE 802.11 std., Jan. 2005.
[5] HTB for Linux, http://luxik.cdi.cz/~devik/qos/htb
[6] L. Bucket and J. Turner, IEEE vol 24, 1986.
[7] AmerHis system, *http://www.alcatel.es/espacio/pdf/amerhis.pdf*
[8] D. Black, S. Blake, M. Carlson, E. Davies, Z. Wong, and W. Weiss, "An architecture for differentiated services," IETF RFC 2475, 1998.
[9] "Satellite Earth Stations and Systems (SES); Broadband Satellite Multimedia; Common air interface specification: Satellite Independent Service Access Point (SI-SAP)", ETSI TS 102 357.
[10] EN 301 192, "Digital Video Broadcasting (DVB); DVB Specifications for Data Broadcasting", European Telecommunications Standards Institute (ETSI).

Chapter 14
Satsix and Recent Standardisation Results in ETSI Broadband Satellite Multimedia (BSM) Networks

R.J. Mort and H. Cruickshank

Abstract This paper describes the recent results of the work in the ETSI BSM standardisation group and its relationship to Satsix.

The ETSI BSM WG focuses on IP-based networks and the associated needs for standardisation when satellites are integrated within them.

The BSM group comprises major organisations from the satellite world and therefore its outputs represent a consensus of the most important techniques and solutions relevant in this sphere of activity. SatSix has had an important influence on the standards produced through the many contributions made and resulting technical discussions.

The BSM protocol architecture used throughout the standards produced is characterised by the separation between common Satellite-Independent (SI) protocol layers and alternative lower Satellite-Dependent (SD) layers.

Introduction

The most recent phase of work of the ETSI BSM WG has resulted in specifications for generic broadband satellite systems aimed at supporting optimal IP-based services integrated within Next Generation Networks. The existence of a set of relevant open standards will enable service providers to identify and support common service platforms, and manufacturers to design competitive solutions.

The main subjects addressed are as follows:

- Address Management at SI-SAP (satellite-independent service access point)
- Multicast Source Management
- QoS Functional Architecture
- Interworking with IntServ QoS
- Interworking with DiffServ QoS
- General Security Architecture
- Multicast Security Architecture
- Transition to IPv6.
- DVB-RCS C2P (connection control protocol) requirements (on-going work).

L. Fan et al. (eds.), *IP Networking over Next-Generation Satellite Systems*
© Springer 2008

The standards resulting from the completed work items can be freely accessed on the ETSI site: portal.etsi.org.

The recent work was founded upon the body of Technical Standards that has been developed in the BSM WG over recent years [1] [12–14]. The existence in these standards of a broadband satellite communications architecture for IP interworking, using the concept of a division between Satellite Independent and Satellite Dependent functions, with a clearly defined boundary at the 'SI-SAP' [2] has enabled the BSM WG to address complex technical issues in a systematic and consistent manner. This interface has also enabled the treatment of issues for the IP layer and above to done for all types of satellite system and independently of the satellite technology-specific layers below the interface.

As a result of the latter approach and also due to the wide industry participation and consensus in the BSM WG, the results are intended to have generic application for IP services to different satellite systems.

The work has been actively supported by inputs from Satsix partners. Satsix has also used the ETSI BSM specifications as a basis for its system designs. Satsix examines and implements specific instances of satellite networks and therefore the ETSI standards have been extrapolated for this purpose, in particular for DVB-RCS implementation.

Papers on the initial QoS and Security work were presented at the IST Mobile Summit 2006 [3, 4].

Address Management at SI-SAP

This subject [5] describes the relationships between IP addresses and lower layer addresses, called BSM_IDs as a generic term for lower layer addresses in different satellite systems (for example in a DVB-RCS system the BSM_ID could relate to a MAC address). It also covers how to create, manage, and query the BSM_IDs for the purpose of sending and receiving user data (in particular IP packets) via the SI-SAP.

The task divides into two parts:

1. address management scenarios and architectures,
2. unicast address resolution at the SI-SAP,

The technical standard document elaborates the details of the address management functions, notably the address resolution function for relating BSM_IDs to IP addresses.

BSM Address Resolution (B-AR) is defined as the function that associates a BSM_ID with the corresponding IP Address. The BSM Architecture (see Fig. 14.1) must provide a service where AR is supported for B-AR clients in all STs by a central B-AR server. In this content, STs includes both the Gateways (typically the Hub for a Star network) and remote STs. Each ST should also have an AR table as part of the B-AR client.

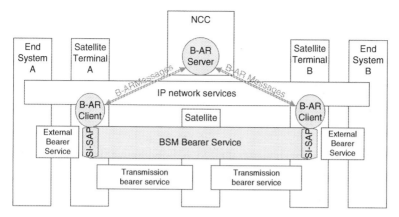

Fig. 14.1 BSM AR Architecture

The B-AR server could, in principle, be located anywhere but it is realistic to assume that is under the control of the BSM operator since it needs knowledge of the BSM address space. Typically, the B-AR Server will be located at a Gateway or at the NCC. Having the B-AR server located at the NCC may be appropriate if the AR function is used to support traffic management, i.e. allowing and denying IP packets access to the BSM network.

B-AR is a C-plane function. However, two distinct processes are required for AR to function. Above the SI-SAP, a BSM_ID must be associated with an IP address. Below the SI-SAP, a BSM_ID is associated with a MAC address. The BSM_ID must be resolved to a MAC address whenever an IP packet has to be transmitted by the lower layers; this is also usually a static or pseudo static process. The process of association can store pairs of values in a table, as is the case with entities using AR over wired networks such as Ethernet. The process of resolution can examine the table, usually stored at the location where resolution occurs. Mechanisms are required to populate and update the local tables that store the associations. Updates to a local table will normally be performed by periodically transferring data from a central, or reference, table. The transferred data will either replace or enhance local data. It is highly desirable to minimise the flow of B-AR data traversing the satellite link.

B-AR may be used whenever an IP packet is to be forwarded to a new destination across a BSM network. The BSM_ID of the next hop must be determined for the packet to be forwarded, and B-AR must be used if the BSM_ID for a given next hop IP address is not already known at that ST. There are three cases:

1. Star network inbound: all IP addresses resolve to the hub gateway BSM_ID. The BSM_ID associated with the hub gateway is either acquired at ST startup as part of Configuration Management by the NCC or it is pre-programmed,
2. Star network outbound: IP addresses resolve to specific BSM_IDs. IP address resolution may require policy decisions in connection with access management. An OBP Satellite that performs layer 2 switching may need an AR table that is at least partially managed by the B-AR server,

3. Mesh network: IP addresses resolve to specific BSM_IDs. IP address resolution may require policy decisions. An OBP Satellite that performs layer 2 switching may manage an AR table itself or it may be managed by the B-AR server.

In all cases a B-AR client at the sending side performs address resolution. In the first instance this should use the cached entries in the local AR table, but if there is no match the ST sends an address resolution request to a B-AR server whose address is acquired dynamically or is pre-configured.

NOTE: In contrast to Ethernet ARP a B-AR client should never send a broadcast message over the satellite link (this may not be possible in any case) and should use the address of the known B-AR server (Fig. 14.1).

Multicast Source Management

This subject [6] defines the architectures and functions required for the interworking of IPv4 multicast protocols, including multicast sources, with the BSM.

The technical standard document firstly considers the BSM network scenarios for IP multicast interworking, including two main aspects:

1. The satellite network architecture
2. Management of multicast sources and data forwarding, either statically or dynamically.

The BSM functional and protocol architectures are then derived for management of:

- IP multicast control messages (group management and routing protocols),
- Multicast access control (including resource management) and
- Multicast address resolution.

The document then defines the detailed functional requirements and interactions of the above three functions with respect to the BSM lower layer interface, the SISAP. The Satellite-Dependent (SD) functions below this interface are system specific and are not treated here.

In the case of multicast routing protocols, the PIM-SM protocol (including the PIM–SSM variant) is taken as the basis for this document since it is almost exclusively used in existing and proposed multicast routing applications today.

IPv6 protocols are not explicitly covered here, though they may be compatible with the architectures described.

To make multicast services effective over the BSM, multicast must take advantage of satellite's native multicast capabilities. Unlike Unicast, where destination IP and link layer addresses are specific to an end host, multicast employs a common IP 'group' address for a given flow to all receivers, and therefore the BSM SISAP should also employ a corresponding common address, or GID (Group ID), for each multicast flow. The way in which these GIDs are controlled and employed is also defined in this document.

The four main network scenarios and their features are summarised in Table 14.1, where the network configuration is either star or mesh.

- Star topology—refers to a star arrangement of links between a central Hub station and remote STs through the satellite. The Hub acts as the sole BSM ingress router and distribution node for BSM multicast. The ST's are all egress routers connected either directly to hosts or via premises networks.
- Mesh topology—refers to a mesh arrangement of links between STs where all ST's can be interconnected directly through the satellite and each ST can act as a multicast distribution node to ST's. ST's can therefore be both ingress and egress routers (Table 14.1).

The above push and pull cases refer to:

- Push—multicast services are configured by the BSM network operator, or similar centralised management entity, in terms of which groups are forwarded over the BSM on a quasi-static basis. The manager may not always know in advance what kind of resources (bandwidth, delay, jitter) will be required for a given multicast flow, but it has to configure BSM resources based, for example, on a service level agreement.
- Pull—multicast services are requested and initiated dynamically (i.e. on demand) by each receiver host issuing a 'join' to an IP multicast group, and therefore by relay though each egress ST, to the Ingress ST using IP multicast protocols. The conditions under which new group membership can be allowed and the associated multicast flows forwarded over the BSM are determined by BSM network policies.

The Functional Architecture shown in Fig. 14.2 is an example for the Mesh Pull scenario. This architecture is focussed on the functional entities involved in the end-to-end BSM multicast control mechanisms that enable multicast flows to be forwarded or removed across the BSM from Ingress to Egress. The architecture must support dynamic control of multicast groups, allowing groups to be added and removed on demand.

Table 14.1 BSM multicast network scenarios

Scenario	Multicast traffic Ingress Point	Multicast traffic Egress Point	BSM network IP multicast control	Ingress IP multicast control	Egress IP multicast control	BSM Access Control	BSM Address Management
STAR PUSH	Hub	ST	None	None/ IGMP/ PIM	None/ IGMP/ PIM	Static	Static/ Dynamic
STAR PULL	Hub	ST	IGMP/ PIM	IGMP/ PIM	IGMP/ PIM	Dynamic	Dynamic
MESH PUSH	ST	ST	None	None/ IGMP/ PIM	None/ IGMP/ PIM	Static	Static/ Dynamic
MESH PULL	ST	ST	IGMP/ PIM	IGMP/ PIM	IGMP/ PIM	Dynamic	Dynamic

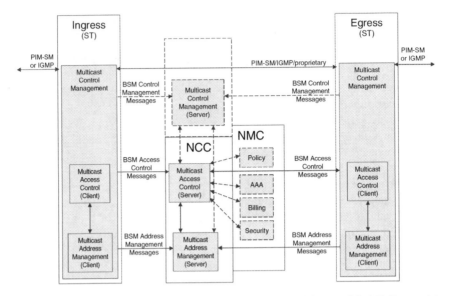

Fig. 14.2 BSM Multicast Source Management Control Plane Architecture (Mesh Pull example)

BSM Multicast Source Management refers to the combination of Control Plane functions needed to create, maintain and remove BSM multicast distribution trees, and which includes Multicast Control Management (using PIM and IGMP), Multicast Access Control, and Multicast Address Management.

The NCC is concerned with BSM SISAP and SD layer functions. The NMC is considered closely related to, or part of, the NCC, whose actions are performed under the aegis of the NMC for aspects such as policy, security and authentication.

Figure 14.3 shows the message flows for the Mesh Pull scenario.

The detailed Multicast Source Management: architecture considers the three constituent functions:

1. Control Management (MCM)
2. Access Control (BMAC)
3. Address Management (MAM).

The protocol architecture differs between the Ingress or Egress ST's, as shown in Figs. 14.4 and 14.5.

In the above figure the IGMP/PIM protocol entity (in the MCMC) establishes the IP group membership list (under the aegis of the BMAC) for each of the Ingress ST BSM interfaces. Whenever there is a change of aggregate group membership over all of these interfaces, and/or periodically as necessary, the MCMC sends a resolution request for any new groups to the lower layers of the attached network in order to obtain associated link layer addresses. It also sends a resolution request for the groups to the SISAP to obtain associated GIDs on the BSM side. Reception and forwarding of multicast groups is controlled by the MCMC, and having obtained the BSM resources necessary, the MCMC issues a 'Listen' command to the attached network interface together with the binding of relevant IP groups and

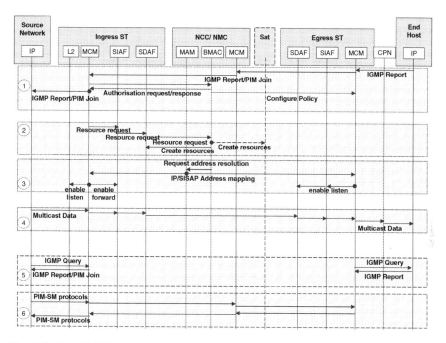

Fig. 14.3 Mesh Pull Scenario Messages

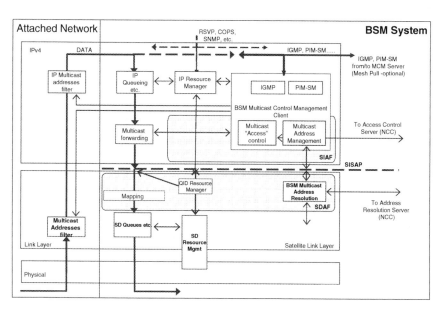

Fig. 14.4 Detailed BSM Multicast Ingress ST protocol Stack

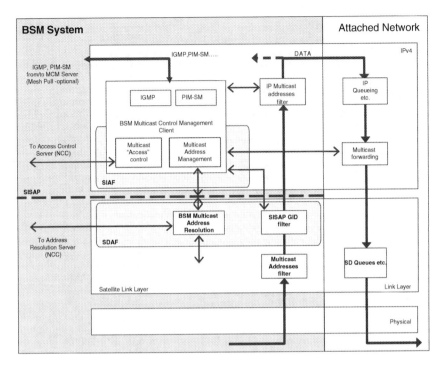

Fig. 14.5 Detailed BSM Multicast Egress ST protocol stack

multicast link layer addresses. The MCMC also issues a 'forward' command to the IP forwarding engine together with the binding of the groups to GIDs (Fig. 14.5).

In the above figure the IGMP/PIM protocol entity (in the MCMC) establishes the IP group membership list (under the aegis of the BMAC). Whenever there is a change of group membership it issues a join request to the upstream router.

The MCMC also sends a resolution request for any new groups to the SISAP to obtain associated GIDs on the BSM side. It also sends a resolution request for the groups to the lower layers of the attached network in order to obtain associated link layer addresses. Reception and forwarding of multicast groups is controlled by the MCMC, and it issues a 'Listen' command to the IP forwarding engine together with the binding of the groups to GIDs. The MCMC also issues a 'forward' command to the attached network interface together with the binding of relevant IP groups and multicast link layer addresses.

QoS Functional Architecture

QoS is a network feature which will be increasingly valuable for service differentiation and support of more QoS-sensitive applications. In contrast to wired or optical networks where over-provisioning of capacity is often used to ensure QoS for packet-based transport, satellite systems, as for other wireless networks, allocate capacity efficiently and carefully. This requires more sophisticated QoS methods that are closely linked to resource provision and control at lower protocol layers than IP.

No standardised or common approach to network architecture for end-to-end QoS provision to applications exists at present.

Various approaches to QoS provision can be proposed based on varying complexity and performance. A modular architecture is therefore required which can be adapted to meet different needs.

QoS provision within ETSI BSM systems is one of the first aims, but since BSM systems are intended to access the Internet, end-to-end QoS across integrated networks including satellites is also important.

A BSM QoS functional architecture has been defined for IP-based applications [7]. Compatibility with QoS requirements for generic internetworking including Next Generation Networks (NGN's [8]) are taken into account.

The BSM architecture is characterised by the separation between common Satellite-Independent (SI) protocol layers and alternative lower Satellite-Dependent (SD) layers. At the SI layers, several methods of ensuring end-to-end QoS over integrated networks are foreseen, by means of signalling protocols (e.g. based on SIP) at the session (or application) layers and DiffServ, RSVP/IntServ, or NSIS at the IP layer. At the SD Layers alternative lower protocol layers offer different QoS characteristics. The focus of the architecture definition here is on maintaining compatibility with these alternative methods and approaches by addressing the generic BSM QoS functions required in the SI layers (including IP). These functions will provide interfaces where appropriate with higher-layer and lower-layer QoS functions, and with external networks and customer equipment.

The BSM Global QoS functional architecture, including the relationship of the BSM with QoS protocol layering and with the rest of the network, is illustrated in Fig. 14.6. The figure also shows the range of possible functions involved in QoS and their functional partition between Control and User Planes (Management functions are not shown for clarity since they are more implementation-specific).

Two main kinds of message flows between functional blocks are indicated in the diagram: primitives between protocol layers, and secondly peer-to-peer protocols. Note that the peer-to-peer protocols are shown as horizontal lines for clarity, though in reality they are transported via the user plane.

The BSM QoS architecture is based on centralised control and management of the BSM subnetwork through a Server entity called the BSM QoS Manager (BQM). Like typical servers the BQM can consist of several physical entities. The ST's, as network edge devices, are responsible for traffic processing and policy enforcement at the ingress and egress, but they should be controlled from the BQM. The BQM should contain all the necessary functions to manage QoS for all layers above the SISAP in both Management and Control Planes. The BQM interacts with equivalent local functions in the ST's.

The control and management functions below the SISAP (for connection control, bearer set up, BSM QoS etc.) are usually also centralised in the NCC, which may be closely associated with the BQM.

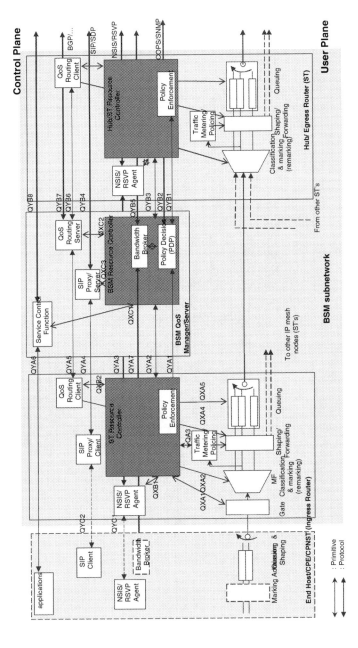

Fig. 14.6 BSM QoS Functions in the IP layer and higher layers (one data direction shown)

Many of the functions in the BQM are standardised functions such as those for signalling (RSVP/NSIS or SIP Proxy/SDP), but others specific to the BSM, such as those for managing the BSM's global IP and SIAF layer resources, are allocated to a functional entity called the BSM Resource Controller (BRC) (Fig. 14.6).

Central to the QoS capability of the BSM is the interface of the IP layer with the lower SD layers at the SISAP. To abstract the User Plane QoS interface at the SISAP the concept of QID's (Queue Identifiers) has been introduced. These represent abstract queues available at the SISAP, each with a defined class of service for transfer of IP packets to the SD layers.

The satellite dependent lower layers are responsible for assigning satellite capacity to these abstract queues according to the specified queue properties (e.g. QoS). The QID is not limited to a capacity allocation class; it relates also to forwarding behaviour with defined properties.

A QID is only required for submitting (sending) data via the SISAP and the QID is assigned when the associated queue is opened. An open queue is uniquely identified

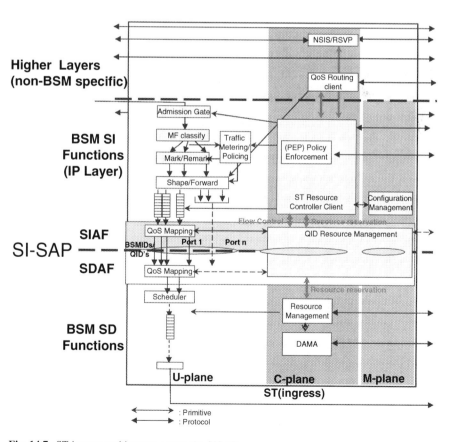

Fig. 14.7 ST ingress architecture across the SISAP

by the associated QID: in particular, the QID is used to label all subsequent data transfers via that queue.

The way in which QIDs are mapped to the IP layer queues is an important consideration for overall QoS.

The different cases of interaction between QoS requests and the BSM involve not only the User Plane containing the QIDs, but also the Control and Management Planes that influence the way the QIDs are used. The interaction between the IP layer QoS and the SD layer QoS takes place across the SISAP and is thus the major issue for the BSM.

An architecture for the ST ingress is shown in Fig. 14.7.

Interworking with IntServ QoS

The key to providing real-time multimedia services such as those offered by the Intserv model is the interaction of a resource reservation protocol like RSVP with lower layer (i.e. link layer) resource reservation. For IntServ provision in a BSM network the concept of QID's (Queue Identifiers) at the SISAP is the concept used to provide this interaction with alternative link layers. QIDs represent abstract queues, each with a defined class of service, for transfer of IP packets to the SD layers. The satellite dependent lower layers are responsible for assigning satellite capacity and/or particular forwarding behaviour to these abstract queues according to defined properties.

Two main scenarios for the use of BSM resources in an IntServ network can be foreseen [9]:

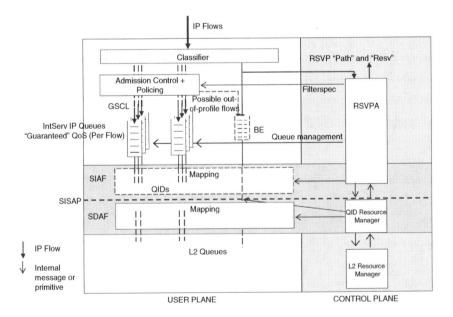

Fig. 14.8 Ingress Architecture (Dynamic SD Resources)

1. **Static SD resources**: BSM SD resources for IntServ (i.e. high priority SD class) are provisioned and managed quasi-statically, and no interaction between RSVP and the SD resource control is available. A range of QIDs is however assumed to be available for specific use of IntServ and they may be of a range of data rates within the total SD resources.
2. **Dynamic SD resources**: BSM SD resources for IntServ are requested dynamically, and an interaction between RSVP and the SD resource control is available.

The functional architecture of the Ingress ST (either a remote ST or a Gateway ST, in the case of a star network) for dynamic SD resources is shown in Fig. 14.8.

Interworking with DiffServ QoS

This subject aims at an open specification for enabling QoS for IP-based multimedia satellite systems, based on the DiffServ model. If IP packets entering the BSM network require a particular QoS treatment, they have to be mapped onto QIDs. The choice of the QID to be used inside the BSM network is thus particularly important. So the present document specifies the allocation of the QIDs and their mapping to IP QoS classes, when DiffServ is used to provide QoS at IP layer.

The technical standard document [10] describes in detail how QIDs are defined, how they are allocated and handled by the BSM network, and the requirements needed by sending and receiving Satellite Terminals (STs) in a BSM network to provide QID management functionalities. The document also defines the primitives that shall be used across the SI-SAP when allocating QIDs, when mapping DiffServ Code Points (DSCPs) and IP services to QIDs, when mapping QIDs to SD queues. The functional architecture of the Ingress ST for DiffServ is shown in Fig 14.9.

General Security Architecture

This section presents the detailed security system in various architectural cases [11]. These security cases are focused on the positioning of security functions above or below the SI-SAP. For example the security key management and data encryption entities can both be above or below the SI-SAP or one above and one below. In addition, the concept of BSM Security association Identity (SID) is presented. For example, if there is a secure connection between an ST and a Gateway, then SID is the reference number that is used to convey security information between **BSM Local and Network** security managers such as encryption keys, digital signature methods and security policy exchanges.

If there is only one single BSM **Network** security manager, then SID will be unique for the whole BSM network. If there are several **Network** security managers (for example one for each ISP), then SID must be used in conjunction with BSM-ID

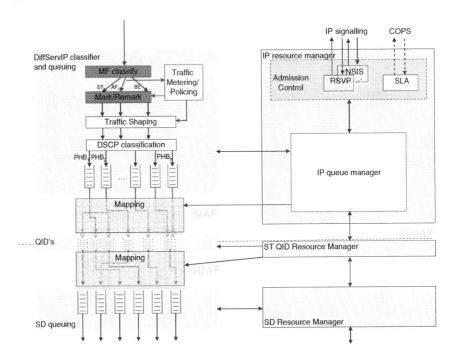

Fig. 14.9 Detailed architecture of an ingress DiffServ-aware ST

of the source and destination entities, in order to identify a security association between two BSM entities.

BSM Security Architecture Cases

The security cases presented here apply to both BSM star and mesh topologies. For a mesh topology with no On-Board Processor (OBP), STs communicate with each other through a BSM gateway (hub). For a mesh topology with OBP, STs communicate directly with each other without the need for the BSM Gateway (Hub). With respect to the security cases presented here, the star and mesh (no OBP) are the same, where the BSM **Network** security manager function is likely to be located at the BSM Gateway (Hub). However, for a mesh topology with OBP, the main difference is that BSM **Network** security manager function can be located at any BSM ST.

Case 1: IPsec and Security Entities in BSM

As shown in Fig. 14.10, this case illustrates the use of IPsec over BSM network in a security gateway-to-gateway configuration such as VPN over satellites scenario. IPsec protocol operates above the SI-SAP.

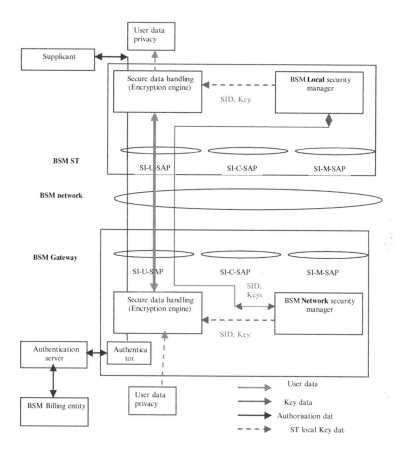

Fig. 14.10 Case 1 IPsec and BSM security entities

Security is provided between a security gateways (that can be co-located with BSM ST or Gateway). The security gateway consists of two functional entities:

Secure data handling entity (privacy/integrity engine): IPsec must operate in tunnel mode.

Key management entity: In a star topology, there will a **Network** security manager for the whole BSM network (co-located with BSM gateway/hub). In addition there is a **Local** security manager in each ST.

Figure 14.10 shows all security entities are above SI-SAP. The diagram also shows that the SI-U-SAP (the user interface) ONLY is used to communicate all secure information (user data and key management messages).

The client authentication process (supplicant, authenticator and Authentication server entities) is shown here as well, where IPsec is used to carry authentication information (such as user name and password) between Supplicant and authentication server.

Both the authentication server and the BSM network manager communicate with the BSM NCC regarding security and authorisation. These interactions are not shown here in order to simplify the diagram. Registration and re-key security association must be established between the **BSM Network** security manager and **Local** security managers in each ST. In the case of IPsec, the IETF Internet Key Exchange (IKE) protocol (RFC 4109) must be used to establish all security associations. This will ensure the mutual authentication between all security entities, establishing the keys used subsequently to secure the user data. Using IKE will also ensure compatibility between BSM and the general Internet (terrestrial) security systems.

The Security association identity SID must be used in all security management message exchanges.

However IPsec for multicast (star topology) is a challenge because IPsec tunnels must be set from the BSM gateways per ST. This is effectively a unicast configuration and the benefits of IP multicast are lost.

Draft-ietf-msec-ipsec-extensions-02.txt is work in progress in defining the extra detail needed for IPsec to work efficiently with multicast. The Security Architecture for the Internet Protocol security architecture document (RFC 4301) describes security services for traffic at the IP layer. That architecture primarily defines services for Internet Protocol (IP) unicast packets, as well as manually configured IP multicast packets. The draft-ietf-msec-ipsec-extensions-02.txt further defines the security services for manually and dynamically keyed IP multicast packets within that Security Architecture.

Case 2: Mixed Link Layer Security Entities in BSM (Security Manager Above SI-SAP and Security Engine Below SI-SAP)

As shown in Fig. 14.11, this case illustrates the use of link layer security (below SI-SAP) with the key management (security manager) as an application (above the SI-SAP in a star topology with a centralised security **Network** manager (can be co-located with the BSM gateway/hub). Typical examples of such system are DVB-RCS with MPE or Unidirectional Lightweight Encapsulation (ULE) RFC 4326 IP encapsulation.

Like case 1, the security is provided between security gateways (can be co-located with BSM ST or Gateway). The security gateway consists of two functional entities:

Secure data handling entity (privacy/integrity engine): e.g. is DVB-RCS security which performs data encryption below SI-SAP

Key management entity: In a star topology, there is a **Network** security manager for the whole BSM network

(co-located with BSM gateway/hub). In addition there is a **local** security manager in each ST.

The client authentication process (supplicant, authenticator and Authentication server entities) is shown here as well, where secure link layer is used to carry authentication information (such as user name and password) between supplicant and authentication server.

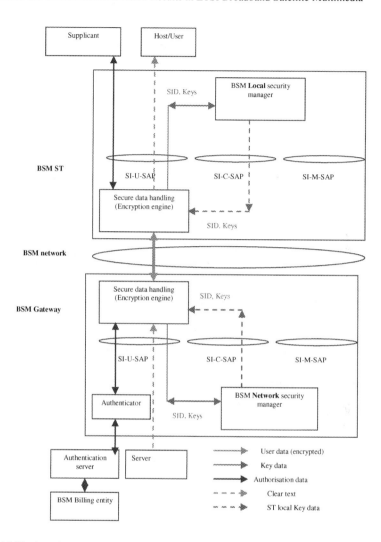

Fig. 14.11 Case 2 Mixed link layer BSM security entities

Figure 14.11 shows security entities above and below the SI-SAP. The diagram also shows that the SI-U-SAP (the user interface) is used to communicate secure user information, while the key management secure information is passed through the SI-C-SAP interface. The client authentication messages use the SI-U-SAP interface.

Both authentication server and the BSM **Network** manager communicate with the BSM NCC regarding security and authorisation. These interactions are not shown here in order to simplify the diagram. Registration and re-key security association must be established between the **BSM Network** security manager and **Local** security manager in each ST. In the case of link layer security, the specific satellite systems security must be used. For example, for DVB-RCS satellite systems, the logon and key exchanges procedures of DVB-RCS recommendations [3] must be used to establish all security associations. For BSM systems operating with ULE, then the ULE specific key management procedures must be used (RFC 4326).

This will ensure the mutual authentication between all security entities, establishing the keys used subsequently to secure the user data. Using link layer security will also authenticate BSM terminals (STs and gateways), which is not possible with using IPsec (case 1).

The Security association identity SID must be used in all security management message exchanges.

Case 3: End-to-End Security

This case is applicable to IPsec, TLS/SSL and application layer security (Fig. 14.12). This is useful for end-to-end and remote access scenarios and is transparent to BSM network. If cases 1, 2 or 4 are used simultaneously with case 3, then a careful consideration must be paid to the BSM network performance degradation due to the dual security processing.

Case 4: Pure Link Layer Security

This case (Fig. 14.13) is applicable to ATM, DVB-RCS and ULE security systems that are implemented in the BSM network in the satellite link layer only. This case is transparent to BSM network. However, the BSM **Local** and **Network** security managers must be able to enforce the BSM security policy rules in this case such communication must use the SI-M-SAP interface. The Security association identity SID must be used in all security management message exchanges.

Generalised Interactions Between Security and Other BSM Entities

This subsection addresses interaction and interworking with BSM QoS, address resolution management.

If QoS is used, then key management messages must use the high priority QoS classes to ensure fast and reliable key exchanges. This imply assigning QIDs with high class of service to security message exchanges. This applies to security cases 1, 2 and 3.

Figure 14.14 illustrates the use of BSM security to encrypt/authenticate QoS and Address resolution requests/responses between ST/Gateway and NCC. SI-SAP interfaces are not shown here because the focus of this diagram is securing message exchanges, over BSM network, between the BSM **Network** managers (QoS and address-resolution) and the **Local** manager in BSM ST/Gateway. The encryption engine can be below or above the SI-SAP.

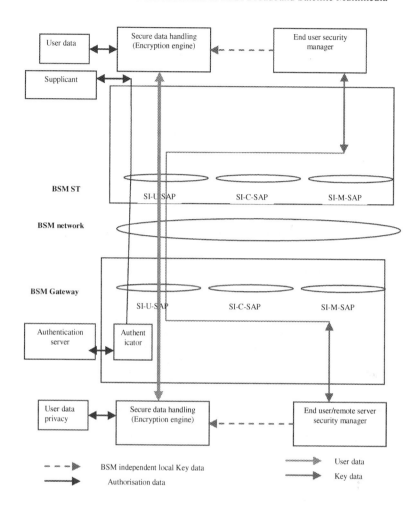

Fig. 14.12 Case 3 End-to-end security, transparent to BSM

Interactions Between Security and QoS Entities

Security of QoS Signalling in BSM Network

The QoS functional architecture document (TS 102 462) presents QoS cases. In all these cases, it is assumed that the BSM system provides different levels of bearer QoS through a certain number of QIDs, which determine the nature of the QoS offered at the SI-SAP. It is the way in which the QIDs are accessed or modified by the IP layer and above that changes between cases. Security issues are the same in all these cases.

User and management planes are not addressed here. In the control plane, communications between the resource management in the ST/GW and the NCC must

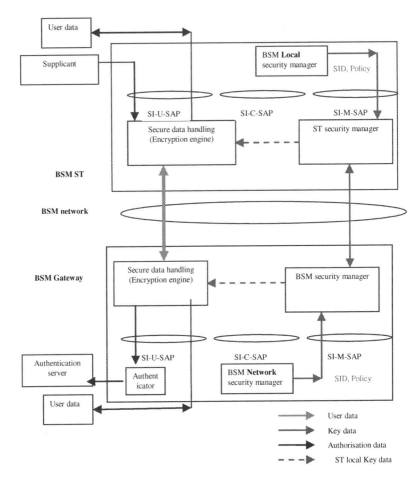

Fig. 14.13 Case 4 link layer security, transparent to BSM

be secured. These QoS messages between the ST/GW and the NCC must be authenticated and optionally may be encrypted (this depends on the security policy for the BSM network).

In Figure 14.15 (copied from the QoS functional architecture document (TS 102 462); QoS case 3), if security is implemented below the SI-SAP, then link number 1 must be secured, using link layer such as DVB-RCS security procedures. If security is implemented above SI-SAP, then link number 2 must be secured, using IPsec or TLS security procedures. Either links 1 or 2 must be secured. However, it is possible to secure both links 1 and 2 at the same time, but the impact of security processing on BSM network performance must be assessed carefully in this situation.

Also Fig. 14.15 shows link number 3 between NSIS/SIP entities in the ST/GW and the NCC. The security issues for these entities are out of scope for BSM networks. However, if SIP or NSIS signalling is used in BSM, then the IETF security

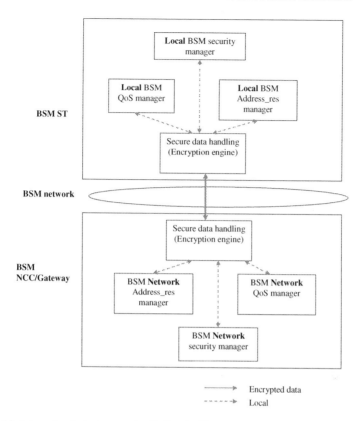

Fig. 14.14 Interaction between security, QoS and address resolution entities

recommendations for both protocol must be observed (such as RFC 4081 for NSIS and RFC 3893 and RFC 3329 for SIP security).

Using COPS Protocol for Security Policy Provisioning

In BSM networks, the Common Open Policy Service(COPS) protocol can used to carry QoS or security information between BSM management entities and satellite terminals (gateways/ST) (RFC 2748). In addition, if COPS is used for QoS provisioning, then COPS Policy Provisioning protocol (COPS-PR) can be used for security policy transfer (RFC 3084).

Figure 14.16 (from the QoS functional architecture document [7]; QoS case 3) shows the interaction between COPS entities to carry QoS and security related information. In the ST/Gateway, the Policy Enforcement Point (Policy-PEP) interacts with the **Local** security manager. In the NCC, the Policy Decision Point (Policy-PDP) will interact with BSM **Network** security manager. These interactions are not shown in the diagram for clarity.

Satsix Workshop, Budapest

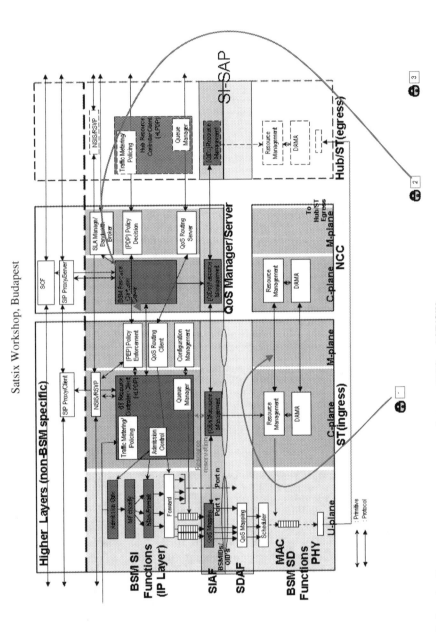

Fig. 14.15 Securing Resource management messages between NCC and ST/GW

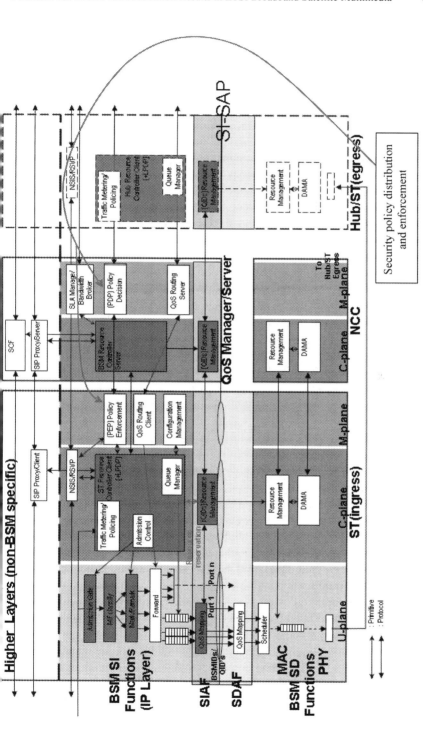

Fig. 14.16 Securing policy distribution using COPS

The management plane is used to carry security policy related communications. Such communications do not need any special QoS treatment unless specified in the QoS or security policy rules.

Using Reliable Transfer Mechanisms (QoS) to Transfer Key Management Messages

In security cases 1 and 3 the security management messages are transferred in the user plane through the SI-SAP interface. Therefore, the queues for security information are managed in the same way as any other user data. However, security management messages must be allocated a relatively high priority. Such allocation can be static and decided by the security policy of the BSM network or it can be dynamic depending on the nature of QoS offered at the SI-SAP.

Case 2 is similar to case 1 and 3, except that security key management messages are passed in the control plane through the SI-SAP interface. Therefore, a similar QoS management is needed in this plane for the security messages.

In case 4, all security management messages are below the SI-SAP. Therefore, there is no need for QoS management above the SI-SAP for these security messages.

Interactions Between Security and Address Resolution Entities

Security of Address Resolution Signalling in BSM Network

BSM address resolution is defined in the SI-SAP spec (TS 102 357) and the Address Management at the SI-SAP document (TS 102 460). The basic issues are how to map IP addresses to BSM-IDs and then to satellite specific MAC addresses.

A generalised model is shown in Fig. 20. Regarding security, any address resolution signalling across the SI-SAP interface within a single ST/Gateway or the NCC has no security implications.

However, communications between the address resolution entities (in ST/GW and the NCC) must be secured between ISPs, customers, network access providers and satellite network operators (as shown in Fig. 14.17). These address resolution messages between the ST/GW and the NCC must be authenticated and optionally may be encrypted (this depends on the security policy for the BSM network).

Using RADIUS with DHCP Servers

If DHCP is used in BSM, then the RADIUS Attributes sub-option enables a network element to pass identification and authorisation attributes received during RADIUS authentication to a DHCP server (RFC 4014). When the DHCP server

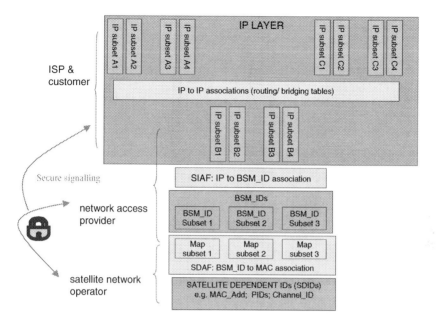

Fig. 14.17 Generalized Address management model in BSM network

receives a message from a relay agent (Network Access Server, NAS) containing a RADIUS Attributes sub option, it extracts the contents of the sub option and uses that information in selecting configuration parameters for the client.

Transition to IPv6

This Technical report studied the key issues for IPv6 introduction and produced recommendations for work on specific topics and standards in this area as follows:

- Support for 'native' IPv6 packets at the SI-SAP and in the Satellite dependent layers.
- Use of dual-stack architecture above the SI-SAP.
- IPv6 Header Compression.
- IPv6 address mobility and any security implications arising.
- Management aspects and IPv6-aware Management Information Bases.
- IPv6 stateless autoconfiguration (consequences of Unreachability Detection and Neighbour Discovery)
- Revision of TS 102 460 (Address management at the SI-SAP) to take account of IPv6 issues.
- Multicast Source Management with a suitable IPv6-to-GID mapping scheme
- Trade-offs for implementing and modifying the MLDvx protocols over BSM

Conclusion

Satsix has helped to establish a range of ETSI technical standards covering a range of issues focussed on IP services provision over networks with integrated satellite sub-networks. These generic standards have been further employed in Satsix as a basis for its own specific solutions to network implementation.

References

[1] ETSI TR 101 985: "Satellite Earth Stations and Systems (SES); Broadband Satellite Multimedia; IP over Satellite".

[2] ETSI TS 102 357 "Satellite Earth Stations and Systems (SES); Broadband Satellite Multimedia; Common air interface specification: Satellite Independent Service Access Point (SI-SAP)".

[3] "QoS architecture for Broadband Satellite Multimedia (BSM) networks", R.J. Mort et al., IST Mobile Summit 2006, Myknonos.

[4] "A security architecture for Broadband Satellite Multimedia (BSM) networks", H. Cruickshank et al., IST Mobile Summit 2006, Myknonos.

[5] TS 102 460: "Satellite Earth Station and systems (SES); Broadband Satellite Multimedia; Address Management at the SI-SAP".

[6] TS 102 461: "Satellite Earth Station and systems (SES); Broadband Satellite Multimedia; Multicast Source Management".

[7] TS 102 462: "Satellite Earth Station and systems (SES); Broadband Satellite Multimedia; QoS Functional Architecture".

[8] ETSI TS 185 001 "NGN; QoS Framework and Requirements".

[9] TS 102 463: "Satellite Earth Station and systems (SES); Broadband Satellite Multimedia; Interworking with IntServ QoS".

[10] TS 102 464: "Satellite Earth Station and systems (SES); Broadband Satellite Multimedia; Interworking with DiffServ QoS".

[11] TS 102 465: "Satellite Earth Station and systems (SES); Broadband Satellite Multimedia; General Security Architecture".

[12] TS 102 466: "Satellite Earth Station and systems (SES); Broadband Satellite Multimedia; Multicast Security Architecture".

[13] TR 102 467: "Satellite Earth Station and systems (SES); Broadband Satellite Multimedia; IPv6 Transition".

[14] ETSI home page: http://portal.etsi.org/Portal_Common/home.asp

Chapter 15
IPv6 Networking over Satellite for Mobile User Groups

Àngels Via Estrem and Axel Jahn

Abstract This paper deals with the support of IPv6 for mobile user groups using the scenario of passengers in an aircraft connected via a satellite link to ground Internet. Several solutions are discussed to support user laptops having IPv6. The peculiarity of the scenario lies in the fact that satellite modems usually offer single IPv4 addresses only. We address IPv6 over IPv4 tunnelling and Network Address and Port Translation (NAPT) with dual stack routers as possible solutions. The NAPT solution is optimized to save overhead over the satellite link, and required signalling to allow ground-to-air traffic is addressed. Finally, a performance estimation is given.

Introduction

Satellite networks play a vital role in serving passengers in aircrafts, ships or trains with Internet and mobile telephony [7]. The passengers in such mobile platforms represent a collective user group which is connected by satellite with ground networks. Today, Internet Protocol version 4 (IPv4) is practically the only version used in such environments. However, IP in version 6 is now available, and it can be expected that users will soon like to use the new IPv6. In this paper we address the support of IPv6 for moving platforms connected by satellite, e.g., in aeroplanes, trains and other vehicles.

In terrestrial networks for fixed IPv4/IPv6, migration is done in several ways, such as Tunnelling, NAPT-PT, or automatic tunnels such as 6over4 (RFC2529 [3]), which uses multicast in IPv4. The ISATAP protocol (RFC4214 [9]) works similar to 6over4, except for the multicast addresses. The protocol 6to4 (RFC3056 [2]) is designed for isolated IPv6 networks. Teredo (RFC4380 [6]) addresses isolated nodes behind NAT, whereas Tunnel Broker (RFC3053 [4]) controls tunnels configured by a new element.

We address the case when we have the transition between IPv6 and IPv4 in the satellite link connecting the aeroplanes to the Internet on-ground. Our goal is to connect the IPv6 users in aeroplanes minimizing the resources on the satellite.

L. Fan et al. (eds.), *IP Networking over Next-Generation Satellite Systems* 217
© Springer 2008

Mobile User Groups Scenario

A scenario is considered with IPv6 users in aeroplanes. The IPv6 devices are connected to the Internet through satellites, which only support IPv4, as shown in Fig. 15.1.

The system under consideration is INMARSAT BGAN (Broadband Global Area Network) [1]. The terminal bit rate is up to approximately 432 kbps, though other transparent Ku-band satellite systems can provide up to 1 Mbps or higher. Typically, satellite systems serve IPv4 only by providing a routed IP interface.

In our scenario, we assume that each aircraft has a multitude of IPv6 users on board. An onboard router serves as gateway to the satellite transport; a corresponding router is placed on ground. The aircraft router must have, at least, one IPv4 interface with a public IPv4 address connected to the satellite. This address may change every time the router changes its point of attachment to the satellite. As there are not enough addresses to be assigned, nodes inside the network can not have public IPv4 addresses. However, the IPv6 nodes inside the aircraft have at least one public IPv6 address each; those IPv6 addresses may change with the router's IPv4 public address.

A solution is wanted to send the IPv6 packets through the IPv4 satellite link. The main changes from IPv4 to IPv6 packets are summarized in the following list:

- The address space in IPv6 has been expanded from 32 bits to 128 bits and new auto configuration mechanisms have been added.
- The IPv6 header format has been simplified: fixed length and optional Extension Headers between the IP header and the upper layer header. Apart from that, five fields have been removed (Header Length, Identification Flags, Fragment Offset and Header Checksum).
- Authentication and privacy options have been added in IPv6.
- All flows can be labelled so the processing is faster in forwarding nodes.
- Mobility aspects have been improved: all nodes must have these capabilities, Foreign Agents are no longer necessary and Route Optimization is performed instead of Triangular Routing.

Possible Networking Solutions

In the following, 2 solutions are selected and explained in detail: Tunnelling and NAPT-PT. After that, other solutions are explained briefly.

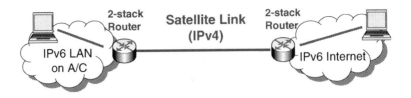

Fig. 15.1 Mobile user groups scenario

Manual Tunnelling

Manual Tunnelling is a mechanism to enable IPv6 hosts to communicate to other IPv6 host without an IPv6 connection between them. It uses a tunnel, which means that the packets are added an IPv4 header so they can traverse an IPv4 network. Unlike the other tunnelling solutions, it needs to know the other end of the tunnel to work.

In tunnels, the encapsulator (the entry node of the tunnel) creates an encapsulating IPv4 header and transmits the encapsulated packet. At the other end of the tunnel, the decapsulator receives the encapsulated packet, reassembles it (if it has been fragmented), removes the encapsulating header and processes the IPv6 packet. Note that the encapsulator records information for each tunnel, such as the Maximum Transmission Unit (MTU), in order to process IPv6 packets forwarded into the tunnel.

Tunnels can be used in a variety of ways [5]:

- *Router to Router*: the tunnel is between two routers, so the communicating nodes do not know that it exists.
- *Router to Host*: the same explained before, but the receiving node is the one to decapsulate, instead of a router in the middle of the path.
- *Host to Router*: the contrary to the previous way.
- *Host to Host*: the two end points of the connection are directly the two end points of the tunnel.

In this paper only router to router encapsulation will be used, as the hosts inside the aeroplane have connection to an IPv6 network and it is only the satellite link which does not support the new version.

Encapsulation and decapsulation are explained below, and also the problems with MTU and Internet Control Message Protocol (ICMP) translation between both versions of IP.

How It Works

As said before, an IPv4 header is added to the IPv6 packet [5], as shown in Fig. 15.2.

In this scenario, packets generated inside the aeroplane are sent to the edge router using IPv6. There, they are encapsulated with an IPv4 header and forwarded to the gateway on ground through the satellite. Then, they are decapsulated and forwarded to the destination through IPv6-Internet, as shown in Fig. 15.3.

The header fields of the IPv4 header are used in the same way as a normal IPv4 packet except the following:

- *Protocol*: 41 (Assigned payload type number for IPv6).
- *IP Header Length*: 5 (in words of 32 bits). There are not any options.
- *Total Length*: Payload length from IPv6 packet plus length of IPv6 and IPv4 headers.
- *Flags*: The Don't Fragment flag is set when needed. The More Fragments bit is set as necessary if fragmenting.
- *Fragment Offset*: Set as necessary if fragmenting.
- *Source Address*: An IPv4 address of the encapsulator: either configured by the administrator or an address of the outgoing interface.

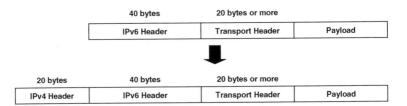

Fig. 15.2 Addition of a new header when a tunnel is used

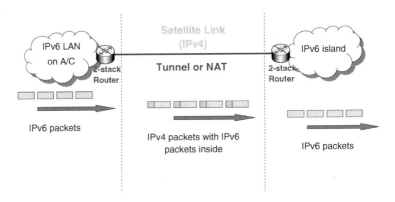

Fig. 15.3 Operation scheme for tunnelling

- *Destination Address*: IPv4 address of the tunnel endpoint.

Unlike in IPv4, where the node fragments the packet when it is too big, in IPv6 only the generator of the packet can fragment it, so a 'Packet Too Big' ICMP message is sent to the source node when necessary.

However, in tunnelling, sending ICMP messages would not be efficient. Then, the encapsulator must decide, statically or dynamically, when to send an error message or when to fragment the packet. Note that in any case the 'Don't Fragment' bit is never set.

A node using static tunnel MTU treats the tunnel interface as having a fixed-interface MTU, whose values are between 1280 and 1480 bytes, being 1280 bytes the most used. When the packet length is bigger, the node sends a 'Packet Too Big' message and discards the packet.

On the other hand, a node using dynamical tunnel MTU uses the IPv4 Path MTU Discovery Protocol. Then, if a packet is larger than the MTU, the node sends a 'Packet Too Big' ICMP message and drops it. However, IPv6 do not permit the packet length to be less than 1280 bytes, then, the node sends it anyway and the following routers will fragment the packet if it is necessary (as the 'Don't Fragment' bit is not set) and it will be reassembled again at the end of the tunnel.

When the decapsulator receives a packet with 41 in the protocol field, it verifies if the packet belongs to one of the configured tunnels and if it does, it reassembles it, removes the IPv4 header and processes it. It must be noted that the IPv6 header is not modified at this point; only the hop limit is decremented by one if the packet is then forwarded. An important issue is that all packets coming from multicast, loop-back, IPv4-compatible or IPv4-mapped IPv6 addresses will be discarded. In addition, the node should be able to perform ingress filtering on the IPv6 address.

When ICMPv4 messages are generated inside the tunnel, the encapsulator (the source address in the IPv4 packet) decides if an ICMPv6 message must be generated or not. Old routers send only 8 bytes of data, so not enough information inside the ICMPv4 message is sent and the encapsulator does not know the IPv6 source address. If sufficient data is available, the ICMPv6 message is generated. The 'Packet too big' messages are translated to v6 and sent. The other types are logged as an error related to the tunnel and if sufficient headers are available, the encapsulator can send the error message with 'address unreachable' code.

Advantages and Disadvantages

The tunnelling mechanism is easy to implement and to process packets at each end of the tunnel. Moreover, it is widely implemented and used.

However, tunnelling adds a new header, which means 20 bytes more. In some terrestrial networks, this fact is not important, but in the satellite links, adding some bytes is expensive due to the scarce satellite spectrum.

Another drawback is that each tunnel must be manually configured, as if it did not, it would not know where to send the tunnelled packets. Nevertheless, the tunnel in the router can be configured knowing the coverage area of each spot beam and the configuration could be automated.

Translation: NAPT-PT

Network Address Translation – Protocol Translation is a mechanism to allow IPv6 hosts to communicate with IPv4 hosts or vice versa. There are two types of translation: Address Translation (NAT) and Address and Port Translation (NAPT). Unlike the first, where one IPv4 public address is needed for each IPv6 node, in the second, as ports are also used, 64 k IPv6 nodes can fit in only one IPv4 public address. This is the reason why NAPT-PT was investigated in this paper.

The following explains this mechanism, as defined in RFC2766 [10] and the modifications which permit it fitting to this system.

How NAPT-PT Works

This method uses a pool of IPv4 public addresses to assign them to IPv6 nodes that want to communicate with IPv4 nodes as shown in Fig. 15.4.

As we have said before, NAPT-PT also uses port translation. Moreover, not only it translates the addresses and ports, but the entire header, as shown in Fig. 15.5.

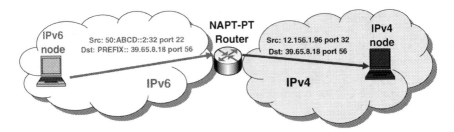

Fig. 15.4 General scheme for NAPT-PT

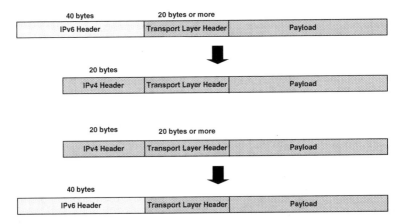

Fig. 15.5 (a) Translation of an IPv6 header into an IPv4 header. (b) Translation of an IPv4 header into an IPv6 header

Like in IPv4 NAT, the router has a table with the relation between the inner addresses and their correspondent external addresses, but in this case the IPv4 addresses with ports match the IPv6 ones.

An IPv6 node establishes a connection with an IPv4 node as follows:

1. Sends the packet with destination PREFIX::IPv4address, where PREFIX is advertised by the NAPT-PT and all packets with this prefix will be routed to the NAPT-PT box.
2. The NAPT-PT receives it and assigns a port and an address to this connection.
3. It sends the packet, and all returning traffic that matches this pair (Destination Address, Destination Port) will be translated again and forwarded to the IPv6 end node.

When translating from IPv6 to IPv4, the new header has not any options and all the Extension Headers in IPv6 are not translated, so the Protocol field and the Total Length field must be adjusted [8].

All other fields are translated directly to their correspondent fields in IPv4. The addresses are taken from the table and the destination address:

1 PREFIX::IPv4address.

In the other way, from IPv4 to IPv6, it must be noted that, in IPv6, packets are fragmented in the origin, so if the packet is bigger than the MTU (if the node has MTU Discovery capabilities) or 1280 bytes (if it has not), the packet must be fragmented.

Like in the other way, if there are optional headers in IPv4, they are not translated to IPv6 and all the other fields are just kept as in IPv6.

The problems with this transition mechanism are the following:

- ICMP messages must be translated, not only the header but the payload. This is a bit complicated as these versions have different parameters.
- If there are some addresses in the payload, then they must be translated also. This is what Application Level Gateway does, which is out of the scope of this document.
- It needs special features in Domain Name System (DNS), so they can resolve IPv6 names when asked from IPv4 hosts or vice versa.

Modifications from the Standard to Fulfil the Requirements

In the previous section, standard NAPT-PT has been explained. However, with the scenario in Fig. 15.1, this scheme does not fit, as the two nodes have the same IP version. Thus, some modifications are necessary:

- Two translations between IPv6 and IPv4 and vice versa are required, one on the aeroplane and one on the router on ground.
- Application Addresses Translators are no longer necessary, as the destination node has the same version as the source.
- The translation tables must be enlarged so they have entries like the one shown in Table 15.1.
- The destination addresses in the IPv6 domains will be IPv6 addresses of the destination node instead of the IPv4 address with a prefix.
- ICMP messages do not need to be translated, as both ends are IPv6.

All packets entering or exiting the aircraft (A/C) IPv6-network should be translated to version 4 and again to version 6. This means that both translation routers

Table 15.1 Entry for one connection in the translation table for modified NAPT-PT

IPv6 Domain		IPv4 Domain	
	Address		Address
Source		Source	
	Port		Port
	Address		Address
Destination		Destination	
	Port		Port

must have the same translation tables or some information about the source or the destination must be sent with the packet.

Header Translation

As commented before, IPv6 headers and IPv4 headers do not have the same length or the same fields. However, all the fields in v6 must be transmitted to the other end of the link in order to translate the packet to IPv6 again. Before explaining how, the scheme of an IPv4 header is shown.

The header fields (Fig. 15.6) are:

- Version: 4
- Header Length: Counted by 32-bit words.
- Type of Service: Informs the network of the QoS desired. It has the same syntax as IPv6 Traffic Class field.
- Total Length: Length in 32-bit words of the payload plus the Header Length.
- Identification: Identifies each datagram.
- Flags:

  ○ Not used.
  ○ Fragmentation Flag: 1 if the packet should not be fragmented.
  ○ More Fragments: 1 indicates that there are more fragments to come.

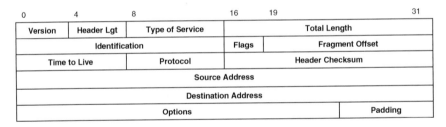

0	4	8	16	19	31
Version	Header Lgt	Type of Service	Total Length		
Identification			Flags	Fragment Offset	
Time to Live		Protocol	Header Checksum		
Source Address					
Destination Address					
Options				Padding	

Fig. 15.6 IPv4 header format

4	8	20	16	8	8
Version	Traffic Class	Flow Label	Payload Length	Next Header	Hop Limit
Source Address					
Destination Address					

Fig. 15.7 IPv6 header format

IPv4 Header		
Flow Label (20 bits)	Next Header (8 bits)	Padding (4 bits)
First Extension Header or Transport Layer Header		

Fig. 15.8 IPv4 header with flow label and next header fields after it

- Fragment Offset: Identifies the fragment inside the datagram.
- Time To Live: Maximum hops permitted till the destination.
- Protocol: Identifies the next layer protocol.
- Header Checksum: It is only used in the header.
- Source Address: 32 bits.
- Destination Address: 32 bits.
- Options

The header will not have any options, that is, the length will be 20 bytes, which is not enough to fit all the IPv6 header fields in. Looking at the Fig. 15.7 and comparing it to Fig. 15.6, only two fields are left: *Flow Label* and *Next Header*.

Both together have 28 bits, so we suggest adding them just after the IPv4 header and before the payload, as shown in Fig. 15.8. Four 0-bit padding has been added in order to have this extended header length multiple of 32 bits. Apart from that, another important issue are the Extension Headers, which will be treated as payload.

Protocol Operation

The protocol explained here assumes that most of the packets are generated inside the aeroplane, thus, these packets are forwarded automatically and the packets generated outside need some signalling before being forwarded. First of all, the protocol for the packets generated inside is explained and then, for the others.

An important issue to design are the NAT tables, which can be static or dynamic. The first option means that all possible addresses must fit inside the table, which is, obviously, impossible. Then, the tables used in this solution will be dynamic, though some extra information will have to be sent with the packets.

The basic operation when the packets are generated inside the A/C is shown in Fig. 15.9 and explained in the following paragraphs:

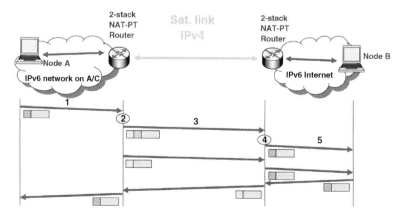

Fig. 15.9 General scheme for operation of this modified NAPT-PT

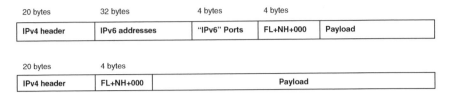

20 bytes	32 bytes	4 bytes	4 bytes	
IPv4 header	IPv6 addresses	"IPv6" Ports	FL+NH+000	Payload

20 bytes	4 bytes	
IPv4 header	FL+NH+000	Payload

Fig. 15.10 (a) Header format for the first packet of the session. (b) Header format for the following packets of the session

1. A node generates the first packet of a session, which is destined to the outside IPv6 world. As its destination is not inside the network, the packet is sent to the NAT router.
2. When the router receives it, it looks for an entry in the translation Table 15.1. As it is the first of a session, it does not find any. Then, it creates a new entry, assigning an address and port pair in this router and also in the router on-ground.
3. When the router has an entry for this packet, it sends it with the IPv6 addresses and original ports, as shown in Fig. 15.10a.
4. After going through the satellite link, the packet arrives at the router on-ground, who looks for an entry in the table, and, as it does not find any, it reads the IPv6 addresses and ports and creates a new one.
5. Afterwards, the packet is translated and sent to the IPv6-Internet.

The following packets of the same session are translated using the table and sent to the other end of the satellite link using the format described in Fig. 15.10b.
On the other hand, for the packets generated outside, the protocol is the following (see Fig. 15.11):

1. A node outside the A/C sends a packet to a node inside.
2. When the on ground station receives it, it looks it up in the table and, as it does not find anything, it looks up the destination address in a table where there are all the addresses that each router has.
3. When it finds the correspondent router, the on ground station sends a signalling packet to the Mobile Router (MR) with the addresses and ports in IPv6 and asking for the IPv4 corresponding addresses.
4. MR assigns IPv4 addresses and ports and creates a new entry in the table.
5. MR sends them back to the on ground station.
6. After receiving the information, the on ground station creates a new entry in the translation table and translates the header. Note that, unlike in the other direction, this first packet does not contain the IPv6 addresses.
7. The on ground station sends the packet.
8. The MR receives the packet and translates it to IPv6 again.
9. The MR sends the packet to the node inside the A/C.

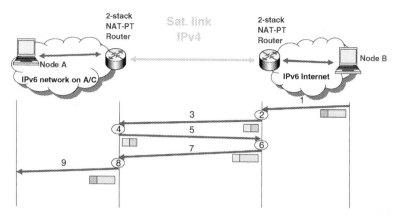

Fig. 15.11 Protocol scheme of this modified NAPT-PT when packets are generated outside

Advantages and Disadvantages

The most important advantage is the fact that no extra headers must be added, which makes the transmission through the satellite link cheaper.

However, it leads to a more complex packet processing at each end of the IPv4 link, which is expensive in terms of delay and also because more specific routers must be used.

Apart from this, the following lists other drawbacks, though less important:

- If packets are fragmented, at the other end of the satellite link they should be reassembled before forwarded, which adds delay and requires more resources. However, it can be solved with a 'Packet Too Big' ICMPv6 message when the packet is bigger than the Satellite Link MTU, which is known by the router.
- It does not support multicast.
- Limited number of connections, depending on how many IPv4 addresses the router has, if too many IPv6 ports are used. Each address has 64 k ports and for each one, 64 k ports of the router on-ground can be used. Thus, this limit is not expected to be achieved.
- It is not compatible with some security systems, as the packet is modified in the path.
- When a packet is generated outside, some signalling is sent before forwarding the packet. However, almost all the packets are generated inside.

Performance Comparison Between Manual Tunnelling and NAPT-PT

First of all, looking at Figs. 15.2 and 15.10, it can be noticed that NAPT-PT has always less bytes of header than tunnelling. On the other hand, encapsulation is

Table 15.2 Mean values of VoIP and TCP sessions

	VoIP	TCP
Payload length (Bytes)	20	710
	UDP 8	
Header length (Bytes)		20
	RTP 12	
Packets/session	9000	86

Table 15.3 Theoretical comparison between manual tunnelling, NAPT-PT and standard IPv6

	Manual tunnelling		NAPT-PT		IPv6	
	TCP	VoIP	TCP	VoIP	TCP	VoIP
Header length (Bytes)	60		60 (1st packet) 24 (others)		40	
Overhead	8.4%	300%	3.4%	120%	5.6%	200%
Delay (ms)	1.87		1.87 (1st packet) 0.75 (others)		1.25	

easier in terms of packet processing and it is not needed to send signalling when the packet is generated neither inside nor outside.

To resume it, a numeric comparison between both solutions and between normal IPv6 is shown in Tables 15.2 and 15.3, where the length of the headers has been taken from Figs. 15.2 and 15.10 and Table 15.2. Note that the packets generated outside are not considered.

Other solutions: Automatic Encapsulation

6over4

This method is known as multicast tunnelling [3], as it uses the IPv4 Multicast Network to connect isolated IPv6 hosts between them or to the native IPv6 network. Thus, IPv4 network is used as a virtual data link layer to IPv6. The requirement is that multicast must be supported in IPv4.

ISATAP

The Intra-Site Automatic Tunnel Addressing Protocol (ISATAP) [9] enables tunnelling from nodes behind a NAT router. It uses IPv4 as a virtual Non-broadcast Multiple Access Network link layer, thus the underlying IPv4 network infrastructure do not need to support multicast.

6to4

This method [2] applies to sites, but it can work as well with a single host. Every site has a 6to4 router, at least with one IPv4 global address, which is used to construct the 48-bit prefix of that site: 2002:x:x::/48 where x:x is the hexadecimal notation of the 32-bit public IPv4 address. This prefix is announced by the router within the site, so every node can configure a globally unique IPv6 address.

Teredo

Teredo [6] encapsulates IPv6 packets within UDP/IPv4 datagram, which is useful when the tunnel traverses a NAT because most NATs can forward UDP easier than TCP encapsulated within IPv4. Using UDP/IPv4 means that 2 headers should be added to the original IPv6 packets: 20 bytes of IPv4 header and 8 bytes of UDP header.

Tunnel Broker

Defined in RFC 3053 [4], the Tunnel Broker configures the tunnels from outside. The scheme is similar to the other solutions: there is a node with only IPv4 connections who wants to communicate to IPv6 nodes; the tunnel broker, in the IPv4 network and the Tunnel Server in the first router in the IPv6 network.

NAPT-PT Solution and Results

In order to simulate the scenario, we implemented a test bed, which is implemented in a *XEN* virtual environment. There are 4 virtual machines: a client, which is a user inside the aircraft, the mobile router, which connects the aircraft to the satellite, the ground station and a server, which simulates a remote user/server on the Internet. It is shown in Fig. 15.12.

The available module from iproute2 for Linux ip tunnel is used, configuring each end of the tunnel.

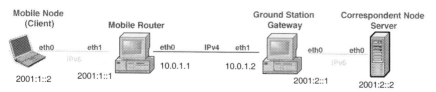

Fig. 15.12 Test bed scheme

Table 15.4 Results for different applications with tunnelling, NAPT-PT and standard IPv6

		HTTP	FTP	VoIP
	Header length (Bytes)	25.44	24.03	24.002
NAPT-PT				
	Overhead	2.50%	2.26%	86%
	Delay (ms)	0.79	0.75	0.75
	Header length (Bytes)	60	60	60
Tunnel				
	Overhead	5.90%	5.60%	215%
	Delay (ms)	1.25	1.25	1.25
	Header length (Bytes)	40	40	40
IPv6				
	Overhead	3.90%	3.70%	142%
	Delay (ms)	1.25	1.25	1.25

NAPT-PT, unlike tunnelling, is not implemented in Linux. There are only some proprietary implementations, which not fulfil our system. Thus, we have developed a protocol converter in C which runs in Linux user space using libraries for taking packets from kernel queues or sending them again to kernel.

Table 15.4 shows the results for 3 different applications: HTTP, FTP and VoIP using NAPT-PT (the mean has been calculated, as the first packet has more bytes in the header), Tunnelling and standard IPv6.

It is easy to see that NAPT-PT adds fewer bits to each packet on average than the tunnelling. Moreover, in some cases, such as the transfer of big files, it can even reduce the delay due to the headers to the half.

On the other hand, as said before, tunnelling is easier to implement as there are available modules, does not need extra memory space to store the translation table nor needs to pre-process the packet before sending it.

As overhead is more important than complexity when talking about satellite communications, the best choice is NAPT-PT.

Conclusions

This paper dealt with IPv6 transport over IPv4 satellites in aeronautical mobile environments. IPv6 over IPv4 tunnelling and network address and port translation (NAPT) with dual stack routers have been investigated. A NAP-PT solution was presented which can exchange IPv6 addresses and ports with IPv4 headers, both at aircraft and ground routers. It could be shown that the performance of the NAPT-PT outperforms tunnelling.

References

[1] BGAN service specification March 2006. INMARSAT. www.inmarsat.com/bgan

[2] B. Carpenter (2001). "Connection of IPv6 Domains via IPv4 Clouds", Internet Engineering Task Force (IETF) RFC 3056.

[3] B. Carpenter and C. Jung (1999). "Transmission of IPv6 over IPv4 Domains without Explicit Tunnels", Internet Engineering Task Force (IETF) RFC 2529.

[4] A. Durand, P. Fasano, I. Guardini and D. Lento (2001). "IPv6 Tunnel Broker", Internet Engineering Task Force (IETF) RFC 3053.

[5] R. Gilligan and E. Nordmark (2005). "Basic IPv6 Transition Mechanisms", Internet Engineering Task Force (IETF) RFC 4213.

[6] C. Huitema (2006). "Teredo: Tunnelling IPv6 over UDP through Network Address Translations (NATs)", Internet Engineering Task Force (IETF) RFC 4380.

[7] A. Jahn, M. Holzbock, J. Müller et al. (2003). "Evolution of Aeronautical Communications for Personal and Multimedia ervices", IEEE Communication Magazine, vol. July, pp. 36–43.

[8] E. Nordmark (2000). "Stateless IP/ICMP Translation Algorithm (SIIT)", Internet Engineering Task Force (IETF) RFC 2765.

[9] F. Templin, T. Gleeson, M. Talwar and D. Thaler (2005). "Intra-Site Automatic Tunnel Addressing Protocol (ISATAP)", Internet Engineering Task Force (IETF) RFC 4214.

[10] G. Tsirtsis and P. Srisuresh (2000). "Network Address Translation - Protocol Translation (NAT-PT)", Internet Engineering Task Force (IETF) RFC 2766.

Chapter 16
Multicast Architecture for IPv6 over DVB-RCS Satellite Networks

A. Yun, D. Elkouss, E. Callejo, L. Liang, L. Fan, and Z. Sun

Abstract As one of the key elements in an IPv6 supported DVB-RCS satellite network, the multicast architecture design is facing the challenges of interworking between the IPv6 multicast protocols and the satellite signaling while efficiently using the satellite bandwidth.

How to enable the multicast group management functions for satellite end users with and without direct IPv6 MLD router support and how to translate the IP multicast routing protocol messages to the satellite lower layer signaling to establish the satellite channels between two spot beams need to be answered.

With the regenerating satellite space segment, dynamic multicast routing is possible and making an efficient use of the satellite bandwidth has to be considered. This paper presents the SATSIX approaches of a multicast architecture involving both the IP multicast and the C2P protocols to address and solve these questions.

Introduction

The new generation of satellites challenges the traditionally efficient implementation of broadcasting and multicasting. Equipped with on board processing (OBP) satellites are capable of switching the different flows between multiple beams and enable complex network topologies. These satellite networks, based on the DVB-S(2)/DVB-RCS standards, feature multiple gateways/feeders instead of a single hub; interconnection with terrestrial networks may involve (multiple) multicast sources external to the satellite network thus making the problem of efficient provision of multicast services much more complex that in the limited, traditional scenario.

DVB-RCS systems are classified in two different families depending on the network topology:

1. Star/mesh transparent (no processing on board).
2. Star/mesh regenerative (based on OBP technologies).

Depending on the management of multicast groups, the following multicast implementations are possible:

L. Fan et al. (eds.), *IP Networking over Next-Generation Satellite Systems*   233
© Springer 2008

1. Static multicast. There is no control on the members of multicast groups. A station wishing to transmit multicast requests a point to multipoint connection to the destinations, without knowing if there are interested receivers attached to them. Multicasting is implemented by broadcasting, filtering out data in the terminals without interested receivers attached to them. This implementation is simple but inefficient in terms of bandwidth.
2. Dynamic multicast. The NCC (Network Control Center) is notified about interested receivers through join/leave messages, thus the NCC configures dynamically the OBP in order to multicast the information only to where there are interested receivers.

Multicast implementation in any network interconnection may involve replication to reach all intended destinations. The integration with a satellite network arises some issues concerning replication. Replication outside the satellite network may not optimize satellite resource usage. As for replication within the satellite network, the typical scenario in transparent satellite systems is replication on the gateway, with one copy for each beam with a receiver. With OBP satellites, on board replication is possible, yielding the most efficient spectrum usage.

This paper will address the most relevant issues regarding the design of an IPv6 over DVB-RCS multicast architecture and will illustrate on this issue on a case study: SATSIX [7–9].

In the second section we review IPv6 multicast for satellite networks: Issues and solutions on IP multicast over IPv6 networks, i.e. MLD approach, address resolution, etc.

The third section of the paper briefly describes the particularities of C2P (Connection Control Descriptor) the connection control protocol over DVB-RCS, then presents two specific solutions for dynamic multicast compared with the AmerHis case (see [12]). The first solution translates the IGMP/MLD concept to the satellite scenario using C2P messages. The second one, more complex, based on the first solution also relies on IGMP adapted for satellite networks in order to reduce the complexity on the NCC and improve the responsiveness of the overall network.

The third section describes some results comparing both solutions and the fourth section offers a critical analysis of them.

IPv6 Multicast for Satellite Networks

IP Multicast in IPv6

There have been several changes in IP multicasting from version 4 to version 6. One of the most significant modifications in the general addressing model in IPv6 is a change to the basic types of addresses and how they are used. Broadcast as a specific addressing type has been eliminated. Instead, support for multicast addressing has been improved and made a required part of. Under IPv6, multicast addresses are allocated from the multicast address space, which consists of all addresses beginning with prefix "1111 1111."

An IPv6 multicast address is an identifier for a set of interfaces that typically belong to different nodes. A packet sent to a multicast address is delivered to all interfaces identified by the multicast address. The second octet following the prefix defines the lifetime and scope of the multicast address (see Fig. 16.1). The lifetime parameter is set to 0, when the multicast address is permanent whereas, the lifetime parameter is set to 1, when the multicast address is temporary. A multicast address that has the scope of a node, link, site, organization or a global scope has a scope parameter of 1, 2, 5, 8, or E, respectively.

IPv6 multicast requires the following protocols to be implemented.

- MLD for IPv6 is used to discover multicast listeners (hosts that wish to receive multicast packets destined for specific multicast addresses) on directly attached links.
- A multicast routing protocol used between routers so that they know which multicast packets to forward to each other and to their directly connected LANs. There are several options such as PIM-SM or PIM-SSM.

MLD for Satellite Networks

The IPv6 Multicast Listener Discovery (MLD) Protocol provides a mechanism for IPv6 Multicast routers to discover on their interfaces hosts interested in receiving IPv6 Multicast packets. MLD (Multicast Listener Discovery) enables each IPv6 router to discover the presence of multicast listeners on its directly attached links and to determine specifically which multicast addresses are of interest to those nodes. MLDv1 is defined in RFC 2710. MLD is IPv6 specific and is equivalent to Internet Group Management Protocol (IGMP) in IPv4. However, there is a new version of MLD. MLDv2 adds the support of source specific IP Multicast as it has

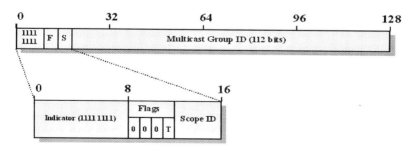

Fig. 16.1 IPv6 multicast address format

been standardized. MLDv2 is equivalent to IGMPv3 [1] for IPv4. There has been
some previous work (in the GEOCAST project) in adapting IGMPv2 for satellite
environment for the scenario DVB-S(2)/DVB-RCS transparent source in the internet.
This work has been standardized under [2, 3]. However, still MLDv2 must be
adapted to the satellite environment, where the new features necessitate a different
approach to adaptation.

In addition, multicast routers build a multicast tree for distribution of data in the
multicast network, and multicast routing protocols, like DVMRP, MOSPF and PIM
(DM and SM), perform this function. In a DVB-RCS network, gateways will behave
as multicast routers and thus they will feature a multicast routing protocol. Terminals
will have multicast end nodes, and in these cases IGMP/MLD proxying is enough to
provide multicast services to the attached networks. In other cases the organization
of networks attached to terminals may be more complex and involve also multicast
routing protocols. But finally, terrestrial networks attached to terminals behave as
"leaf nodes" in the multicast tree, and a solution based on IGMP at the terminal can
be used to signal group membership to the satellite enabled multicast network.

Functional Architecture

In many practical applications, participants in the multicast group (MCG) change.
During a session new receivers join or old ones leave. The group membership is
said to be dynamic when a receiver may join or leave the group at any instant of
time during the session. A MCG algorithm should be able to allow changes in the
multicast group without disrupting communications between the source and the
existing members of the multicast group.

Dynamic multicast implies that any RCST is capable of dynamically handling
multicast flows (without any static configuration).

Two multicast architectures are foreseen in the frame of SATSIX depending on
the topology. The first one focuses on the star topology matching the transparent
payload systems, and the second one is dedicated to mesh topology (but can also
address star topology).

For both architectures two approaches have been considered:

- IGMPv2 adaptation over the satellite environment for IPv4 satellite networks
  and PIM-SM for inter routers multicast communication.
- MLDv2 adaptation over the satellite environment for IPv6 satellite networks and
  PIM-SM for inter routers multicast communication.

Star Multicast Functional Architecture

The following figure gives the network topology to offer star dynamic multicast
service over an IPv6 satellite network (Fig. 16.2)

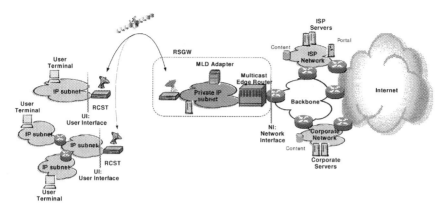

Fig. 16.2 Star multicast functional architecture

We introduce a brief overview of the involved network entities:

User Terminal (UT)—A User Terminal is on the IP subnet connected to the RCST through the User Interface. This UT has an MLD v2 host function to subscribe/de-subscribe to a multicast group in case of IPv6, or a IGMPv2 host function in case of IPv4.

Router on the IP subnet of a RCST—If there is a router on the path between the RCST and User Terminal, this router is an MLD Proxy in case of IPv6 (MLD version 2 is supported) or an IGMP Proxy (IGMP version 2 is supported) in case of IPv4, based on [6]. The upstream interface is the interface towards the RCST and the downstream interface is the interface towards User Terminals.

RCST—RCST have the same MLD Proxying function for IPv6 or IGMP Proxying function for IPv4. An MLD Proxy based on [6] is implemented in a RCST. The MLD V2 is supported. The RCST acts as an MLD Router and Querier on its User interface and it acts as an MLD Host on the satellite air interface. The MLD proxy avoids the RSGW receiving request message for an IP multicast flow from a UT, on the LAN of a RCST, which has already been requested by another UT on the same LAN. The MLD Proxy knows that the IP multicast flow is already transmitted on the TDM and provides it to the new UT without asking the RSGW. This enables to reduce the number of MLD messages sent on the air interface. The MLD proxy forwards multicast data flow received from the satellite air interface to its User Interface according to its group membership table.

RSGW—The RSGW is composed of a private subnet and several entities such as an MLD Adapter and a Multicast Edge Router.

MLD Adapter—Based on an MLD Proxy and having some specific features optimized to satellite network. It has an interface towards the RSGW (and then RCSTs) and an interface towards a Multicast Edge Router.

MER—The Multicast Edge Router is a multicast router with an MLD Router/ Querier function on its interface towards the MLD adapter and a multicast routing function on its other interface. The multicast routing function of the Multicast Edge Router is commonly based on PIM-SM. It is in charge of joining or pruning to the group-spanning tree.

Backbone network, ISP or MSP network, Internet, Corporate Network—These networks have Multicast Core Routers with multicast routing protocol such as PIM-SM, MBGP, MDSP.

Mesh Multicast Functional Architecture

The following figure gives the network topology to offer mesh dynamic multicast service over an IPv6 satellite network (Fig. 16.3).

The main difference between the previous architecture is that all RCST/RSGW have multicast router capabilities, and use multicast routing information over C2P (Connection Control Protocol).

In the network configuration considered, the satellite terminals (RCSTs) are configured as routers and implement a multicast routing protocol. MLD over the terrestrial interface and over the satellite interface a long term solution could consider PIM-SM over the satellite interface. The short term possible solutions are detailed on the next section.

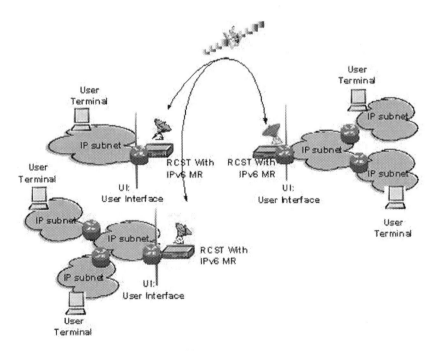

Fig. 16.3 Mesh multicast functional architecture

A Multicast Approach over DVB-RCS Satellite Networks

This section will develop two different scenarios that address dynamic multicast group management over regenerative networks. They are introduced by an example of the current implemented situation represented by AmerHis [12]. This example analyses the star multicast scenario, where dynamic multicast group management is solved thanks to IGMP satellite adaptation combined with C2P (Connection Control Protocol) for multicast connection establishment/release. The two solutions proposed afterwards, highly rely on C2P for multicast connection control and dynamic multicast group management and can be seen as an enhancement of the AmerHis star multicast scenario.

C2P for DVB-RCS Satellite Networks

In meshed DVB-RCS systems RCSTs can have a direct (one-hop) communication. This is possible thanks to a Connection Control Protocol (C2P) capable of setting up MAC connections to convey any traffic transmission among RCSTs.

The first C2P initiative was identified in the frame of the DVB-RCS technical activities in 2001. It was accepted by the DVB-RCS group and included as Annex J of DVB-RCS Guidelines. This proposal was completed in 2004 based on the experience with AmerHis, the first multi-beam DVB-RCS/DVB-S regenerative satellite system, capable to provide simultaneous star and mesh communications.

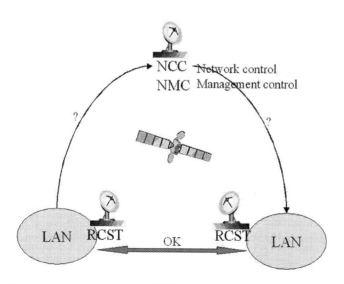

Fig. 16.4 The connection control protocol (C2P) allows one hop satellite communications

C2P is necessary in any mesh scenario, but C2P can also be extended to other DVB-RCS scenarios. Within the frame of SATSIX C2P has been extended to accomplish a star transparent DVB-RCS scenario, IPv6, DVB-S2 support, full dynamic multicast, DVB-RCS security and multicast group management (Fig. 16.4).

The RCSTs, RSGWs and NCC of the regenerative platform already rely on C2P. To adapt C2P to the transparent platform the transparent equipment, the transparent NCC (in transparent mode) and transparent RCSTs must support also the protocol. Summarizing, C2P is a protocol that links the NCC/(TSGW) with the RCST/RSGW.

Case Study: AmerHis

AmerHis IP Multicast

Current RCS systems use transparent satellites, efficient for a centralized star network, e.g. for access to broadband terrestrial internet services, but incurring a double-hop satellite delay, making it unattractive for multimedia communication between user terminals.

AmerHis, a joint project of the European Space Agency (ESA), CDTI and Hispasat [10, 4], represents a response to the growing demand in multimedia broadband services and the adaptation of real time services to the satellite world. The Hispasat Amazonas Satellite, launched in August 2004, carries the AmerHis OBP. This OBP is a multibeam regenerative payload with DVB-RCS MPEG profile inputs and DVB-S outputs, allowing the multicross connectivity, switching, data replication and remultiplexing MPEG packets on board. It represents the first operational regenerative, on-board processing (OBP), DVB-RCS/DVB-S satellite switching. This OBP greatly shortens the satellite delay by reducing the double hop nature of the current bi-directional RCS systems into a single hop architecture for both star and mesh topologies.

AmerHis IP multicast solution is based on IGMPv2 proxying at terminals and an IGMPv2 adapter on the regenerative gateway. As mostly mesh multicast services are expected to be multiconference type (controlled number of sources), and star multicast could be any source from terrestrial networks, only dynamic multicast group management is offered for star multicast.

Star Dynamic Multicast

The star dynamic multicast service allows any RCST that has access to a RSGW, to join any multicast source from terrestrial networks. Star multicast services are one to many only, such as NVoD, Broadcast TV/Radio and file transfer. These services are provided with a single hop satellite delay. For star dynamic multicast Fig. 16.5, multicast flows are dynamically forwarded from a RSGW to several RCSTs. The RSGW will forward into the satellite network the terrestrial network multicast sources under RCSTs request. The multicast forwarding is dynamic: the

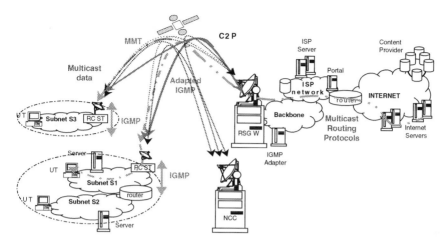

Fig. 16.5 AmerHis Star Dynamic Multicast

IGMPv2 protocol is running between the default RSGW and RCSTs to discover multicast group membership. A RSGW forwards only a multicast flow on the uplink assuming that at least one RCST has joined this flow, and the OBP replicates this flow to downlink TDMs covered by the RSGW. A RSGW does not forward multicast flow on the uplink if no RCST has joined it.

AmerHis IGMPv2 Adaptation

AmerHis IPv4 star multicast group management is based on IGMPv2 adaptation. The network entities considered in AmerHis star multicast topology are: end user hosts, RCSTs supporting IGMP proxy functionality and a RSGW that includes an IGMP adapter and a multicast edge router compliant to PIM-SM.

The IGMP proxy included in the RCST has two interfaces:

- The upstream interface: the satellite interface (SI)
- The downstream interface: the user interface (UI)

The IGMP proxy functionality is compliant to [4, 5] The IGMP querier function on the UI is configured to be elected as Querier. To receive IGMP messages and multicast traffic sent by the RSGW, the RCST filters the multicast PID associated to this data, which is given in the MMT.

The IGMP Adapter included in the RSGW is an IGMP Proxy optimized to satellite environment. The IGMP host function is standard function as defined in [6] but the IGMP querier function is optimized to the satellite environment.

AmerHis IP multicast solution confirms an efficient IP multicast over satellite architecture. This solution is based on the OBP replication on board, the usage of C2P messages to dynamically request multicast resources, the MMT (Multicast Map Table), to dynamically inform all RCSTs in the system all active multicast sessions (mapping multicast session to PIDs) and IGMP adaptation. This IGMP adaptation introduces certain modifications to IGMP to improve its performance when applied over satellite link, but not affecting the system interoperability with IP. Remote satellite users may join Internet multicast sessions (external and from the satellite network), and simultaneously send IP multicast flows to other users.

Still two aspects need to be solved to ensure a full IP multicast: mesh dynamic IP multicast group management (not only star) and dynamic multicast routing on the OBP depending on the locations of the members that join or leave a multicast session. These two aspects are solved thanks to C2P in the following sections.

Pure C2P

The first option would be to rely only on C2P messages (join request/response, leave request/response). In this way the NCC is allowed to directly control the multicast sessions as all C2P messages are exchanged between the RCST/RSGWs and the NCC.

In terrestrial networks a router takes the role of querier and periodically requests the hosts in the network of multicast groups interests, a satellite network, in comparison, needs to set up a connection between every pair of hosts connecting (in mesh scenarios.) A direct translation of a multicast protocol to a satellite network implies setting up multiple connections just to transmit the multicast connection signaling information. It might even happen that a host is interested in forwarding a certain group but has no bandwidth available to set up the connection to request it. Or, if there were several feeds the querier would need to exchange messages with all the feeds multiplying the final number of messages.

All management functions are centralized by the NCC: resource assignment and the decisions over the routes to use for the multicast traffic and also the Query functionality. The NCC acts as a multicast router querying, controlling and managing the multicast groups and multicast specific addresses active within the satellite network.

The objective of the multicast protocol is to decide if a flow must or must not be sent through a given branch of a network. In terms of satellite networks this translates to set or to not set up a point-to-multipoint connection between the feed and the hosts. Thus, the NCC is in any case heavily involved in the process. Moving the multicast control towards the NCC and to the control plane heavily reduces the amount of connections and messages needed at the price of putting more complexity on it.

The new set of C2P messages maps IGMP/MLD messages to C2P. With these messages the following procedures define the pure C2P multicast scenario:

- *Add*/drop IP multicast addresses of "queried" multicast hosts in the multicast group memberships list contained in the RCST the host is connected to, and in the NCC.

- *Add/drop* parties in a multicast connection, i.e. aiming at adding/dropping IP multicast addresses of "unsolicited" multicast hosts in the multicast group memberships list contained in the RCST the host is connected to, and in the NCC.

This approach goes one step further than AmerHis. AmerHis uses IGMP adaptation over the satellite network, IGMP is treated like any other data and connections are established to carry it (i.e. connections between the terminal RCST and the gateway RSGW). Now it is proposed to send this multicast connection control messages as C2P signaling, relieving the system from this overhead and reducing the number of messages exchanged.

One additional advantage is that being the NCC aware of multicast it can dynamically inform the OBP of the switching, optimizing bandwidth use.

The solution is basically a translation of IGMP/MLD to C2P messages, thus the IGMP/MLD messages and actor names have been kept to show the parallelism between both.

General Query/Report Procedure

The querier is uses the General Query, in this case played by the NCC, to learn which groups have members on an attached network. The NCC periodically sends a General Query on the satellite interface, to solicit membership information (**step 1** Fig. 16.6).

On the RCST subnet the process is similar, being the main difference that on the terrestrial network the messages are IGMP/MLD and on the satellite network the messages are C2P messages. The RCST as explained in "Star Multicast Functional Architecture," previous section, acts both as IGMP/MLD router with the role of querier on its local network and as an IGMP/MLD host on the satellite network (**steps 4 and 5** Fig. 16.6).

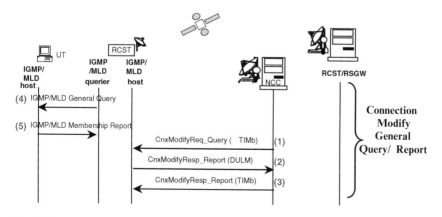

Fig. 16.6 Pure C2P scenario. General query/report procedure

When a host receives a General Query, it sets delay timers for each group of which it is a member on the interface from which it received the query. Each timer is set to a different random value, selected from the range (0, Max Response Time]. When a group's timer expires, the host sends a membership report to the NCC (**step 2** Fig. 16.6). If the host receives another host's Report while it has a timer running, it stops its timer for the specified group and does not send a Report, in order to suppress duplicate Reports. In the satellite network because of the technical implications of sending broadcast messages by the RCSTs the proposed solution is that the NCC after reception of a unicast report answers the report with a broadcast message. This message acknowledges the sender and at the same time informs the remaining RCSTs that a report has been sent (**step 3** Fig. 16.6).

After receiving an IGMP/MLD general query the RCST whose timer expires sends the Membership Report. In the same way the RCST is aware of which groups the local machines are interested on by the Membership Reports the local machines send (**step 5** Fig. 16.6).

Join Procedure

When a host joins a multicast group and it is the first member of that group on the network, it should immediately transmit an unsolicited Membership Report for that group, **step 1** and **step 2** for the terrestrial and satellite network respectively (see Fig. 16.7).

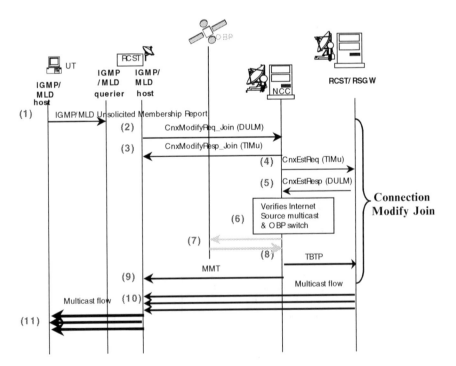

Fig. 16.7 Pure C2P scenario. Join procedure

In the satellite network for reliability the join message is acknowledged by the NCC (see **step 3** Fig. 16.7). Once the NCC is aware of the interest on a specific group the NCC checks its routing table. Then it decides if the feed resides on the satellite network or if it rather resides on the outside, correspondingly the NCC initiates a connection with a RCST or RSGW (**steps 4 and 5** Fig. 16.7).

Finally the NCC allocates timeslots for retransmission on the TBTP (**step 8**) to the feed and notifies the RCSTs of the PID to be listened on the MMT (**step 9**).

Leave and Group Specific Query/Report Procedure

When a host leaves a multicast group, if it was the last host to reply to a query with a Membership Report for that group or if it has no way of knowing that, it should send a Leave Group message (**step 1 and 4** Fig. 16.8). The querier replies the leave message (**step 2 and 5**). It sends a specific group query asking if there is any remaining member of the group. The querier does not decide that there is no remaining member until the specific group query times out (**step 3**). Thus the RCST does not send a leave message towards the NCC until its local specific group query has timed out.

If the NCC receives a response (**step 6**) before the specific group query times out it replies with a broadcast group query response (**step 7**). This mechanism is the same explained for the general query in "General Query/Report Procedure," this section, its objective is to avoid the other RCSTs to unnecessarily send a response.

In case that the Group query times out on the satellite network (**step 6′** Fig. 16.9), the NCC releases the connection (**steps 8′ and 10′**).

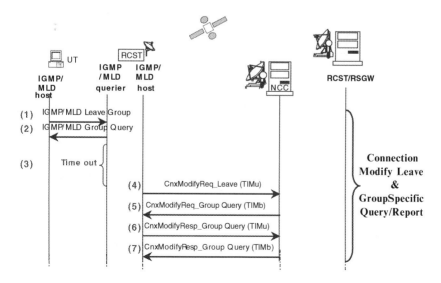

Fig. 16.8 Pure C2P scenario. Group specific query/report procedure

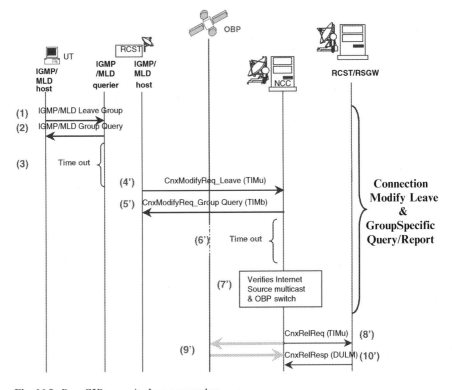

Fig. 16.9 Pure C2P scenario. Leave procedure

This pure C2P solution would solve the dynamic IP multicast mesh scenario. The Hybrid solution, mix of C2P signaling and IGMP/MLD will enhance the IGMP adaptation solution, to be able to switch the OBP in accordance to the members of the multicast session.

Hybrid Solution: C2P and IGMP/MLD

In this scenario the management of the multicast connections is shared between the NCC and the RSGW, the NCC exchanges C2P messages with the RCSTs receiving the flow to process the join and leave messages from the RCSTs for the different multicast groups. However, the querier is in this case the RSGW; IGMP/MLD messages will be exchanged between the RCSTs and its associated RSGW. For that reason this scenario is only applicable to star connectivity and it is not valid for a mesh multicast connection as the RCSTs can direct their IGMP/MLD unsolicited reports to only one destination, their associated RSGW. Putting part of the intelligence in the RSGW reduces the NCC processing. Another advantage is that there is no delay between the multicast messages and the response is faster, i.e. after a leave message the RSGW can immediately stop the multicast flow.

The procedures are going to be introduced explaining the differences between the pure and the hybrid C2P solutions.

General Query/Report Procedure

The general query/report procedure message exchange differs from that in the previous subsection ("General Query/Report Procedure") in the duplication of the report message. The host sends one C2P report message to the NCC (**step 2** Fig. 16.10) and a normal IGMP/MLD message to the RSGW (**step 3**).

Join Procedure

The join procedure in the satellite network is different from the terrestrial one. In this case the unsolicited report is sent twice, the first is a C2P message (**step 2** Fig. 16.11) towards the NCC, the second a normal IGMP/MLD message towards the RSGW (**step 3**). As soon as the message arrives to the RSGW it checks the availability of the feed, if it is correct the RSGW starts the connection establishment procedure (**steps 7 and 8**). This is the inverse situation to the pure C2P and the complexity is carried by the RSGW.

Once the connection is established the NCC allocates timeslots for the RSGW and informs the RCSTs about the PID to be listened on the MMT.

Group Specific Query/Report Procedure

The leave process is the same as explained in "Leave and Group Specific Query/ Report Procedure," this section. The difference relies in the IGMP/MLD messages

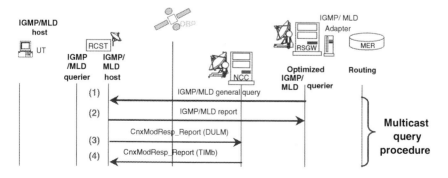

Fig. 16.10 Hybrid scenario. General query/report procedure

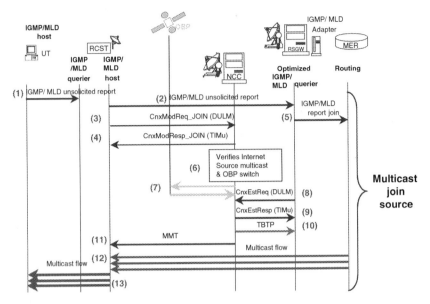

Fig. 16.11 Hybrid scenario. Join procedure

sent to the RSGW (**steps 2 and 5** Fig. 16.12) and the group specific query is not sent as a C2P message but as an IGMP/MLD message by the RSGW (**step 4**). If the group specific query is not answered the connection is released but now the release is initiated by the RSGW (**steps 9 and 10**) as the NCC is unaware of the multicast connection.

Conclusions

New generation satellites based on DVB-RCS/DVB-S(2) multibeam OBPs have opened the way to a more complex problem for efficient provision of multicast services. Certain issues must be taken into account when applying IP multicast over satellite, such as the satellite network topology, the management of the multicast (static or dynamic), IP encapsulation, address resolution or the interaction with IP multicast protocols, even more to cope with IPv6 multicast needs.

AmerHis DVB-RCS/DVB-S multibeam system already introduced a first approach to dynamic multicast group management based on IGMP satellite adaptation. This IGMP adaptation introduces certain modifications to IGMP to improve its performance when applied over satellite link, but not affecting the system interoperability with IP. Remote satellite users may join Internet multicast sessions (external and from the satellite network), and simultaneously send IP multicast flows to other users.

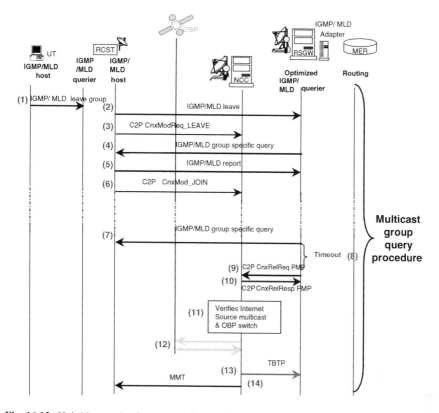

Fig. 16.12 Hybrid scenario. Group specific query/report and leave procedure

This paper studies different ways to extend this solution to a more efficient implementation based on an extended used of Connection Control Protocol (C2P). The DVB-RCS C2P protocol already solves basic connectivity issues for mesh DVB-RCS scenarios, but can be extended to solve more complex issues as dynamic multicast group management and multicast routing. The scenarios described show how C2P signaling may allow the NCC to handle dynamically all mesh (not only star) multicast groups from the satellite network and solve the problem of behaving as IGMP querier and IGMP proxy on the same interface. As a second step, a more complex solution allows a dynamic multicast routing of the OBP depending on the number and dynamic location of the members of the mesh multicast group.

References

[1] B. Cain, S. Deering, I. Kouvelas, B. Fenner and A. Thyagarajan; IETF RFC 3376—"Internet Group Management Protocol, Version 3"; October 2002.

[2] ETSI TS 102 293 "IGMP Adaptation".

[3] ETSI TS 102 294 "IP Interworking Via Satellite: Multicast Functional Architecture".

[4] ETSI TS 102 429-1,2,3,4: SES BSM Regenerative Satellite Mesh—B (RSM-B); DVB-S/ DVB-RCS Family for Regenerative Satellites; "Part 1: System Overview"; "Part 2: Satellite Link Control Layer"; "Part 3: C2P".

[5] G. Fairhurst and M.J. Montpetit; IETF draft-ietf-ipdvb-ar-06.txt "Address Resolution Mechanisms for IP Datagrams over MPEG-2 Networks"; March 2007.

[6] B. Fenner, H. He, B. Haberman and H. Sandick; IETF RFC 4605—"IGMP/MLD-Based Multicast Forwarding (IGMP/MLD Proxying)"; August 2006.

[7] SatSix. "Network Architecture".

[8] SatSix. "Satellite Access Architecture".

[9] SatSix. "Satellite Network Requirements".

[10] F. Vallejo and A. Yun; "AmerHis: Triple Play over an OBP-based DVB-RCS Satellite Platform"; IEEE International Conference on Satellite Systems Communications; September 2006.

[11] R. Vida and L. Costa; IETF RFC 3810—"Multicast Listener Discovery Version 2 (MLDv2) for IPv6"; June 2004.

[12] A. Yun, I. Moreno and F. Ruiz; "IP Multicast over New Generation Satellite Networks. A Case Study: AmerHis"; 23rd AIAA International Workshop on Satellite and Space Communications (ICSSC 2005) and 11th Ka Broadband Communications Conference; September 2005.

Chapter 17
PLATINE: DVB-S2/RCS Enhanced Testbed for Next Generation Satellite Networks

C. Baudoin, M. Dervin, P. Berthou, T. Gayraud, F. Nivor, B. Jacquemin, D. Barvaux, and J. Nicol

Abstract Emulation is a cost effective and efficient tool to perform performances evaluation and innovative access and network techniques validation. Its ability to interconnect real equipments with real applications provides excellent demonstrations means. The main problem is to overcome the emulation weakness which is the accuracy of the model reproducing the systems to be evaluated. Owing to its modular design and implementation, the PLATINE satellite emulation platform, presented in this paper, is able to emulate a complete DVB-RCS (Digital Video Broadcasting – Return Channel via Satellite)- DVB-S2 (Digital Video Broadcasting – Second) system in a realistic and flexible way. It is possible to configure the platform to emulate a transparent DVB-RCS system dimensioned around a single Hub, or to emulate a system using a regenerative satellite with an onboard switching matrix. Different DVB-RCS protocol stacks are implemented, and the adaptive physical layer is emulated in real time thanks to precalculated DRA schemes and MODCOD files. A DiffServ-like QoS Architecture that couples MAC and IP-Layer QoS mechanisms and Layer 2 security framework are currently under development. At the network side, IPv4 and IPv6 are fully supported as well as IPv6 mobility and dynamic multicast. In this paper, we mainly focus on the emulation platform and the tools developed to help the performance analysis of the emulated system.

Introduction

Evaluating performances over real satellite data links or network is often costly, even impossible for systems in development phase. Simulation and emulation both provide the opportunity to evaluate performances, at low cost, on more or less realistic systems. When simulation needs a complete modeling of the systems from applications to physical network and operates in virtual time, emulation is more demonstrative since real applications can be deployed over the model describing transfer characteristics, delay and error behavior for instance.

The IST SATSIX [1] project relies on these different methods of evaluation and validation, with the design and development of simulations, emulation and trials. The emulation testbed relies on the satellite emulation platform (PLATINE) [7]

L. Fan et al. (eds.), *IP Networking over Next-Generation Satellite Systems*
© Springer 2008

formerly developed within the frame of the IST SATIP6 [2] project. It aims to demonstrate the network and application services integration on next generation satellite systems and the possibility to interoperate with terrestrial networks. With regards to the previous release, PLATINE includes a DVB-RCS and DVB-S2 emulation, with ULE/MPEG2-TS and AAL5/ATM stacks together with the adaptive physical layer simulation and the associated radio resources management (RRM). A complete QoS architecture [5–6] mixing SIP proxies, IP/MAC scheduling and cross layer techniques is available. Finally, a layer 2 security framework will be implemented. Moreover, various network techniques, such as IPv6 mobility, dynamic multicast and its interaction with mobility thanks to MLD proxy are carried on in the frame of SATSIX, making our platform a good opportunity to test new schemes, protocols and services for next generation satellite networks.

This paper first briefly sums up the main functionalities of DVB-RCS satellite systems. The second section depicts the design and the implementation of our full DVB-S2/RCS emulated satellite platform. It mainly focuses on the satellite Physical and Data Link layers techniques and the final QoS architecture that was retained to provide differentiated QoS on a satellite network. The final section presents the environment tools used to collect statistics, events and errors and to control the testbed in a distributed way.

Introduction to DVB-RCS Systems

PLATINE Objectives

The Satellite network emulation testbed (PLATINE) has been initially developed within the frame of the SATIP6 IST project. It is compliant with the architecture adopted within the ETSI BSM [4] group and the DVB-RCS and DVB-S2 standards. Different system scenarios can be handled by the emulation testbed. The first scenario, shown in Fig. 17.1, consists in a geostationary satellite network with onboard switching capabilities and DVB-RCS uplinks and DVB-S2 downlinks. The satellite is regenerative meaning that only a single hop is needed to interconnect two end users. Satellite Terminals (RCST) provide single PC or LANs with the access to the network, while Gateways (GWs) allow the connection with Internet core networks. The second scenario is related to a more classical transparent satellite providing network access to users. RCST can access a gateway. STs and GWs are boundary devices between the satellite and terrestrial links and play an important role in access to satellite resources and hence in QoS provisioning. Both devices implement IP routing and have an IP interface on the satellite segment, as IP serves as a common denominator between the satellite and terrestrial networks. That is to say that Satellite Network is considered as a special link from a classical network point of view.

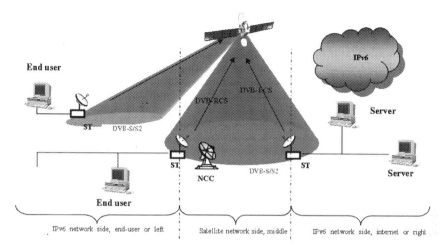

Fig. 17.1 DVB-S2/RCS architecture

Figure 17.1 depicts the satellite network architecture emulated by the PLATINE platform. On the left is represented the end-user side of the platform. On the right is shown the provider/enterprise/Internet side of the platform. We distinguish also between the satellite network side (in the middle) and the IP network sides (on left and right ends), interconnected by RCSTs.

Three main components have to be distinguished in the satellite network side (middle): the Satellite, the Return Channel Satellite Terminals (RCST) and the Network Control Center (NCC).

The role of the different network elements (NCC, RCST) is explained in a brief introduction to the main ETSI recommendations related to QoS in satellite.

Introduction to DVB-RCS Architecture

Initiated in 1993, the international European DVB Project published, in the end-nineties, a family of digital transmission specifications, based upon MPEG-2 (Motion Picture Expert Group) video compression and transmission techniques. In each specification, data are thus transported within MPEG-2 transport streams (MPEG2-TS) which are identified through DVB Service Information Tables. Adapted for satellite systems, DVB-S defines one of the most widespread formats used for Digital TV over the last years and still nowadays. However, DVB-S Satellite Terminals can then only receive frames from the satellite. The need for a return link rapidly becomes essential so as to support emerging Internet services via satellite. DVB-RCS, published in March 2000, provides full bidirectional satellite

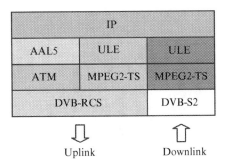

Fig. 17.2 PLATINE protocol stack (star topology)

architecture. Concerning IP encapsulation, various schemes already exist and are still being specified for the forward and return link.

The latest version of DVB-RCS and DVB-S2 introduces Fade Mitigation Techniques (FMT) to mitigate propagation effects on communication links. It consists in adapting the waveform, either by adapting the coding and/or modulation scheme or by reducing the data rate (on the uplink) to cope with systems constraints (see Fig. 17.2).

The return link access scheme in DVB-S/RCS systems is MF-TDMA. The return link is segmented into portions of time and frequency ("superframes"), each of which is divided into timeslots ("bursts") of either fixed or variable durations and bandwidths during which STs are able to transmit MPEG2-TS packets or ATM cells.

The entire satellite system control, especially STs synchronization and resource allocation, is performed by the NCC. It periodically broadcasts a signaling frame, the TBTP (Terminal Burst Time Plan), which updates the timeslot allocation within a superframe between every competing ST. This allocation can be dynamically modified on STs demand thanks to a bandwidth on demand protocol called Demand Assignment Multiple Access (DAMA). It supplements the STs with the ability to frequently request capacities that fit their current respective traffic load to the NCC. However the DAMA request/assignment cycle exhibits a non negligible latency and additional delays that cannot always match interactivity requirements of multimedia services. In order to maximize satellite resource use and meet multimedia requirements, the DVB-RCS norm discriminates RCST capacity requests into 4 categories:

- Continuous Rate Assignment (CRA): fixed slots are assigned in each MF-TDMA frame for the whole duration of a RCST connection
- Rate-Based Dynamic Capacity (RBDC): a dynamic rate capacity (in slots/frame) granted in response to explicit RCST requests
- Volume-Based Dynamic Capacity (VBDC): a dynamic cumulative volume capacity (in slots), granted in response to explicit RCST requests

- Free Capacity Allocation (FCA), which is assigned to STs on an "as available" basis from unused capacity.

The standard, after defining separate MAC traffic priority queues (Real-Time, Variable Rate and Jitter-Tolerant priorities), suggests a requesting strategy for each of them, that is to say a relevant mapping between traffic and request categories. Any given RCST can be assigned one or a mix of the four capacity types. In general, higher priority classes of service (e.g. IP DiffServ EF and AF classes) are associated with guaranteed capacity (CRA, RBDC), while lower priority classes (e.g. Best Effort) are predominantly given best effort capacity (VBDC, FCA).

Experimental Platform

The testbed we have built is able to emulate a complex scenario for next generation satellite network as presented in Fig. 17.1. Each network element involved in the satellite network is emulated in our platform on a dedicated node. In fact, 3 users LAN composed of two nodes (standard Linux systems) are connected to the emulated satellite network with 3 RCST that implement an almost complete DVB-S2/RCS stack. The satellite core network is emulated thanks to the Satellite Emulator (SE) as link emulator and the Network Control Center (NCC) for bandwidth management (DAMA). Nine computers are used as described in Fig. 17.3.

We first describe the protocol stacks implementation, the satellite carrier emulation, the MAC layer implementation and then the QoS architecture.

The PLATINE Architecture Implementation

The Satellite System Emulation

In order to have the most modular platform and so preserve room for future evolution (GSE), stringent requirements were fixed before the development phase.

At first, the emulation testbed takes advantage of a linux system (Fedora Core 5) which natively supports IPv6 and a wide panel of IPv6 applications (Apache as HTTP Server, Mozilla as HTTP Client, Vsftpd as FTP Server, Gnomemeeting for Videoconferencing, VideoLanClient for Videostreaming), as well as advanced network and QoS features.

The core architecture of PLATINE relies on a derived GPL C + + runtime called Margouilla [3], that provides platform independent messaging and synchronization toolkit, bloc management, a set of common blocs ready to use (IP/ATM/Ethernet layers...) and various utilities packages such as configuration file and logging mechanisms for error and debug messages. The final PLATINE protocol stack is detailed in Fig. 17.4.

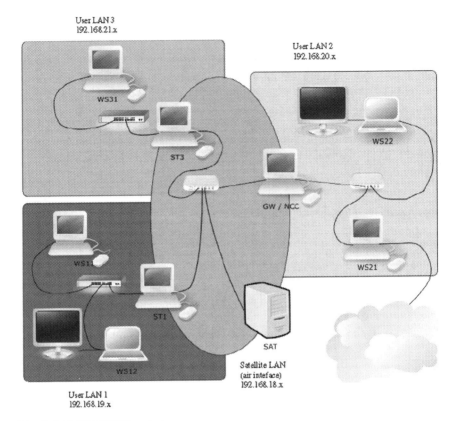

Fig. 17.3 Full PLATINE testbed

The blocks that were developed within the Margouilla runtime are:

- The *satellite carrier package* is responsible for the different satellite carriers emulation on top of Ethernet (DVB-RCS, DVB-S2 and Signaling Channels) and the simulation of typical satellite bit errors and delay.
- The *DVB-S2/RCS package* implements a framing structure compliant with the DVB-S2/RCS standards. and fills DVB-RCS frames with upper layer packets (ATM or MPEG2-TS) coming from the ENCAP bloc layer. In order to achieve proper QoS, this layer manages synchronization and queues according to the authorizations a DAMA algorithm delivers.
- *The DAMA package* implements the DAMA algorithms used to manage the satellite resources allocation at layer 2 taking into account adaptive physical layer information.
- The *ENCAP package* implements AAL5 and ULE encapsulation schemes, and is in charge of the segmentation and reassembly functionalities (ATM or MPEG2-TS).

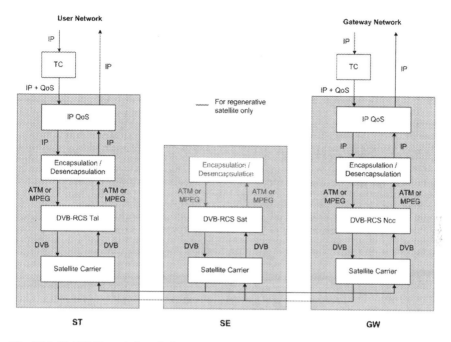

Fig. 17.4 PLATINE emulation platform architecture

- The *IP QoS package* implements common mechanisms to enable differentiation at this level. It mostly relies on QoS services offered by Linux kernel through the Traffic Conditioning (TC) tool, retrieves incoming packets from IP network with their associated tag and forward them to the lower layer.

A detailed description of the way the different layers are emulated is done through the following paragraphs.

The Satellite Carrier Emulation

The satellite carrier emulation is designed to operate on top of Ethernet frames and for each satellite channel corresponds an Ethernet multicast address. Ethernet was chosen for its native broadcast abilities (like a spot) and also for its high bandwidth capacities. For each spot, different logical channels are defined to transport data on the air interface depending on their type (data, logon, control,...). This component is also in charge of dropping frame not belonging to the node or received with a MODCOD more efficient than it can receive.

The Satellite Link Emulation

The satellite emulator (SE) can act as a transparent or a regenerative satellite. It is able to emulate spot switching and signal format conversion. Indeed, the regenerative satellite with an onboard switching matrix processes DVB-RCS frames, switches ATM cells or MPEG2-TS received from the DVB-RCS frames, encapsulates ATM cells into MPEG2-TS frames (in the ATM case) and finally forward them within DVB-S2 frames. The switching mechanism is modular and can manage either ATM cells or MPEG2-TS frames. The switching tables could be updated by appropriated control message. Currently, the ATM and MPEG tables are allocated statically.

What is more, the main functionality of the SE is to emulate its satellite link, modeling configurable delay and jitter. In addition to simulate delay, error model can be introduced, either using precalculated BER files or statistical laws.

Medium Access Control

Due to hardware limitation and OS constraints, RCST and NCC components are synchronized on uplink frame and thus do not handle time slot synchronization. The precision of the emulation is therefore at best equal to one frame duration.

The emulation of the adaptive physical layer is done thanks to corresponding simulations performed in SATSIX the aim of which is to generate spatial rain patterns.

First of all, rain cells are spatially modeled by circular patterns regarding the rain rate parameter. Then, the circular hypothesis is extended to the rain attenuation, and to the allocated DRA schemes. Finally, Spatial rain patterns are generated from the DRA sequences time, issued from FMT simulations taking as an input simulated rain attenuation time series for an uplink availability of 99.9% of the time.

The result example for a spotbeam is shown in Figs. 17.5 and 17.6 with 20 rain cells with 5 dB max. attenuation, and 4 cells with 11 dB max. attenuation moving across the beam and their impact on DRA schemes allowed for RCST.

The *frame synchronization* is carried out by emitting Ethernet frames at fixed instant on all RCST (superframe tick) and then internally by awaking processes each frame ticks (50 ms) to send already queued packet. The DVB-RCS allocation scheme is repeated for all the frames composing the superframe.

The *allocation procedure (DAMA algorithm)* implementation in the demonstrator is described below:

- At log-on RCSTs request a fixed bandwidth (CRA) for MAC "real-time (RT)" traffic. This allocation can be updated thanks to the dynamic QoS architecture using SIP proxy information.
- Then RCSTs is mainly in charge of the capacity request calculation that aims at achieving full link utilization thanks to a multi-MRC approach.

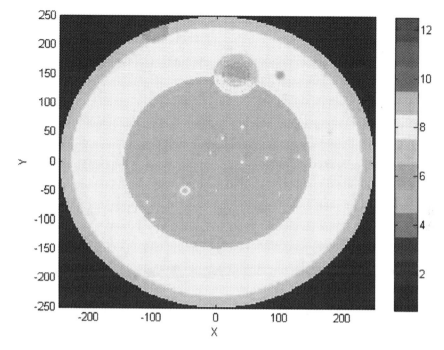

Fig. 17.5 Return link scenario with individual terminals: spatial repartition of the DRA schemes in the spot at the beginning of the simulation

- Each Capacity Request (CR) is sent on a SYNC slot basis. Each CR is then delayed by the satellite emulator and reach NCC $250\,\text{ms} \pm 10\,\text{ms}$ after. CRs are processed by NCC upon arrival.
- NCC computes allocations at the beginning of each superframe according to an internal SACT table based on CR received. The DAMA algorithm takes into account DRA scheme of each RCST to constitute the frame structure. FCA based on weighted fair queuing scheduler can also been applied. Authorizations are sent back to RCSTs using a TBTP table.
- Upon reception of a TBTP, allocations are stored by RCST and used in the next frame.

IP Encapsulation

This section deals with the transport of IP datagrams over the PLATINE emulation platform. The following paragraphs describe how IP encapsulation is handled by the platform components.

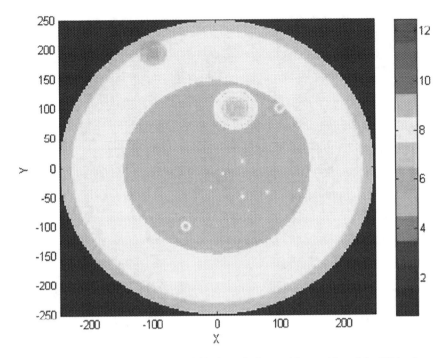

Fig. 17.6 Return link scenario with individual terminals: spatial repartition of the DRA schemes in the spot at the end of the simulation

The ENCAP package is in charge of the IP encapsulation and segmentation/reassembly functionality. As presented in Fig. 17.4, the ENCAP package is used by the RCST and GW components to send or receive IP datagrams through the satellite link. When the SE emulates a regenerative satellite, the OBP can manage two different encapsulation schemes for return and forward links thanks to the ENCAP package.

Various IP encapsulation schemes already exist or are still being specified for the forward and return link of DVB-RCS satellite system. The ENCAP package is designed to be modular, so that new encapsulation schemes can be easily added to the emulation testbed. The encapsulation schemes for forward and return links are chosen at simulation startup.

The ENCAP package currently implements AAL5 and ULE encapsulation schemes, as well as ATM and MPEG2-TS segmentation/reassembly mechanisms. The layer 2 security framework extends the ULE encapsulation scheme to add integrity and encryption functionalities.

The AAL5/ATM encapsulation and segmentation/reassembly scheme is presented in Fig. 17.7. The ULE/MPEG2-TS mechanism defined in [9] is briefly described in Fig. 17.8.

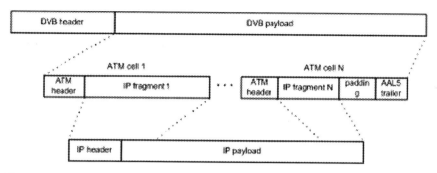

Fig. 17.7 AAL5/ATM encapsulation scheme

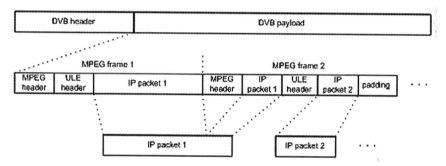

Fig. 17.8 ULE/MPEG-2 TS encapsulation scheme

The QoS Architecture

As presented in the first section, our platform implements a QoS architecture [5–6] compliant with the ETSI Broadband Satellite Multimedia (BSM) group. This architecture links the MAC layer with the IP Layer to ensure the QoS continuity in the satellite protocol stack.

MAC Layer

Three MAC queues are available in the PLATINE system, one to satisfy a "real-time (RT)" service, one to provide a "non real time (nRT)" service and the last one to handle "best effort (BE)" service. The association between MAC queues and allocation categories is fully configurable. So the RT queue can be only served by Continuous Rate Assignment (CRA) or can also request additional capacity using RBDC requests. In the same way, the non real-time and best effort queues can be served either using RBDC or VBDC requests. Anyhow, allocation provided by the

NCC (using CRA, RBDC, VBDC or FCA) is not affected to any specific queues, and is used to empty MAC queues in priority order.

IP Layer

The core of the IP QoS processing is performed by the HDLB scheduler which is based on the Linux HTB policy [8], but uses the Dual Leaky Bucket Filter algorithm to shape traffic (see Fig. 17.9). It is mainly designed as a kernel module. It can be controlled in user space through the "TC" tool. The link between TC and the HDLB module is provided by a rtnetlink socket.

Rtnetlink allows the kernel's routing tables to be read and altered. It is used within the kernel to communicate between various subsystems, and for communication with user-space programs. Network routes, IP addresses, link parameters, neighbor setups, queueing disciplines, traffic classes and packet classifiers may all be controlled through NETLINK_ROUTE sockets. It is based on netlink messages.

TC presents interfaces to attach or remove disciplines to qdiscs, and create or destroy classes of traffic. It creates messages to be sent to the rtnetlink socket. Platine uses a special version of TC adding HDLB classes compatibility. On the other side, the kernel module retrieves TC's commands and parameters and acts consequently. To get along, both TC and HDLB module use identical data structures. When HDLB module is first loaded into the kernel, the module registers the qdisc and class operations it is able to perform and that can be called from the user space. Fig. 17.10 depicts an overview of the HDLB scheme.

The IP Layer uses the following hierarchy, with basically 3 DiffServ categories (EF, AF31 to AF33, and BE) the parameters of which are dynamically updated by the QoS server according to QoS agent or SIP proxy information.

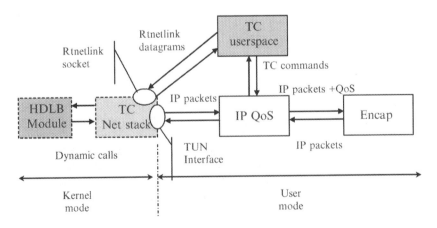

Fig. 17.9 IP QoS architecture

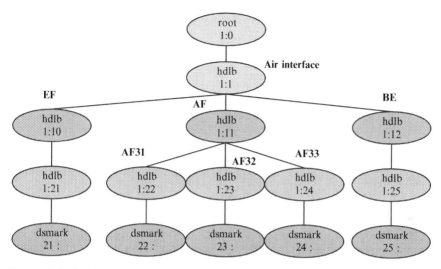

Fig. 17.10 HDLB overview

Cross Layer Mechanism

The BSM group recommends cross layer mechanisms to enhance the adaptation between the IP and the MAC layers. The output throughput of the IP scheduler is adapted to the MAC output throughput thanks to feedback information. This scheme allows to optimize the overall scheduling process since the MAC scheduler does not perturb the scheduling decision made at IP level, and to limit or to prevent the loss in the MAC FIFOs.

QoS Tools

In order to configure properly the QoS for every kind of application, different tools have been developed. Firstly, a QoS Server running in the satellite terminals is in charge of collecting QoS information on traffic flows and configuring accordingly the IP and MAC layers. Those information are provided by a QoS Agent for non-QoS aware applications or an enhanced SIP Proxy for QoS aware applications.

QoS Agent

Very few applications currently implemented are aware of the QoS provided by the underlying network. As the applications are rarely able to define their own requirements, a user-oriented solution, called « QoS Agent », has been defined in the SatIP6 project [2]. Implemented on the user terminal, the QoS Agent detects the

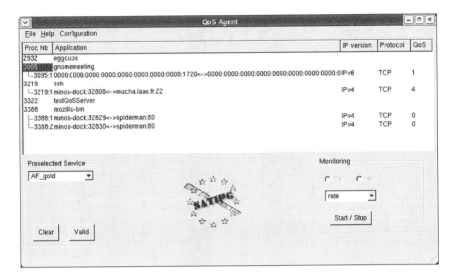

Fig. 17.11 QoS Agent graphical interface

outgoing traffic flows and thanks to a graphical interface listing them (cf. Fig. 17.11), the user is able to assign one of the available QoS services and, through interactions with the QoS Server, to remotely configure the multi-field classifier in the RCST, for only one or for all the streams the application uses. In fact, each time a service is selected by the user for a given application, the QoS Agent sends to the QoS Server the concerned connections' list with the reference of the associated service using a dedicated transactional protocol. Using the received information, the QoS Server is able to tag and redirect these packets coming from the user terminal toward the appropriate requested IP service. If resources are not available, the QoS Agent is immediately informed.

Enhanced SIP Proxy

An alternative solution to the QoS Agent is to automate the resources reservation and make it transparent to users who wouldn't be able to choose the most appropriate classes of service to the different flows (audio and video for example). To make that, the Session Initiation Protocol (SIP) has been chosen because of its growing success in both public (open source code) and private (company) fields. This mechanism allows to configure the QoS transparently by the analysis of the SIP messages exchanged between the applications. This analysis is made by an enhanced SIP Proxy (or QoS-aware SIP Proxy).

As shown in the Fig. 17.12, the enhanced SIP Proxy intercepts the session descriptors included in SIP messages (INVITE and OK), deduces the characteris-

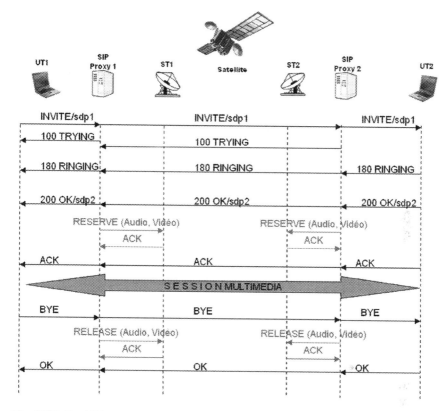

Fig. 17.12 Establishment of a SIP session with QoS

tics of each media involved in the session and realize the reservations and releases
with the QoS Server.

Some additional functionalities are needed to realize that:

- An SDP analyser making the Proxy able to analyze the session descriptions
- A table of medias updated during the session establishment. The medias negoti-
  ated between the caller and the callee are identified by a call-ID.
- An SDP/Diffserv mapping
- A QoS module which realizes the resources reservation associated to each media
  with the QoS Server.

A QoS-Aware SIP proxy is deployed in each user LAN. This distributed architecture is well suited for an access or mesh topology based on a regenerative satellite because it answers the following two concerns:

- scalability concerning flow QoS management in user LAN;
- session establishment delays: the number of round trips of session and QoS signaling on the satellite link are minimized.

Interface with the QoS Server

In both case, the exchange of resources reservation messages and resources release messages is performed through a TCP connection over IPv4/IPv6 to assure a reliable and ordered message delivery. The protocol used to communicate with the QoS Server is based on XML.

The structure of the messages exchanged by the QoS Agent or the enhanced SIP Proxy with the QoS Server is quite similar. In both case, it indicates the type of message (RSV, FREE) and the source address, the source port, the destination address and the destination port which are required to the identification of a specific connection. Then, in the QoS Agent case, it is necessary to precise the application, the PID of the application and the wished class of service and in the SIP Proxy case, the type of media for example.

PLATINE Outputs and Control

Statistics, Events and Errors

Additionally to aforementioned advantages, emulation provides interesting characteristics such as measurement points which would not be so easily accessible in real systems. Regarding our satellite emulator, statistics are computed at each superframe at the NCC/GW side, and at each frame at the RCST side (for both MAC layer and IP QoS layers).

These measurement points, as well events or errors, are handled by a distributed framework. Each emulation component embeds 3 agents in charge of collecting and gathering all these information with basic treatments (such as mean, max, …) and sends them on a regular basis to centralized statistic, event and error controllers. These processes aggregate the data with optional processing (sliding window for example), store them in files according to the simulation and run numbers, and send them to a real time graphical display.

Statistics includes delay measurements, load measurements, queues, DAMA and congestion control information.

Figure 17.13 shows several measurement points at different levels on the RCST return link channel. Measurement points are located at ingress points, at egress points as well as inside of IP_QoS and MAC layers of the satellite network emulator. As a result, the DAMA algorithm effects can be observed at different levels of the system. So, when higher layers transmit traffic, we are able to measure the throughput per traffic category at IP layer ingress points and egress points. The throughput crossing the IP_QoS layer can be therefore characterized. Similarly, relevant IP QoS information extracted from the HDLB scheduler are available.

The MAC layer behavior can also be analyzed through the study of the throughput at ingress and egress measurement points. Inside the MAC layer, the DAMA

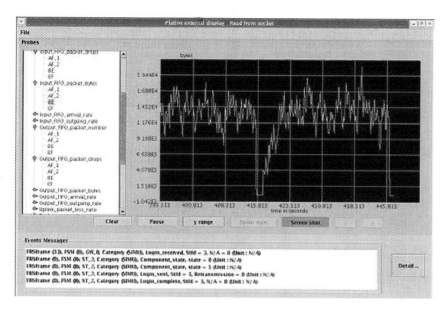

Fig. 17.13 PLATINE statistics tool

algorithm behavior can be evaluated. The evolution of capacity requests can be obtained for the NRT MAC service class in term of kbits/s and time-slot number. The DAMA efficiency can be directly obtained through the ratio between used and allocated time-slots.

We also benefit from statistics on the downlink channel, like the downlink traffic throughput received on the AIR interface by the RCST and the uplink traffic throughput sent on the AIR interface by the RCST.

What's more, in order to study more thoroughly the DAMA algorithm efficiency in a configuration comprising numerous RCSTs, the NCC is supplemented with the ability to emulate virtual RCSTs. To do so, the MAC scheduler is able to replay virtual CRs and takes them into account in addition to CRs sent by individual emulated RCSTs. These virtual CRs may be obtained through past recorded logs or created from scratch. Moreover, this tool allows to generate background traffic to study congestion for instance.

Managing the Emulation Testbed

One of the classical problem of emulation testbeds is their poor user interface resulting to difficult configuration and management, especially in a distributed environment. PLATINE includes a new control interface (see Fig. 17.14) to start

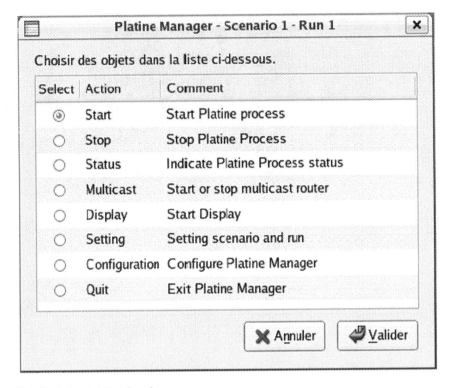

Fig. 17.14 Control User Interface

and stop the emulation process on a configurable network topology. Moreover, it offers various services to retrieve the component status, to start network components such as multicast routers and to execute the JAVA display. Finally, it can distribute the complete software on a group of nodes in a LAN to simplify maintenance and tests.

Conclusion & Further Works

This testbed is an important tool to evaluate DVB-S2/RCS system performance and in particular its impact on innovative network techniques. Providing the user with several kinds of measurement, traffic flows and results analysis tools, it is an interesting way to investigate how to configure and use such systems.

This testbed is efficient enough to support various kinds of experimentation, dealing with multimedia user applications for instance. It may be enhanced in order to implement mechanisms such GSE. Thanks to the software design of the testbed, enhancements may be done easily. Any interested user is clearly welcome.

Acknowledgments This work is supported by the IST FP6 SATSIX project, funded by European Commission (EC). The financial contribution of the EC toward this project is greatly appreciated.

References

[1] IST SATSIX Project (Contract IST-2004-26950), http://www.ist-satsix.org

[2] IST SATIP6 Project (Contract IST-2001-34344), http://satip6.tilab.com

[3] Margouilla c + + Runtime : http://cqsoftware.free.fr/margouilla

[4] ETSI TR102157 v1.1.1. ETSI TC SES; Broadband Satellite Multimedia; IP Interworking over satellite; Performance, Availability and Quality of Service, July 2003

[5] A. Pietrabissa, T. Inzerilli, O. Alphand, P. Berthou, E. Fromentin, T. Gayraud, F. Lucas. Validation of a QoS architecture for DVB/RCS satellite networks via a hardware demonstration platform, Computer Networks Journal, Issue 6, Volume 49, pp. 797–815, December 2005.

[6] O. Alphand, P. Berthou, T. Gayraud, S. Combes, QoS Architecture over DVB-RCS satellite networks in a NGN framework, GLOBECOM 2005, St-Louis USA, December 2005.

[7] O. Alphand, P. Berthou, T. Gayraud, F. Nivor, S. Combes, C. Baudoin, PLATINE: DVB-S/RCS Testbed for Next Generation Satellite Networks, ASMS 2006, Munich Germany, May 2006.

[8] http://luxik.cdi.cz/~devik/qos/htb/manual/theory.htm

[9] G. Fairhurst, B. Collini-Nocker, Unidirectional Lightweight Encapsulation (ULE) for Transmission of IP Datagrams over an MPEG-2 Transport Stream (TS), RFC 4326, December 2005.

Chapter 18
Overview of the SATSIX Trials

A. Ramos, D. Pérez, R. Muñoz, F. Vallejo, B. de la Cuesta, J. Aguiar,
B. Carro, N. Hennion, and P. Zautasvili

Abstract SATSIX contributes to enhance the positioning of satellite access systems in the Broadband Access Market and capitalises the development and deployment of NGN and IPv6 through satellite. Through live demonstrators, SATSIX will prove the capability of the satellite to provide affordable services to a large customer range. The SATSIX Work package/Activity 3200 aims at validating this concept, demonstrating its feasibility by means of demonstration scenarios.

Introduction

This paper summarises the tests that will be carried out over four real Satellite Systems: the Regenerative AMERHIS system of HSA and the Transparent DVB-RCS systems of HSA, HDT and TASF. It includes the tests procedures for validating each key SATSIX feature selected, the subsequent recording of results and the methods of evaluating the results. These tasks will be carried out for each of the following trial scenario:

- Scenario 1: Corporate applications on DVB-RCS transparent access platform.
- Scenario 2: Collective Access Terminal on DVB-RCS transparent access platform.
- Scenario 3: Corporate applications on DVB-RCS regenerative access platform.
- Scenario 4: Residential applications on DVB-RCS transparent access platform.

The Corporate applications on TASF DVB-RCS transparent access platform (scenario 1) will allow to assess and to validate the IPv6 application scenarios defined in the project over a DVB-RCS star network and operational architecture. It includes the required adaptation for coupling the hub and the applications which will have to be studied and then implemented. For doing that, the current platform will be adapted to host SATSIX key features, and hence will be able to assess in that context the Corporate Applications. Also, innovative collaborative working applications will be evaluated on the same platform. This will allow a benchmarking between commercial applications and applications developed internally from open source modules.

L. Fan et al. (eds.), *IP Networking over Next-Generation Satellite Systems*
© Springer 2008

The Collective Access Terminal on DVB-RCS transparent access platform (scenario 2) is developed and evaluated at Hungaro Digitel HDT premises with the participation of HDT for the Collective Access Terminal part. HDT's DVB-RCS HUB will be used for this scenario. Wireless Local Loop (WLL) is able to share satellite connection and its costs among the users. Therefore this combination of access technologies can reduce the costs of a user and can make satellite connection more affordable. This live trial is able to show the real user's behaviour and can help to evaluate not only the measured parameters but also the feeling of the users.

Scenario 3 is developed and evaluated on HSA's pilot AMERHIS system. It will allow to assess and to validate the IPv6 application scenarios defined in the project over a DVB-RCS mesh network and operational architecture. It includes the required adaptation for coupling the platform and the applications which will have to be studied and then implemented.

The residential applications on DVB-RCS transparent access platform (scenario 4) is developed and evaluated on HSA's pilot DVB-RCS system. The principal residential application of satellite networks is the provision of Digital TV services which includes audio, video and data broadcasting to final users.

For each scenario the list of functions to be demonstrated are specified, as well as the systems characteristics and validation scenarios are defined.

Corporate Applications on DVB-RCS Transparent Access Platform

Functions to be Validated

The objectives of this scenario are to test corporate applications on a DVB-RCS transparent and an applications platform modified to be compliant with IPv6. The TAS infrastructures hosted in Cannes (France) will be used during this trial. They are represented in Fig. 18.1.

What we mean by corporate scenario is the test of applications used in the scope of a corporate. For example, we will make end-to-end tests with:

- Mail server (SMTP/POP3)
- Web browsing (HTTP/HTTPS)
- Unicast Files transfers (FTP)
- Multicast file transfers (MPUSH based on multicast UDP)
- Collaborative working server (based on the LASS server)
- Voice over IP (SIP)

Before testing the corporate applications, we will make some basics test based on network testing tools (like ping, iperf, sjitter...).

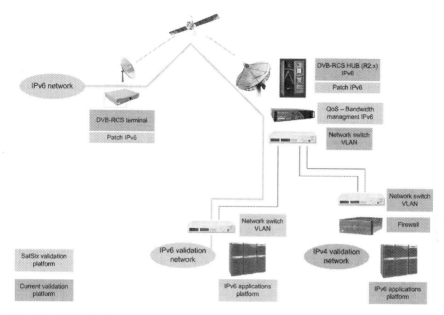

Fig. 18.1 TASF platform

Platform Architecture Overview

Figure 18.1 represents the diagram of the TAS platform:

Demonstration Instruments

Demonstrations instruments are composed of three parts:

- IPv6 applications platform (Satsix platform)
- DVB-RCS satellite gateway version R2.x with Ipv6 patches
- 3 satellites terminals with 3 PC where we can install the applications to test

The IPv6 Satsix applications platform will be connected to the existing one (IPv4) through a central router/firewall (based on a FreeBSD operating system). This router will hosts an IPv6 to IPv4 gateway in order to have access to the Renater terrestrial network. The IPv6 part of the platform will hosts the following services:

- Network testing tools (Iperf and Sjitter)
- Mail server (Postfix)
- Web server (Apache)
- FTP server (Proftpd)
- Multicast files transfers server (TAS Mpush server)

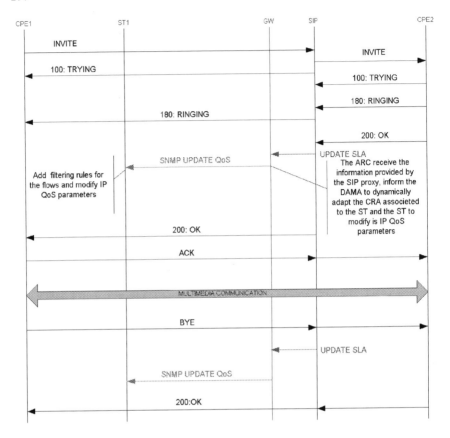

Fig. 18.2 IP QoS Dynamic mechanism

- Collaborative working server (LASS server)
- SIP server (Asterisk)

It is planed to install those services on a Linux operating system (Fedora Core 7).

The DVB-RCS satellite gateway will be patched to be able to manage IPv6 addresses. Only the core of the system will be modified (not the interface of the manager). Additionally, the IP dynamics QoS layer will be also added to the gateway (Fig. 18.2).

On the other side of the satellite link, 3 Satsix terminals (with one PC per terminal) will be installed with all the needed applications:

- Network testing tools (Iperf and Sjitter)
- Mail client (Thunderbird)
- Web browser (Firefox)
- FTP client (Filezilla)
- Multicast files transfers client (TAS Mpush client)
- Collaborative working client (LASS client)
- SIP client (X-lite)

Trials and Validation

We will start with basic end-to-end test between clients (behind a Satsix satellite terminal) and servers hosted on the Satsix platform.

- Ping (ICMP packet) between a first client and a server hosted on the platform
- Ping (ICMP packet) between two clients
- Iperf/Sjitter between a first client and a server hosted on the platform

  ○ TCP with different packet sizes
  ○ UDP with fixed bandwidth (from 64 Kbps to 512 Kbps)
  ○ Network Jitter is compute using the Sjitter software (developed by TAS)

⇒ Outputs: Packets delay, packets lost, bandwidth, jitter…

Then we will focus on the corporate applications:

- Mail

  ○ Send an email
  ○ Receive an email

- Web browsing

  ○ Display HTML page (static and dynamic)
  ○ Download file using HTTP protocol

- Unicast files transfer

  ○ Download a file from the FTP server
  ○ Upload a file to the FTP server

- Multicast files transfer

  ○ Receive a file from the Mpush server
  ○ Acknowledgment of the file through the return link

- Collaborative working server

  ○ Connection to a session
  ○ Chat
  ○ Audio
  ○ Video
  ○ Application sharing

- Voice over IP

  ○ Register software phones on the SIP server
  ○ Register hardware phones (for example hardware Wifi phone) on the SIP server
  ○ Make a call between two clients (software/software, software/hardware, hardware/hardware)

Collective Access Terminal on DVB-RCS Transparent Access Platform

Functions to be Validated

In central and eastern part of the European Union broadband network service is not available in many rural places. Because of the high costs of terrestrial communication network development, broadband two-way satellite connection is the most affordable technology and particularly, WLL is able to share satellite connection and its costs among the users.

In this scenario, educational, governmental, home and small office users will be connected to the Internet and the functions to be validated will be the following:

- IPv6 WLL access validation of the connection between Windows XP/Linux PCs and a server on the Central LAN segment via Wireless Local Loop (WLL) using IPv6 protocol.
- WLL security methods (WEP, WPA and WPA2) validation using 802.1x authentication over an IPv6 satellite connection. One of the most vulnerable parts of the network is the Wireless Local Loop. Therefore, the validation of the WLL security is one of the most important parts of the trial.
- Application/Network security validation of several commonly used applications. A dual stack (IPv4 and IPv6) network is going to be configured in order to enable operating systems and applications, which are not supporting IPv6 protocol or IPv6 DNS queries, to communicate through the network.
- IPv6 Mobility validation as a way to extend the network environment of Application Security with IPv6 mobility functionality. The IPv6 mobility will be tested between two test-WLLs.
- Dynamic Multicast validation over IPv6 network, using DVB-RCS system and WLL. In this scenario the ability to join and leave the multicast group will be tested.

Platform Architecture Overview

This scenario will be developed and evaluated in Szigetszentmiklos-Lakihegy (Hungary) at HDT premises. HDT's DVB-RCS HUB will be used for this scenario. The HUB uses AMOS1 satellite, which has an excellent coverage in the CEE countries.

In this scenario, the access network is made of a connection of WLL (WiFi and WiMax) to the satellite network (Fig. 18.3).

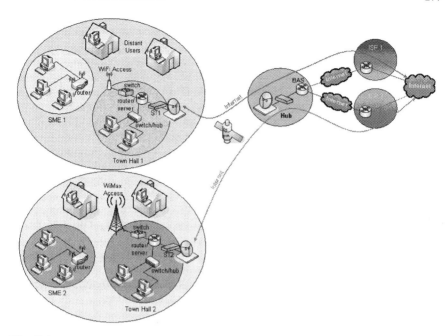

Fig. 18.3 Collective Access Terminal on DVB-RCS transparent access platform

Demonstration Instruments

The network architecture composition varies according with the functions to be validated and the sum of the equipment and software used is as follows:

- Additional Application Servers on the Central LAN;
- Application Server on the Protected IPv6 LAN;
- Application servers for Home Network and Corresponding Node;
- Central IPv6 LAN with FREERADIUS and CA server;
- Central IPv6 LAN with FREERADIUS server;
- Cisco 871 for the IPSEC connection to the Protected LAN;
- Cisco IOS to terminate the IPSEC connection;
- DNS Server service: Bind9 on Linux DEBIAN;
- Home Agent for Mobility (Linux PC);
- IPSEC:

  - Standard IPSEC engine for Windows;
  - Standard IPSEC package for Linux;

- Local DNS server on the remote site to serve the DNS requests from Windows XP OS;
- Mail client application: Thunderbird for Windows and Linux;
- NERA DVB-RCS system (supports IPv4 only, therefore we install routers for IPv6 over IPv4 tunneling):

Central router Cisco 3745;
Site Access router Cisco 871;
- PC with DEBIAN Linux;
- PC with Windows XP;
- POP3 Server service: Courier pop3 and AUTHDAEMON;
- Server with DEBIAN Linux;
- SMTP server service: POSTFIX mail server;
- SSH service:

  ᴐ Putty, WinSCP for Windows;
  ᴐ OpenSSH for L.inux;

- VLC for client and server side application;
- Web access:

  ᴐ Internet Explorer 7.0 for Windows;
  ᴐ Mozilla Firefox for Linux;

- Web and secure web Server services: Apache2 on Linux DEBIAN;
- Wireless Access Point of Cisco 871 W router or Linux HostAP.

Trials and Validation

For the IPv6 WLL access validation the PCs will be connected to the WLL by the Access Point of Cisco 871 W and during the trial it will be applied the normal WEP 128bit encryption method only. The wireless clients will be able to: connect to the wireless segment; acquire IPv6 address, through the DHCP, from the AP; and establish an IPv6 ICMP session to the server, through the satellite connection.

Regarding the WLL security validation there will be a single hop IPv6 connection via satellite between the remote site router and the RADIUS server. This way it should be expected: a successful IPv6 RADIUS authentication sessions established between the AP and the RADIUS server; that the wireless clients are able to authenticate themselves through the router's RADIUS connection, via IPv6 over DVB-RCS system; a successful key exchange on the WLL; that the wireless clients are able to communicate with each other; and that the wireless clients are able to establish IPv6 ICMP sessions with the server (RADIUS), through the satellite connection.

As for the application/network security validation, the hosts in the WLL will be connected to the IPv6 Central LAN segment for accessing the Application servers and the connection to the WLL loop will use the appropriate WLL security authentication (802.1x, RADIUS). This way, it's anticipated: a successful file transfer over the network; the availability of Email services (sending, receiving); an operational Web browsing; a successful establishment of secure shell connections; a successful forwarding of packets through the IPSec connection.

Concerning the IPv6 Mobility validation, in this scenario the IPv6 mobility feature will be set up on a Mobile Node (MN). The MN will be able to communicate to the Corresponding Node on the Outside IPv6 LAN segment with the IPv6 Home

Address from the Home network pool. Following this the Mobile Node will be able to connect with the Home Agent for registering its new CoA and the Corresponding Node will be able to communicate with the Mobile Node without interruption during its roaming between the two WLLs.

Finally, during the dynamic multicast validation over IPv6 network the clients on the IPv6 Wireless Local Loop should be able to play the multimedia stream using the VLC application. On the server side, the multimedia content will be encoded by the VLC application and the multimedia stream is transferred via UDP Multicast technology to be played out at the client site. Thus, by joining the multicast, the routers will be instructed successfully to forward the multicast stream; there will be a successful transfer of the multicast stream over the IPv6 satellite connection; the multimedia content will be played at the client site; and starting and stopping actions from the clients (joining and leaving the Multicast Group) will transfer or stop the traffic over satellite, respectively.

Corporate Applications on DVB-RCS Regenerative Access Platform

Functions to be Validated

The objective of this scenario consists of demonstrating that the regenerative platform is able to use IPv6 protocol. Furthermore, ULE Encapsulator protocol will be used since a complete integration between IPv6/ULE will represent a first step for a future move towards a GSE integration.

SATLIFE IST project is currently bringing many enhancements to the AMERHIS system such as: enhanced QoS and multicast functionality, NAT and security enhancement in ST, new service applications and multiple SP support in NCC. All these improvements will be validated once more thanks to the use of the regenerative platform of Hispasat.

Platform Architecture Overview

In Fig. 18.4 it is shown a complete description of all the elements involved in the trials.

In the transmission side, a key point is the correct selection of the right terminal. Thus, two different options have been considered:

- SP-RCST: Thanks to SATLIFE project, the regenerative platform is able to allocate resources to a particular DVB-RCS terminal, the SP-RCST, without the need of the C2P protocol.
- This terminal, normally used for native MPEG2 video transmission, receives from a VSP (Video Service Provider) the information to be sent which could be IPv6/ULE/MPEG-2 always through an Ethernet interface.

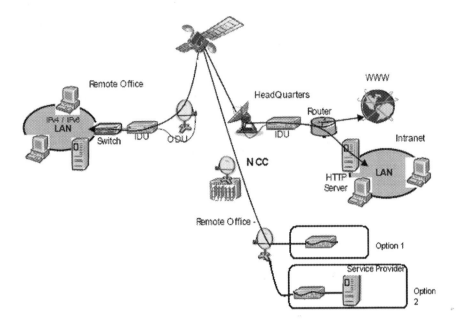

Fig. 18.4 Elements in the Regenerative platform

- ASI-RCST: The other possibility consists of using a specific RCS terminal with an ASI interface. It will allow to transmit native MPEG2 packets onto the air interface. However, an extra development is necessary in order to integrate it within the AmerHis platform.

The reception is the critical part of the demonstration and an ULE de-Encapsulator is required in the project.

A field trial involving the AmerHis payload with all the elements described above can be seen in the next picture (Fig. 18.5):

The Video Service Provider injects the IPv6/ULE data into the terminal using the appropriate interface.

The first option considered is the use of an OPAL IP Encapsulator which can mix sources in IPv4 or MPEG2. The issue for this trial would be to verify if it is possible to update easily the existing OPAL IP Encapsulator existing for ULE encapsulation.

On the other hand, an ULE Encapsulator could be used. It includes an ASI board to connect it to a terminal.

Trials and Validation

Initially the tests would be based on unidirectional flows that could be evolved depending on the constraints of the different elements.

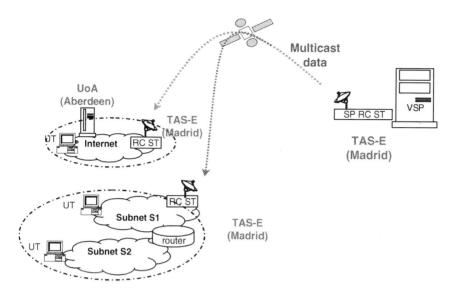

Fig. 18.5 Field trial involving Amerhis payload

The regenerative platform is connection oriented. The connections between RCSTs or GW-RCSTs are established thanks to the C2P protocol. As the SATSIX terminal does not support C2P, other alternatives have been presented in order to permit testing the support of IPv6 over the regenerative platform.

The applications that are interesting for this scenario are based on corporate services, such as VPN connectivity, internet and Intranet access from remote offices to the HeadQuarters and Videoconference.

Residential Applications on DVB-RCS Transparent Access Platform

Functions to be Validated

Nowadays the principal residential application of satellite networks is the provision of Digital TV services, and even more, the provision of triple-play services which includes audio, video and data broadcasting to final users by satellite in one-way communication.

Therefore, the applications validated in this scenario are the following:

* Digital TV is the most classic service in the satellite networks and it is possible to provide audio, video and data broadcasting to a wide range of users without a great deployment of infrastructure. This scenario is based on a broadcast

network via a transparent satellite and the objective is the verification in orbit of all Digital TV services (HDTV, SDTV) working over DVB-S2.

- Interactive TV: Thanks to DVB-S2/DVB-RCS standards will be possible to provide interactive services with a typical star topology.

Through this set of test procedures, the Digital TV and also HDTV functionality will be checked between the Service Provider and the final users using a DVB-S2 connection via satellite in transparent mode.

Platform Architecture Overview

Scenario 4 will be developed and evaluated on HSA's pilot DVB-RCS system.

Hispasat Multimedia Platform is based on the Satlink System, fully DVB-RCS standard compliant. The network has a typical star topology with a central node or Hub which contains the NMS, the satellite front-end, IP infrastructure and terrestrial networks interfaces.

The platform offers bidirectional IP connectivity between terminals and terrestrial networks, always through the Hub, allowing different services such Internet access, Intranets/VPNs, VoIP and multicast. On the other hand, MPEG2/DVB-S2 unidirectional services are available from Hub to the terminals (Fig. 18.6).

Demonstration Instruments

The test platform is composed by several servers that send the multimedia contents and interactive applications to the Service Provider and to the RCS terminal.
The equipment that will be used is composed by:

- A UNIX machine with Solaris O.S. to install the flow server.
- A High Speed MPEG-2 Interface Board:
- A Service Provider Unit.
- The HSA transparent satellite with DVB-S2 platform.
- A DVB-S2 MHP set-top-box IPv6.
- A high definition TV
- PC with a software tool that will measure the quality of the video received (Fig. 18.7).

Trials and Validation

The digital TV service capabilities to be tested in the real system tests are organised as:

- Broadcast Digital TV
- Broadcast High Definition Digital TV

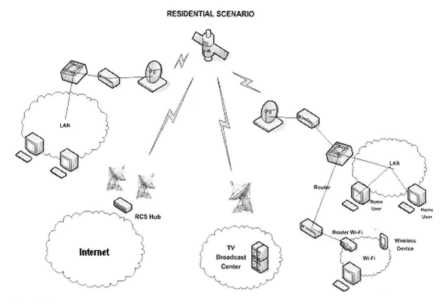

Fig. 18.6 Residential applications on DVB-RCS transparent access platform

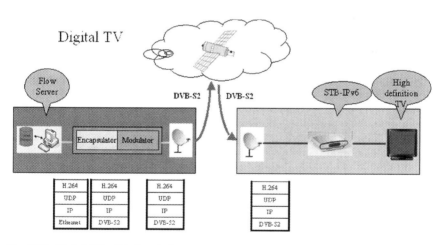

Fig. 18.7 Digital TV validation platform

The transmission of digital TV contents from server to clients and their reception will be checked, and also subjective image and audio quality will be measured means of a tool that gives a score for the video quality comparing the video transmitted and the video received. This will allow making a comparison between the qualities obtained for DTV and HDTV.

The main innovative issue in this scenario is the use of DVB-S2 standard vs DVB-S, which represents a technological advance in relation to previous projects in the same field.

Video Quality Measurements

Digital video data, stored in video databases and distributed through communication networks, is subject to various kinds of distortions during acquisition, compression, processing, transmission, and reproduction. It is imperative for a video service system to be able to realise and quantify the video quality degradations that occur in the system, so that it can maintain, control and possibly enhance the quality of the video data [1].

There are two ways of assessing the quality of an image or video: subjective evaluation and objective image and video quality assessment. In the subjective evaluation, like mean opinion score (MOS), to obtain such scores for the one directional transmission of video sequences, several human observer test methods are described in [2]. The goal of objective methods is to design quality metrics that can predict perceived image and video quality automatically.

Some of them have been considered:

- Subjective evaluation methods:

  - Double Stimulus Impairment Scale (DSIS) [3].
  - Double Stimulus Continous Quality Scale (DSCQS) [3].
  - Single Stimulus Continuous Quality Evaluation (SSCQE) [3].
  - Simultaneous Double Stimulus for Continuous Evaluation (SDSCE) [3].
  - Subjective Assessment Methodology for Video Quality by EBU (SAMVIQ) [4].

- Objective evaluation methods:

  - Peak Signal to Noise ratio.
  - EPFL (École Polytechnique Fédérale de Lausanne) Objective Video Quality Measures [5].
  - TS (Institute for Telecommunication Sciences) Video Quality Measure and the VQM Method [6].
  - Structural distortion based method (SDM) [7].

In order to measure video quality in this scenario, VQM method will be used. This method is one of the most reliable result gives [1].

Conclusions

This paper has presented the definition of the trials that will be carried out inside Satsix project context. For this purpose four real Satellite Systems will be used: the Regenerative AMERHIS system of HSA and the Transparent DVB-RCS systems of HSA, HDT and AASF. Each scenario considers different applications in order to demonstrate the provision of affordable services.

To sum up, at network level, these live trials will mainly demonstrate innovative IPv6 features such as:

- WLL interworking
- Dynamic multicast
- Dynamic QoS
- Application/Network security
- Mobility
- PEP/RRM interaction

And finally, the main features tested at the access network will be:

- WLL security
- DVB-S2 scheduling
- C2P

Acknowledgements This work is supported by the IST FP6 SATSIX project, funded by European Commission (EC). The financial contribution of the EC towards this project is greatly appreciated.

References

[1] Johnny, Biström 2005. Comparing Video Codec Evaluation Methods for Handheld Digital TV. Helsinki University of Technology.

[2] ITU-T Recommendation P.910, "Subjective video quality assessment methods for multimedia applications", September 1999.

[3] ITU-R Recommendation BT.500-11, "Methodology for the subjective assessment of the quality of television pictures", International Telecommuncation Union, Geneva, Switzerland, 2002.

[4] Kozamernik, Franc (EBU Technical Department), Sunna, Paola (RAI CRIT), Wyckens, Emmanuel (France Telecom R&D) and Pettersen, Dag Inge (NRK) 2005. Subjective Quality of Internet Video Codecs—Phase 2 Evaluations Using SAMVIQ. EBU TECHNICAL REVIEW, January 2005 1/22.

[5] Branden, Lambrecht and van den, Christian 1996. Perceptual Models and Architectures for Video Coding Applications. PhD thesis. EPFL. Lausanne Swiss. 1996.

[6] Wolf, Stephen and Pinson, Margaret 2002. "Video quality measurement techniques", National Telecommunications and Information Administration (NTIA) Report 02–392, June 2002.

[7] Wang, Zhou and Bovik, A. C. 2002. A Universal Image Quality Index. IEEE Signal Processing Letters, vol. 9, pp. 81–84, March 2002.

Chapter 19
ULE Link Layer Security for DVB Networks

S. Iyengar, H. Cruickshank, L. Duquerroy, Z. Sun, and C. Baudoin

Abstract The MPEG-2 standard supports a range of transmission methods for a range of services. This document provides a security solution at the link layer using the Unidirectional Lightweight Encapsulation. It describes the SATIPSec key management protocol and its adaptation for DVB-RCS networks to provide security at the ULE level. Also the forward link signalling risks are analyzed and a possible security solution has been proposed.

Introduction

MPEG-2 Transport Stream (TS) has been widely accepted not only for providing digital TV services, but also as a subnetwork technology for building IP networks. RFC 4326 describes the Unidirectional Lightweight Encapsulation (ULE) mechanism for the transport of IPv4 and IPv6 Datagrams and other network protocol packets directly over the ISO MPEG-2 Transport Stream as TS Private Data. ULE specifies a base encapsulation format and supports an extension format that allows it to carry additional header information to assist in network/ Receiver processing.

In ULE, it is suggested that this may be provided in a flexible way using Extension Headers. This requires the definition of a mandatory header extension, but has the advantage that it decouples specification of the security functions from the encapsulation functions. Moreover the use of extension headers for securing the ULE link can also be used for security of Generic Stream Encapsulation (GSE) to be used for DVB-S2 systems.

This document describes a key management protocol and how it is adapted for DVB-RCS networks. A security solution based on extension headers is then described and how it provides ULE level security. Finally the forward link signalling risk analysis is presented and a security solution proposed to protect the signalling traffic.

L. Fan et al. (eds.), *IP Networking over Next-Generation Satellite Systems*
© Springer 2008

SatIPSec Key Management Protocol at IP Layer

SatIPSec [1] is a key management protocol at the IP layer which offers a new way of transparently and efficiently securing unicast and multicast satellite transmissions, on forward and return links, in Digital Video Broadcasting Return Channel systems (DVB-RCS) mesh and star topologies. SatIPSec protocol is derived from IP Security (IPSec) standard protocols [2].

The SatIPSec solution is based on two different entities entitled SatIPSec client and SatIPSec Group Controller & Key Server (SatIPSec GCKS). The SatIPSec architecture is thus formed of several SatIPSec clients and of one GCKS:

- SatIPSec clients are in charge of securing IP traffic flows during their transmission over the satellite links. For that purpose, they apply some security treatments (ciphering/deciphering, computation and checking of authentication values…) to IP datagrams.
- The GCKS centrally ensures the management of SatIPSec clients and the establishment of the security architecture. Its function is to configure SatIPSec clients, by supplying them with the necessary information for securing data transmissions according to the security policies.

Figure 19.1 presents the SatIPSec architecture: SatIPSec clients and GCKS are implemented in external boxes (independently of the other Satellite communication equipments). There is one SatIPSec client box behind each Satellite Terminal (ST); however SatIPSec mechanisms could also be implemented directly inside the IP stack of satellite terminals. The GCKS is preferably located at the gateway side in star topology.

The location of SatIPSec clients allows to apply security mechanisms to IP traffic (if requested) so as to ensure its protection during its transmission over the satellite network. Before transmission on satellite system, a SatIPSec client can cipher each IP packet and can compute an authentication value. In reception, the SatIPSec client(s) can decipher it and check the authentication value, before transmitting it on the terrestrial network it (they) is (are) connected to. SatIPSec allows to protect

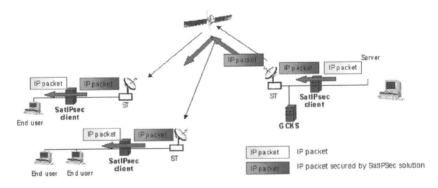

Fig. 19.1 SatIPSec architecture

any unicast or multicast IP flows transmitted on satellite links, from Hub to ST, from ST to Hub and from ST to ST, in Star and Mesh topologies.

Security Architecture

The SatIPSec solution allows establishing:

- Virtual Private Networks (VPN): A VPN is established when two distant terrestrial sites wish to communicate securely together through the satellite network. A unicast secure tunnel is established between the both SatIPSec clients connected to each site (i.e. end-points). This tunnel can be configured to secure unicast bi-directional exchanges between for instance two end-users or subnets belonging to these sites, or all unicast traffic exchanged between these two sites (Fig. 19.2).
- Secure Multicast Groups: A Secure Multicast Group is composed of members (i.e. SatIPSec clients) sharing a common security configuration allowing to secure IP multicast flows. Only group members can get access to the corresponding traffic (in clear) in reception. Each multicast IP packet is indeed encrypted and encapsulated before its transmission on satellite network by a group member. In reception, only group members can decipher it, and then transmit the packet on the terrestrial network they are connected to. The SatIPSec solution encapsulates multicast packets in new multicast packets, the establishment of a secure multicast group can therefore be considered as the establishment of multicast tunnels (Fig. 19.3).

Security Services

Data Security

- Data confidentiality: achieved by the ciphering of the entire IP packet (data and header).

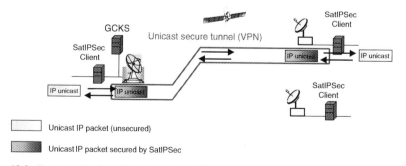

Fig. 19.2 Security of unicast IP packets by a VPN

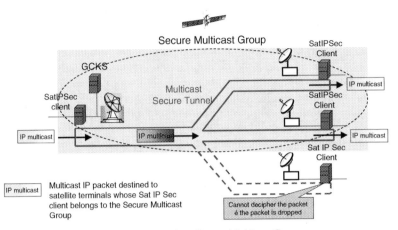

Fig. 19.3 Security of multicast IP packets by a Secure Multicast Group

- Data integrity and Data origin authentication: achieved by the computation of an authentication value depending of the IP packet, before its transmission on satellite network. In reception SatIPSec clients can verify the integrity of the packet (i.e. if it has been modified or not during transmission) and the identity of its source, by checking its authentication value (source has be a member of the multicast group).
- Traffic Processing per data flow: Each traffic flow can be secured independently (with its own keys and security attributes and algorithms), and encapsulated in its own tunnel. An IP data flow can be identified by its IP source and destination addresses (representing equipment, end-user, subnet, multicast IP address) and its source and destination port numbers. Thanks to this service, the access to a flow can easily be limited only to the allowed receivers.
- Replay attack prevention: in order to prevent that a packet is re-sent by an intruder, a unique number is assigned to each packet (ex: counter). Thanks to this sequence number, receivers can detect already received packets and drop them.
- These services can be offered for all types of traffic (unicast and multicast). Confidentiality and origin authentication/integrity services can be combined or used separately according to the required security level.

Access Control—SatIPSec Client Authentication

Access control aims at guaranteeing that only authorized recipients get access to data transmitted on satellite links. With SatIPSec, each IP flow requiring to be protected is encrypted before being transmitted on satellite network. The security parameters allowing to secure it in emission, and to decrypt it in reception, are distributed only to authorized SatIPSec clients (after verification of their identity).

Thus SatIPSec solution enables a strong access control. All security parameters defining how to secure an IP flow (keys, cryptographic algorithms to use, flow identifier...) are contained in a Security Association (SA).

Access control is thus managed by the GCKS. It is based on the authentication of each SatIPSec by the GCKS, and on the selective (and secure) distribution of Security Associations only to authorized SatIPSec clients.

Protocols

In SatIPSec, like in IPSec, data plane and control plane are separated. The control plane implements a new protocol called Flat Multicast Key Exchange protocol (FMKE) [1], in replacement of Internet Key Exchange protocol (IKE) [3]. It is a layer-3 protocol, mainly dedicated to Group Key and Security Associations (SA) Management.

The IPSec Encapsulating Security Payload (ESP) protocol [4] (or Authentication Header (AH) [5]) is the security data plane of SatIPSec at layer 3. The ESP protocol is implemented and used as it is defined in the standard, but allows securing both unicast and multicast data. Only the tunnel mode is used.

For more details regarding SatIPSec protocols, please refer to [1].

Risk Analysis

SatIPSec at IP Layer: Limitations

SatIPSec is a generic solution that can be implemented to protect data at IP layer, but also at the link layer.

A link layer security solution represents several advantages with regards to a layer 3 solution:

- Independence from the type of satellite terminal: A security solution at DVB-RCS [6] level can be used in any types of ST: router or bridge.
- Independence from the type of traffic to protect: A layer 2 solution is a multi-protocol solution, which can secure transparently any types of traffic (IPv4, IPv6, Asynchronous Transfer Mode (ATM), and Ethernet).
- Compatibility with other Internet Service Providers (ISP) or subscriber security functions: Taking into account the role of the different actors (Access Network Operator, Internet Service Provider), it can be possible to have simultaneously different security schemes. For example, the ISP or a subscriber could use its own security system above the data link (may be at the IP or application layers). A security solution at layer 2 would therefore not interfere with the one proposed by the Internet Service provider or by a subscriber.

- Satellite bandwidth optimization: In the satellite context, it is always of a paramount importance to optimize the overhead in order to reduce the waste of the bandwidth. The security solution shall minimize the overhead caused by its services. At IP layer, packets have to be encrypted and encapsulated in new IP packets before their transmission on satellite links. This can generate a significant volume of overhead on satellite links (for e.g. ESP in tunnel mode appends around 44 bytes to every IP packet). A security solution at layer 2 would allow minimizing this overhead (by avoiding packet encapsulation and by defining a security format more optimized to satellite constraints).

Moreover it provides or can provide additional security services which may be considered major requirements in many DVB-S/RCS satellite networks:

- Protection of the complete Protocol Data Unit (PDU), e.g. protection of IPv4 or IPv6 datagrams, including IP addresses.
- Protection of Layer 2 Network Point of Attachment (NPA)/Medium Access Control (MAC) address. In broadcast networks this service would prevent an intruder tracking the identity of receivers and the volume of their traffic.

Link Layer Security

Data Plane Protection

Security Architecture: Overview

In SATSIX project [7], SatIPSec will be adapted to provide link-layer security (with a special focus for the link using Unidirectional Lightweight Encapsulation (ULE) [8]). ULE is an encapsulation mechanism for the transport of IPv4 and IPv6 Datagrams and other network protocol packets directly over the ISO MPEG-2 Transport Stream as TS Private Data.

At the architectural level, SatIPSec is at the link layer and requires one SatIPSec client module to be integrated in each satellite terminal and gateway as shown in Fig. 19.4. The GCKS, which configures SatIPSec clients, is integrated in the DVB-RCS gateway (GW) or Network Control Centre (NCC).

SatIPSec at the link layer will be defined to enable it to provide:

- Data confidentiality at link level, which is the major requirement to mitigate passive threats.
- Data source (ST/Gateway) authentication/data integrity at link level
- Protection against replay attacks
- Protection of Layer 2 NPA/MAC addresses
- ST/Gateway authentication and authorization

The protection of data can be applied either per receiver MAC/NPA address or per IP flow. Concerning involved protocols, key management functions may be decoupled

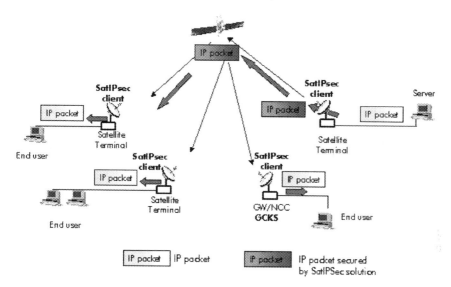

Fig. 19.4 SatIPSec at Link Layer: architecture (DVB-RCS mesh topology)

from data protection functions. Thus it is proposed to re-use the IP key management protocol defined for SatIPSec at IP layer (with some adaptations). Besides SatIPSec security session establishment and management will be integrated in DVB-RCS standard process (in order that security session is inherently part of the DVB-RCS session).

SatIPSec data plane will be located at Encapsulation Layer: ULE (the unit of encryption is ULE PDU) for MPEG-TS streams.

MPEG-TS encryption is not considered because MPEG-TS encryption implies to encrypt all TS Packets sent with a specific PID value. However, MPEG-TS may typically multiplex several IP flows, belonging to different users, using a common PID. Therefore all multiplexed traffic will share the same security keys.

MPEG-TS encryption has the following advantages:

- The bit stream sent on the broadcast network does not expose any L2 or L3 headers, specifically all addresses, type fields, and length fields are encrypted prior to transmission.
- This method does not preclude the use of IPsec, or any other form of higher-layer security.

However it has the following disadvantages:

- Each ULE Receiver needs to decrypt all MPEG-2 TS Packets with a matching PID, possibly including those that are not required to be forwarded. Therefore it does not have the flexibility to separately secure individual IP flows.
- ULE Receivers will have access to private traffic destined to other ULE Receivers, since they share a common PID and key.

Security Data Plane

ULE specifies a base encapsulation format and supports an extension format that allows it to carry additional header information to assist in network/Receiver processing. An extension to the encapsulation is therefore being considered to provide confidentiality (encryption) and, optional source authentication. An extension to the ULE encapsulation is therefore being considered to provide confidentiality (encryption) and, optional source authentication [9]. Moreover the use of extension headers for securing the ULE link can also be used for security of Generic Stream encapsulation (GSE) [10] to be used for DVB-S2 systems.

The security extension aims to secure the transmission of user traffic over MPEG-2 Transport Streams. In order to address the security issues, Fig. 19.5 shows the ULE Sub Network Data Unit (SNDU) format with the security extension header. This security extension is a standard extension header as described in Section 5 of RFC 4326 [8] and it should not affect the ULE base protocol. This security extension header MAY directly follow the ULE base header as shown in Fig. 19.5 or it MAY also follow another specific extension header.

This security extension header is a Mandatory ULE Extension header. This means that a receiver MUST either process this header before it processes the next extension header or the encapsulated PDU, otherwise the entire SNDU should be discarded.

In Fig. 19.5 the Type field in the base header denotes that the mandatory security extension header is present. The receiver destination NPA address is optional. After the base ULE header the security extension header is followed. This header contains the ULE-SID, the optional Sequence Number field and the optional Message

```
0 1 2 3 4 5 6 7 8 9 0 1 2 3 4 5 6 7 8 9 0 1 2 3 4 5 6 7 8 9 0 1
+-+----------------------------+----------------------------+
|D|         Length             |      Type = S-ULE          |
+-+----------------------------+----------------------------+
|           Receiver Destination NPA Address *              |
|                            +----------------------------+
|                            |       ULE_Security_ID       |
+----------------------------+----------------------------+
|       ULE_Security_ID       |   Sequence Num.(Optional)  |
+----------------------------+----------------------------+
| Sequence Number (Optional)  |       MAC (Optional)       |
+----------------------------+----------------------------+
|       MAC (Optional)        |    Type = Type of PDU      |
+----------------------------+----------------------------+
|                                                          |
|                                                          |
=                    Encrypted PDU                         =
|                                                          |
|                                                          |
+----------------------------------------------------------+
|              Cyclic Redundancy Code                      |
+----------------------------------------------------------+
```

Fig. 19.5 General SNDU format with Security extension header (D = 0)

Authentication Code (MAC) field. The next type field denotes the type of the enclosed PDU. The higher layer PDU is encrypted and then encapsulated in the SNDU.

The format of the Destination Address Absent field (D), the Length field the Type field and the Receiver Destination NPA address field directly follow and are used in the same way as defined for standard ULE.

Destination Address Absent (D) Field

The most significant bit of the Length Field carries the value of the Destination Address Absent Field (D) which follows the same definition as in standard ULE [8]. When D is set to 0, it indicates the presence of the Destination Address Field while D set to 1 indicates that a Destination Address Field is not present.

Length Field

A 15-bit Length field denotes the length, in bytes, of the SNDU counted from the byte following the Type field, up to and including the CRC [8].

Type Field

An IANA-assigned 16-bit Type Field indicates that this is a Secure ULE SNDU [8].

Destination NPA Address Field

The SNDU Destination Address Field is optional. This field is MUST be carried when field D is set to 0 and may be omitted when D = 1 [8].

ULE-SID Field

A 32-bit security identifier, the ULE-SID similar to the Security Parameter Index (SPI) used in IPsec has been added to uniquely identify the secure session. This ULE-SID would represent the security association between the MPEG-2 transmitter and receiver for a particular session and will indicate the keys and algorithms used for encrypting the data payload and calculating the MAC. The ULE-SID can be used by a receiver to filter PDUs along with the NPA address.

Sequence Number Field

An optional 32-bit sequence number has been added to the ULE SNDU to prevent replay attacks. The gateway would monotonically increment this number when it sends a packet to the receiver and the receiver would verify the correct sequence number. If an adversary tries to inject or replay old packets the sequence number would not match. This would result in discarding the packet.

Message Authentication Code (MAC) Field

To provide both data origin authentication and data integrity, an optional Message Authentication Code (MAC) is used in the extension header. The MAC is calculated over the security extension header and the encrypted data payload. The receiver would calculate the MAC for the received packet and compare it with the transmitted value. The two would not match in only 2 cases, firstly either there was an error during processing or transmission over the MPEG-2 Network, or secondly the packet has not been sent from an authenticated entity.

In either case, the packet should be discarded. Hence the same MAC can be used for data origin authentication and to provide data integrity for transmission/processing errors.

Type Field

This second type field denotes the type of packet that is encrypted and encapsulated in the Secure ULE SNDU.

Encrypted PDU

To achieve data confidentiality, the traffic between the MPEG-2 TS Transmitter (ULE Encapsulator) and Receiver needs to be encrypted. The network layer PDUs are first encrypted and then encapsulated in the Secure ULE SNDU. The security associations between the two communicating points will describe the algorithms and keys used for encryption purposes. Secure ULE does not impose the use of any specific encryption algorithm and should be able to support the commonly used algorithms like DES, 3DES etc.

ULE Security Databases Interfaces

There are basically two main interfaces between the key Management module (SATIPSec) and the ULE data plane as shown in Fig. 19.6:

- Key management <–> ULE Security databases
- ULE Security databases <—> ULE data plane

Key Management <–> ULE Security Databases

This interface is between the Key Management client (SATIPSec client) and the ULE Security Databases. The Key management client will communicate with the GCKS and then get the relevant security information (keys, cipher mode, security service, ULE_Security_ID and insert this data into the ULE Security database. The ULE Security databases consist of the Security Association Database (SAD) and the Security Policy Database (SPD). The SAD holds the records of all security associations currently used and the SPD all the information for security policy control.

The Key management could be either automated (SATIPSec) or manually inserted using this interface. The following three interface functions are defined:

- Insert_record_database (char * Database, char * record, char * Unique_ID);
- Update_record_database (char * Database, char * record, char * Unique_ID);
- Delete_record_database (char * Database, char * Unique_ID);

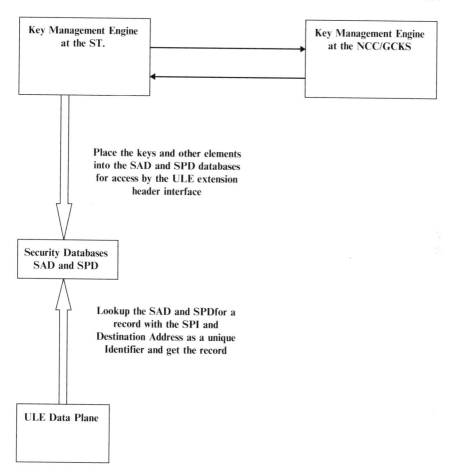

Fig. 19.6 Interfaces between the Key Management Engine and the ULE Data Plane

ULE Security Databases <—> ULE Data Plane

This interface is between the ULE Security Database and the ULE data plane (Satellite Terminal). For Outbound Traffic (Satellite terminal As a Gateway), the ULE data plane will first lookup the ULE Security Databases for the relevant security record passing it the Destination Address and the ULE_Security_ID. It then uses the data to create the ULE security extension header After that this data is passed to the DVB interface. For inbound traffic (Satellite Terminal), the ULE data plane on receiving the ULE packet will first get the record from the Security Databases using the Destination Address and the ULE_Security_ID. It then uses this information to decrypt the ULE extension header and then passes it on to the upper layers. In both cases only one interface is needed:

- Get_record_database (char * Database, char * record, char * Unique_ID);

ULE Transmitter Processing

The following procedure is followed at the encapsulator for processing the security extension header for ULE:

1. Upon reception of the higher layer PDU, the Security Policy Database (SPD) is first queried to check the policy to be applied to the packet. If security is needed then an SA must exist in the SAD (this is done by the key management system). The parameters are retrieved from the Security Association Database (SAD) and it is first encrypted using the key and the algorithm as indicated in the SAD.
2. The header of the base protocol (and other extension headers if present) would be added to the SNDU.
3. The ULE-SID for the security association between the transmitter and the receiver would be added next.
4. The SAD would be used to see if the sequence number has to be added. If yes, then the corresponding sequence number is added to the SNDU.
5. The SAD would also be checked to see if the data origin authentication and data integrity has to be provided. If yes, then the MAC has to be calculated. The MAC is calculated over the encrypted PDU, the Security header i.e. the ULE-SID and the Sequence Number (if present) and the secret key. The MAC is then added to the extension header in the SNDU.
6. Then the encrypted higher layer PDU is encapsulated to the SNDU.
7. Finally, the CRC is calculated as defined in Section 4.6 of RFC4326 [8] and added.

ULE Receiver Processing

The following procedure is followed at the receiver for processing the security extension header for ULE:

1. Upon reception of the Secure ULE SNDU, the receiver may first filter the received packets according to the receiver destination NPA address (if present).
2. The CRC is verified as defined in RFC4326 [8].
3. It would then use the ULE-SID to obtain the security associations between the transmitter and receiver and retrieve the data from the SAD. With this the receiver would know if the sequence number and the MAC are present or not. This would also be used to get the algorithms and keys used for both encryption of the encapsulated PDU and for generation of the message authentication code.
4. It would then use the sequence number for filtering out any out of-sequence packets.

5. The next step would be to check the MAC to verify the authenticity and integrity of the received packet. If the calculated MAC does not match the transmitted MAC, then the packet is discarded.
6. Finally the encapsulated payload will be decrypted.

Security Control Plane

Integration of SatIPSec Key Management in DVB-RCS Standard

It is proposed that FMKE phases (see [1]) are inserted in the DVB-RCS process. FMKE phases 1 and 2 will take place during the DVB-RCS Logon, in replacement of the DVB-RCS optional security scheme. Mutual authentication, establishment of a secure ISAKMP control channel between GCKS and ST (phase 1), and secure transmission of Security Associations (SA) to the ST (phase 2), are achieved.

During DVB-RCS session, FMKE phases 2 and 3 will be used in parallel to configure and update SAs in ST:

- The phase 2 will be used to create, update (i.e. rekeying), or remove unicast and multicast SAs in one ST.
- The phase 3, dedicated to a secure multicast group of terminals, will be used to create, update or remove multicast SAs in its group members. A phase 3 can be launched as soon as one group member is logged on (it receives during logon the multicast control SA which protects the messages of this phase). With the phase 3, the GCKS can configure and update all logged group members of a group.

Thus FMKE phase 1 and 2 messages will replace the Security Sign_On, Security Sign_On Response, EKE (and MKE, QKE for DVB-RCS initial version) requests and responses used during DVB-RCS Logon (Fig. 19.7) and session (Fig. 19.8) for respectively security session establishment and key updates.

This requires the implementation of inter-layer control functions between DVB-RCS and SatIPSec control planes (in ST/GW and GCKS): Indeed the DVB-RCS logon process is implemented at Medium Access Control (MAC) level. As FMKE phases 1 and 2 establishing the security session have to take place during the Logon, an inter-layer control function between DVB-RCS control plane and SatIPSec control plane is required as shown in Fig. 19.9.

Security Session Establishment

As presented in the previous part, the security session establishment, based on FMKE phases 1 and 2, takes place during the Logon process, in replacement of the DVB-RCS security scheme. In the DVB-RCS standard, the CSC burst, sent by the ST, defines a 24 bit ST capability field. The bit number 23 corresponds to the 'Security mechanism' field. If this bit is equal to one, it means that the ST wants to implement security mechanisms for the session duration. The NCC communicates on the

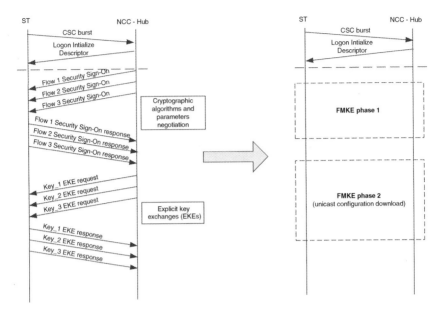

Fig. 19.7 DVB-RCS Logon—security session establishment

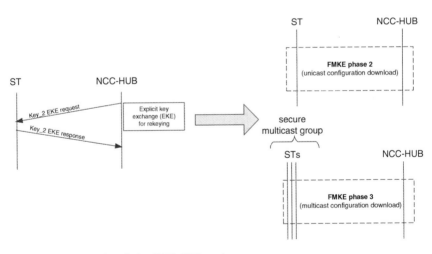

Fig. 19.8 Security update during DVB-RCS session

Logon Initialize Descriptor whether security shall be used. Indeed, the Logon Initialize Descriptor defines a 4 bit '*security handshake required*' field. If the MSB is equal to one, security mechanisms have to be used.

If both CSC and Logon Initialize Descriptor activate security, the messages defined by the DVB-RCS optional security mechanisms are transmitted in order to authenticate the ST and agree on the session keys. In the security solution based on

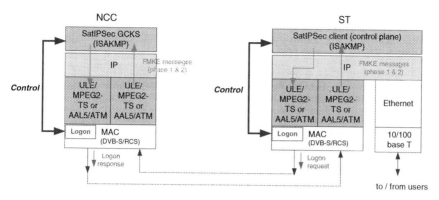

Fig. 19.9 Inter-layer control between DVB-RCS and SatIPSec control planes

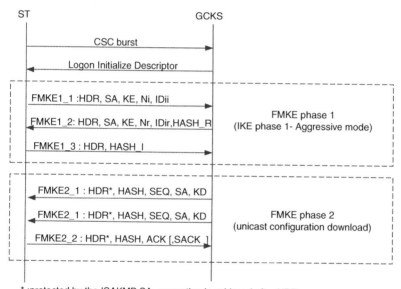

Fig. 19.10 SatIPSec session establishment in DVB-RCS Logon

FMKE, we can therefore re-use this system without any modifications to indicate that security is required. If agreed, the FMKE session establishment is performed.

Figure 19.10 represents the FMKE session establishment implemented during DVB-RCS Logon: FMKE phases 1 and 2 replace the DVB-RCS optional security mechanisms.

The phase 1 is dedicated to the establishment of a secure control channel (i.e. IKE Security Association) between ST and the Group Controller/Key Server (GCKS). This establishment requires that they generate common encryption and authentication keys, and negotiate the cryptographic algorithms and parameters of the secure channel. It is preceded by a mutual authentication. If successful, the control channel is established.

The phase 2 is dedicated to the configuration of the ST at logon. Messages are protected by the IKE SA established in phase 1: they are encrypted and include an authentication/integrity value.

The SAs which are transmitted can concern the protection of:

- unicast data flows
- multicast data flows
- multicast control messages which will be exchanged during phases 3

During DVB-RCS Session

During DVB-RCS session, the FMKE protocol can be used to update the security configuration of logged STs. Updates can be achieved in unicast and in multicast.

Unicast configuration update (phase 2)

The messages of the Phase 2 can also be used during the DVB-RCS session to create, update or delete SAs in the ST (Fig. 19.11). The phase 1 protocol does not allow renewing directly a key. This is realized in fact by the creation and the establishment of new SAs, and the deletion of old SAs. When a SA must be renewed, a new SA is created before the expiration of its lifetime. It has the same attributes, except the keys and the

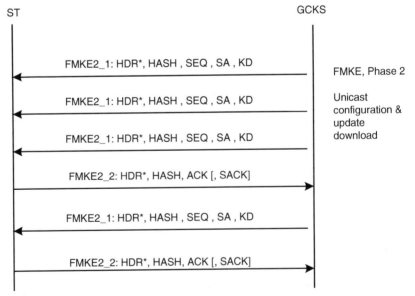

* :protected by the ISAKMP SA, encryption is achieved after HDR

Fig. 19.11 FMKE DVB-RCS Unicast configuration update (phase 2)

SA identifier. The only required modification with regards to the FMKE specification for SatIPSec layer 3 is the adaptation of the SA payload in order to specify encryption/decryption and/or authentication of data flows at MAC level.

Group configuration update (phase 3)

This phase is dedicated to the simultaneous configuration of all (logged) members of a group (Fig. 19.12). It is achieved in multicast.

This phase is protected by a multicast control SA, distributed by the GCKS to group members during logon or during the DVB-RCS session with a Phase 2 message. Like in phase 2, the only required modification with regards to the FMKE specification for SatIPSec layer 3 is the adaptation of the SA payload like in the previous phases.

Adaptation of the SatIPSec Key Management Messages

Changes required in the current definition of the FMKE protocol are:

- Terminal identification must be based on its MAC/NPA address instead of its IP address during security session establishment (like in the current version of the DVB-RCS standard).
- The definition of SA payload fields must be adapted. As a matter of fact, the SA payload defines SA attributes (keys, cryptographic algorithms to use…), but also the traffic to be protected (linked to Security policy). SA payload fields have to be modified in order to identify layer 2 traffic flows (with an IP granularity or not).

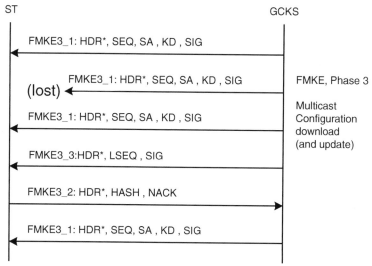

Fig. 19.12 FMKE DVB-RCS Group configuration update (phase 3)

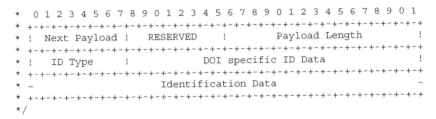

Fig. 19.13 FMKE Identification Payload Format

ID payload adaptation

The only difference with regards to the specification of FMKE for SatIPSec at layer 3 [1], concerns the value contained in the ID (Identification) payload as shown in Fig. 19.13 The Identification Data field must contain the ST's MAC/NPA address. A specific ID type must be defined for MAC address.

The protection of data can be applied either per MAC flow (like the DVB-RCS optional security mechanisms) or per IP flow. With a MAC granularity (flows identified by the MAC/NPA address of the receiver for ULE streams), it is possible:

- For ULE streams: To secure or not, independently, each multicast flow.
- For unicast ULE streams: To secure or not, independently, the unicast ULE traffic issued from one terminal and destined to another unique terminal.

The IP granularity is based on source and destination IP addresses and port numbers. Its advantage is that we can secure only the part of traffic which requires to be secured and thus we do not overload the terminal with unnecessary encryption computations.

If we are based on layer 2 granularity, flow identifiers required is:

- Destination MAC/NPA addresses (ST address or multicast address) for ULE streams.

If we are based on IP granularity, flow identifiers required are:

- Source and Destination IP addresses (and ports), representing IP subnets or hosts.
- And the destination MAC/NPA address for ULE streams.

This information is necessary to allow terminals to build its Security Policies and Security Associations in its databases.

Forward Link Signalling

Risk Analysis

In a two-way satellite interactive network, consisting of a forward link and return link via satellite, the forward link signalling consists of general System Information

(SI) tables, carrying information about the structure of the satellite interactive network, and satellite terminal specific messages sent to individual satellite terminals.

There are 6 general SI tables; they are broadcasted in the network [6]:

- Superframe Composition Table (SCT): This table describes the sub-division of the entire satellite interactive network into superframes and frames.
- Frame Composition Table (FCT): This table describes the partitioning of the frames into time-slots.
- Time-Slot Composition Table (TCT): This table defines the transmission parameters for each time-slot type identified by the time-slot identifier.
- Satellite Position Table (SPT): This table contains the satellite ephemeris data required to update the burst position at regular intervals.
- Correction Message Table (CMT): The CMT provides correction values for burst frequency, timing and amplitude.
- Terminal Burst Time Plan (TBTP): It contains one or more entries for each RCST, with each entry defining an assignment of a contiguous block of times-lots. Each individual terminal is addressed by a logical Logon_ID, notified to the terminal at logon time.

There is also the Terminal Information Message (TIM). This message is sent by the NCC either to an individual RCST addressed by its MAC address (unicast message) or broadcast to all RCSTs using a reserved broadcast MAC address and contains static or quasi static information about the forward link such as configuration. At last, the PCR Insertion TS Packet shall be used for inserting the NCR value used for return link synchronization.

The forward link signalling may be threatened by the following threats:

- Passive attack: An attacker listens to the forward link signalling broadcasts to gain some sensitive information related to the satellite terminals (e.g. the status of satellite terminals in unicast TIM messages) or to the satellite network (e.g. the assignment of time slots to terminals in TBTP). The objective of the attacker is to analyze the status of the network to gain potential interesting information.
- Active attack: An intruder is assumed to be sufficiently sophisticated to over-ride the original transmission from the NCC and deliver a modified version of the forward link signalling transmission. This attack may involve masquerading, modification of messages in an unauthorized manner, or replay attacks, the objective of the attacker being to prevent satellite terminals from accessing the network and sending or receiving data (denial of service). Such a threat is more difficult to implement successfully than the passive threat, requires more sophisticated resources and may require access to the transmitter. Hence, protection against this threat should be used only when such a threat is a real possibility.
- Active attack: An attacker may listen to the forward link signalling broadcasts to collect network information, and thanks to this information may flood the NCC with valid messages. The objective of the attacker is to prevent the NCC to assume its functions: e.g. the satellite terminal attacker listens to all SI tables (NCR (in PCR), SPT, SCT, FCT, and TCT) related to the structure of the satellite

interactive network (initial synchronization procedure), it is then able to flood the NCC with valid Logon requests.

- From the threat analysis above, the following security requirements can be derived:
- Confidentiality of SI tables, TIM and PCR insertion TS packets
- Integrity protection and source authentication (i.e. NCC) of SI tables, TIM and PCR insertion TS packets (optional).
- Protection against replay attacks of SI tables, TIM and PCR insertion TS packets (optional).

Security Solution

The proposed solution in SATSIX project is based on the stack described in Fig. 19.14. GSE encapsulation allows the transport of MPEG-TS SI tables can be encapsulated using GSE and then sent over the DVB-S/S2 interface as shown in the protocol stack in Fig. 19.14. The security extension presented is also compatible with GSE encapsulation and can thus be used to protect general SI tables and satellite terminal specific messages, by providing:

- Confidentiality
- Integrity and source authentication (optional)
- Protection against replay (optional)

Some of these tables must be reachable by all satellite terminals which are authorized to get access to the DVB-RCS satellite network: NCR (in PCR), SPT, SCT, FCT, TCT and broadcast TIM. The control key used to encrypt them before transmission must therefore be shared by all these terminals. Moreover they have to get access to these tables before to be logged (c.f. Fig. 19.15). Thus this key shall be present in each satellite terminals at power-up (either stored in a smart cart inserted in the terminal, either stored during manufacturing in non-volatile storage).

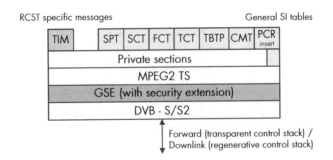

Fig. 19.14 SATSIX forward/downlink signalling secure stack

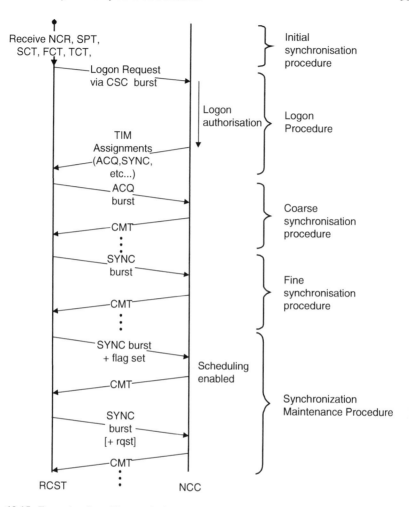

Fig. 19.15 Example of satellite terminal network entry signalling flow

The TBTP table is sent to a group of logged satellite terminals. The group is addressed by a logical Group_ID. Its Group_ID is notified to a terminal at logon time. A unicast TIM message is dedicated to a terminal.

We may envisage using the control key used to encrypt NCR (in PCR), SPT, SCT, FCT, and TCT and broadcast TIM, to encrypt all TBTP tables and TIM messages. However this requires having confidence in all RCST of the system (since they can get access to all TBTP tables and TIM messages).

Otherwise, a group control key (associated to a group of terminals identified by a Group_ID) and an individual control key (associated to a terminal) may be used to respectively encrypt the TBTP addressed to the group and to encrypt unicast

TIM messages addressed to the terminal. These keys may be distributed to each terminal during DVB-RCS security session establishment.

The first unicast TIM message sent to a terminal is transmitted during the logon process, and therefore cannot be encrypted with the individual control key. The proposed solution is to encrypt this message with a key derived from the pre-shared secret of the terminal (the pre-shared key is used then for terminal authentication during the DVB-RCS security session establishment). This key will be then replaced by the individual control key.

Conclusions

This document describes the SATIPSec key management protocol used at the IP layer and its limitations for layer 2 security. It describes how SATIPSec can be adapted to be used as a key management protocol for layer 2 (ULE security). It then explains the ULE Security Header format that will be used in the data plane (i.e. the layer where the security services will be applied). The interfaces between the Key Management application (SATIPSec) and the data plane (ULE) are further described. The risk analysis of the forward link signalling is highlighted and a security solution for control plane security is described.

Acknowledgements The authors gratefully acknowledge the support of the EU Information Society Technologies SATSIX Project [7] 0, the European Commission [11] and the SATNEX II project [12].

References

[1] Duquerroy L, and Josset S, The Flat Multicast Key Exchange protocol, Internet Draft, IETF, Sep. 2004 (status: expired).

[2] Kent S, and Seo K, Security Architecture for the Internet Protocol, IETF RFC 4301, Dec. 2006.

[3] Kaufman C, The Internet Key Exchange v2 (IKEv2), RFC 4306, IETF, Dec. 2005.

[4] Kent S, IP Encapsulating Security Payload (ESP), RFC 4303, IETF, Dec. 2005.

[5] Kent S, IP Authentication Header (AH), RFC 4302, IETF, Dec. 2005.

[6] EN 301 192, Digital Video Broadcasting (DVB) Specifications for Data Broadcasting, European Telecommunications Standards Institute (ETSI).

[7] SATSIX web http://www.ist-satsix.org

[8] Fairhurst G, and Collini-Nocker B, Unidirectional Lightweight Encapsulation (ULE) for Transmission of IP Datagrams over an MPEG-2 Transport Stream (TS), IETF RFC 4326, Dec. 2005.

[9] Cruickshank H, Iyengar S, Security Extension for Unidirectional Lightweight Encapsulation Protocol, IETF IPDVB group, draft-cruickshank-ipdvb-sec-02.txt.

[10] Fairhurst G, and Collini-Nocker B, Extension Formats for Unidirectional Link Encapsulation (ULE) and the Generic Stream Encapsulation (GSE), Work in Progress, draft-ietf-ipdvb-ule-ext-01.txt, Mar. 2007.

[11] European Commission home page http://www.cordis.lu/

[12] SATNEX Project, http://www.satnex.org/

Chapter 20
Implementing VoIP Support in a VSAT Network Based on SoftSwitch Integration

Yosy Hecht

Abstract Satellite communications based on geo-synchronous satellites are characterized by a large delay, and high cost of resources. VoIP applications over satellite networks must thus be handled very carefully so as to provide quality of service, as expected from voice calls, namely: secure resources for a call through-out its duration, regardless of other calls and other data traffic being present in the system. Furthermore, jitter and delay must be kept to a minimum while maintaining as high satellite resource exploitation efficient.

These subjects are well treated in the previous generation of satellite telephony products (Also referred to as 'Native telephony'). Basically, the circuit switching approach dominated these solutions, with specific solutions applied to both the access level and the application level. The media itself, derived from the require-ment of bandwidth efficiency, used the very same CODECs used for VoIP such as G.723 and G.729. The native telephony solution is jitter free, minimal delay, total QoS (regardless of other calls or data) and extremely efficient. The purpose of the current work was to capitalize on the existing infrastructure and its benefits and to port them to the VoIP environment.

The approach described herein was implemented in Gilat's SkyEdge™ VSAT (Very Small Aperture Terminal) product and is operational for about 2 years.

Introduction

Satellite communications based on Geo-synchronous satellites are characterized by the following:

- High delay caused by the propagation path of the information from earth to the satellite and back (about 250[ms] and 500[ms] round trip).
- High cost of BW.
- Often, the inbound path is a FTDMA channel with bursts.

As a result, transporting VoIP over a satellite medium is a demanding and challenging task: QoS must be carefully supplied so as to support the requirements of minimal delay and jitter while consuming minimal bandwidth.

This task may be divided according to the following:

- Identify a call. This task must be achieved relatively fast so as to provide resources for the call before the actual flow of media commences.
- Identify the type of the call: Be it star, mesh or local. Each such type requires a different allocation pattern: A star call requires timeslot allocation and some resources at the hub to deal with the media, a mesh call would need special dedicated mesh resources on both participating VSATs, a double-hop mesh call would require timeslots to both participating VSATs, a local call requires no resources and the media must not travel over the satellite medium.
- Identify the resources required for the call (e.g. codec BW).
- Reject a call for which there are insufficient resources.
- Identify the media flow itself, once it commences.
- Compress the media and send it over the proper satellite resource.

SkyEdge™

SkyEdge™ is Gilat's leading product, a 'one platform does it all' satellite access solution. It provides support for both Internet access as well as other IP applications and legacy protocols. The SkyEdge™ VSAT includes expansion slots where telephony boards and mesh receivers may be installed. The inbound is built on a multi-slot TDM approach on which several access methods may be implemented—RA (Random Access), GA (reservation), DA (Dynamic access) and a special implementation of DA for voice called VDA. The VDA access scheme is unique in that it is dedicated to native telephony and hence requires no flow control reliability or fragmentation facilities enabling use of a very small header structure. A controller called DCAS (Dynamic Call Allocation Server) receives call requests from either VSATs or hub, based on native telephony protocols, analyzes the requests and allocates the resources needed for the calls. The telephony task on the VSATs and hub components are responsible for signalling termination and voice treatment including sampling and coding in one of the popular vocoders such as G729 or G723. The multi-slot structure is designed to fit one of these vocoders, and a compromise is made on minimizing delay and maximizing efficiency such that voice is collected typically in 120[ms] intervals, thereby adding 120[ms] delay, but sharing the same headers with either 12 G729 or 4 G723 frames. The good synchronization between the CODEC requirements and the access setup yields a jitter-less transmission path. Calls for which there are no resources are being rejected.

Implementing VoIP over SkyEdge™

Earlier products as well as the skyEdge™ provided a VoIP solution based on the data and internet features support infrastructure. However, this approach was limited in efficiency due to the fact that it was associated with the DA access mode, it provided no guarantee of reducing the jitter when many calls were present at the same VSAT and more. A new approach was sought, such that the existing native telephony infrastructure with all of its inherent benefits may be leveraged on to gain the same benefits for the VoIP solution. The requirents were:

- Provide 'end-to-end' VoIP solution: At each end of the system (Either VSAT or Hub), the system provides a LAN connector. Not an FXS or E1.
- Support H.323 protocol and SIP at the same time.
- Do not limit the solution to specific endpoints—whichever endpoints that complies with the underlying standards must be supported by the system.
- Provide 'total' QoS—zero jitter, be completely independent of other data types on the same VSAT. Guarantee the required resources throughout the duration of the call.
- Reject the call if there are no resources to satisfy the QoS requirements.
- Use the same vocoders supported by the telephony application, use the same system elements.
- Provide the same BW efficiency as with the telephony application.
- Enable the VoIP and the Native telephony application to coexist within the same network, the same VSAT.

Solution

The main technical decision taken was to rely on the VoIP protocol signalling to achieve the tasks of identifying the call itself, its' type, the participating endpoints and required resources. This would be achieved through either a statefull inspection of the signalling packets in the application layer, and then reporting to the DCAS the relevant information consequently receiving the resource allocation, or through terminating the call and creating a new call leg. It seemed like a major undertaking to implement both SIP and H.323 stacks at each VSAT, and it seemed reasonable that the hub may be assumed to be the place where all signalling will pass through. At the hub, there were 2 possible solutions:

- A passive application that would analyze the signalling and communicate the data to the DCAS.
- A Gatekeeper and a Proxy server jointly referred to as softSwitch.

Due to the requirement that it will be possible to reject the calls in case of resource deficiency, the second option was chosen. The main reason for this selection was that in order to actively reject a call, the rejecting party must be a part of the VoIP network. A semi-passive element that would generate a reject message may lead to unforeseen results and a ping-pong of messages and acknowledgements.

Topology

The topology is described in Fig 20.1. A softSwitch is placed at the hub having an interface to the IP world, where the following entities may exist: Internet network access to remote CPEs, Gateways, other softswitches, CPEs, and access to the CPEs behind the VSATs. The softwitch has another interface to the DCAS. The CPEs behind the VSATs connect to the VSAT over the VSAT's LAN port. There is a dedicated voice packet processor at the hub to handle VoIP media, and a different processor to handle data (and perform TCP spoofing etc).

Signalling

All endpoints must be registered with the softSwitch, and the call models must be of H.323 routed mode and SIP using SIP proxy. When a CPE behind a VSAT initiates a call, it sends the call setup message (SIP INVITE, H.323 ARQ, setup) over the 'regular' data path of the VSAT, as any other IP application, subject to data QoS settings. When the softSwitch receives the setup request, it resolves the source and destination IP addresses of the 2 parties involved in the call, and sends a call request message to the DCAS containing the information. The DCAS looks its tables and classifies the call as either star, mesh, local or hub call, and checks for the availability of resources. If no resources are available, the DCAS rejects the call to the soft Switch and the softSwitch, being a part of the VoIP topology, issues a call clearing command with the proper cause to the CPE that initiated the call.

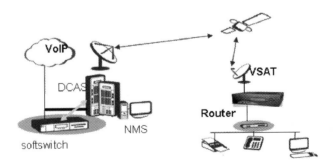

Fig. 20.1 Network Toplology

If the required resources are available, the DCAS allocates the resources and sends assign messages to the voice packet processor at the hub, to the VSATs and to the softswitch which continues the call setup process.

Media path

Once the assign message arrives at the VSAT, it intercepts all RTP packets having the destination IP address and UDP port which it received in the assign message from the DCAS. The VSAT then performs the following:

1. It removes IP, UDP and RTP headers.
2. It collects frames of payload until the transmit timeslot arrives.
3. It adds a 6 Byte header with some RTP fields included that is shared with the VoIP application layer and the satellite access layer.

The VSAT then sends the new packet to the hub, where it is processed by the voice packet processor which reconstructs the original RTP packet structure, with some modifications e.g. the SSRC is not reconstructed, it is rather created a new for the duration of the session. The voice packet processor then forwards the RTP packet over the Ethernet port to its destination.

On the reversed direction (Outbound), the packets arrive at the voice packet processor which strips them of their IP, UDP and RTP headers, constructs a new packet, and sends it to the IP encapsulator, and to the VSAT. At the VSAT the packets are reconstructed to have IP UDP and RTP headers, and the packets are sent to the CPE.

Solution Pros

- Enables total QoS. The voice and data paths are completely separated from each other and there can be no interference caused by data traffic such as FTP upload and the voice traffic.
- There is no interference between different voice calls, each call has its satellite resources reserved for the entire duration of the session.
- The access provides a jitter-less medium for the voice data, which results in a superior voice quality.
- It is possible to reject a call if there are no resources available, and provide a proper cause.
- The bandwidth requirement is known from the signalling session, there is no need to assume bandwidth requirement a-priori, or to spend time in estimating the actual bandwidth.
- Decisions regarding satellite resources are taken where the VoIP decisions are taken: At the softswitch.

- No need to implement complex logics at the VSAT.
- The VSAT implementation is protocol agnostic—it is only aware of RTP.
- The DCAS is protocol agnostic, and only aware of the protocol between itself and the softSwitch.
- Using a softswitch from an external vendor brings the benefits of exploiting the know how of expert—the softSwitch vendor, and concentrate on the satellite-unique issues. It also enables the system to benefit from the VoIP market trends and developments; (to expand in the directions the market is growing in terms of new protocols, new protocol version, redundancy, cost reduction etc).
- Enables potential support for lawful interception, also in mesh modes.
- May enable support for encrypted signalling (because the encrypted tunnel must be between the CPE and the softswitch)—something that 'Transparent' systems can't achieve.

Solution Cons

- Not 'transparent'. If a carrier wants to use its existing softSwitch, it can't do so. Although softSwitch to softswitch communications enables such a topology, usually the carrier will not want some of the subscribers to be managed in one softswitch and others on another.
- It forces the service provider to be a telephony service provider, even if it is not its intention—the service provider must manage the subscribers, provide class 5 services and comply with regulations that otherwise it was exempt of such as lawful interception, emergency calls.
- Does not provide support for protocols such as Skype—that are either proprietary and wouldn't be recognized by the softSwitch, or their topology is not centric, or do not use RTP.
- In case of a VSAT where there are many calls, it cannot benefit from statistical multiplexing based on silence suppression.
- Has problems with VLAN support.
- Forces the network to work in a working point where a certain codec is most efficient. If other codecs are used, efficiency may deteriorate. Other applications may also perform less well.

Adaptation to DVB-RCS Environment

The new DVB-RCS access scheme poses a few challenges to the above described concept—no longer are the existing tools (Such as special satellite resource pool and proprietary headers in the return channel) available, and the question of whether the concept applies in this new environment is a valid one. It can be shown that with some adaptations, the concept is still valid even in the DVB-RCS environment, providing basically the same advantages and possibly more.

The method to be used to request DVB-RCS resources would be CRA, where the NCC analyzes the call needs and allocates the required resources to the RCST without receiving a request from the RCST. A minor adaptation of the standard is required to allow dynamic allocation and de-allocation of CRA resources to an RCST on a per-call basis, according to call characteristics. Providing better efficiency may also be achieved through the a-priori knowledge of the media stream end-to-end characteristics such as UDP ports. This enables header compression to be applied on each media stream, removing even the IP headers, and reconstructing them on the other side. When performing IP header compression, some deviation from the DVB-RCS standard is required due to the fact that the packets are no longer IP—proper modification of the standard in the context of session management should enable this support. Of course, such header compression may require additional information transfer between the two ends of the satellite link—information not currently available within the DVB-RCS specs. C2P, a new protocol currently under development, may be applied to provide such information transfer.

Additional point to consider is the fact that the current approach is built on reducing the VoIP calls back to circuit switching mechanisms. These mechanisms guarantee quality of service equal to that expected by landline telephone subscribers (except for the obvious satellite delay and compression). It must be considered how to provide virtual circuit implementation in the DVB-RCS environment which is geared towards IP traffic. In this respect also arises the question of efficiency caused by the resolution of ATM cells or MPEG frames—the currently available media formats in DVB-RCS.

Last, moving forward with a new access scheme provides an opportunity to resolve existing deficiencies in the model such as the transparency issue, interface to IMS and standardization of the protocol between the NCC and the softSwitch.

Conclusion

Satellite communications based on geo-synchronous satellites are characterized by a large delay, and high cost of resources. VoIP applications over satellite networks must thus be handled very carefully so as to provide quality of service, as expected from voice calls while maintaining as high satellite resource exploitation efficient.

We have described above a circuit-switching approach to overcome the challenges VoIP posed to satellite network vendors and operators. The approach consists of an integrated softswitch included in the network, communicating with the hub and providing information for exact resource allocation such that each call is allocated dedicated resources. The approach had been implemented and tested and is provided by Gilat and is commercially available as a VoIP solution. Finally, some considerations of using this approach in a DVB-RCS network were discussed and required additions to the standard were indicated.

Acknowledgements The work of optimizing VoIP and other session based protocols over the DVB-RCS standard was carried out under the VIVALDI project that is partially funded by the EC within the 6th FP.

Chapter 21
Performance Evaluation of CRTP and Enhanced CRTP Within the DVB/RCS Context

G. Dimitriadis, S. Karapantazis, and F.-N. Pavlidou

Abstract The last years have seen a solid interest in Voice over IP (VoIP), also referred to as Internet or IP Telephony. The interest in this kind of communications is fueled by the multiple advantages that VoIP holds. Nonetheless, several strides remain to be made in order for VoIP systems to be on a par with the well engineered Public Switched Telephone Network (PSTN). From the network's viewpoint, one major downside of this application is the relatively high size of the IP/UDP/RTP headers compared to the packet's payload, which translates in inefficient bandwidth utilization. This work intends to assess the performance of two well-known header compression schemes that is Compressed RTP (CRTP) and Enhanced CRTP, over satellite links. For the needs of this comparison a software platform was implemented which models the behavior of the satellite channel and permits the evaluation of different header compression schemes on real traffic.

Introduction

In recent years, the Internet has experienced an unprecedented growth which brought about significant changes in the telecommunications arena. This explosion, coupled with the growing exigencies for flexible network access, spurred the development of innovative applications. Of the multitude of applications that thrived, Voice over IP (VoIP) is the most prominent one. VoIP, which is also known as Internet or IP Telephony, is the technology that uses the Internet as the transport medium for voice communications.

The trend towards this technology is fueled by the host of advantages that VoIP offers. From the users' viewpoint, VoIP has an edge on Public Switched Telephone Network (PSTN) in terms of monetary savings and flexibility. Making a call over the Internet is usually more economical than a circuit-switched call. In addition, Internet telephony allows end-users to employ the same identity regardless of the point of access. As far as service providers are concerned, VoIP opens up new sources of revenue for them. Moreover, it gives them the opportunity to compete with incumbent operators, since it surmounts the regulatory hindrance.

Notwithstanding the aforementioned advantages, there exist several issues that must be tackled in order for the deployment of a VoIP system to be successful. This

L. Fan et al. (eds.), *IP Networking over Next-Generation Satellite Systems* 317
© Springer 2008

paper focuses on an issue that has much bearing on the efficient use of network resources, namely, header compression. The majority of the narrowband voice codecs that have been developed thus far generate data of a few kbps. Also, since we are dealing with real time communications, the bundling of many voice frames in a single packet is subject to practical constraints, since the mouth-to-ear delay must be kept to as low as possible levels. The combination of the above two remarks, leads to the conclusion that in VoIP the payload of a voice packet is of similar size or even quite smaller than the size of the headers which are added by the lower layers. Specifically, in IPv4 the IP header contributes 20 octets, while the UDP and RTP headers are 8 and 12 octets respectively, for a total of 40 octets. When it comes to IPv6 (IP version 6), the size of the IP header becomes 40 bytes, rising the total headers length to 60 octets. However, the versatility of IP and the one-for-all solution that it represents means that many of its features are not employed in VoIP communications (e.g. fragmentation). Furthermore, the header information shows significant redundancy between consecutive packets, thus rendering its compression a possibility. Except for the obvious advantage of efficient bandwidth utilization, header compression exhibits some other assets as well. It reduces transmission delay and packet error rate (PER). The latter can be explained considering that for a given bit error rate (BER), the smaller the size of the packet, the smaller the number of the received packets containing erroneous bits.

Concerning the compression of the IP/UDP/RTP header train, three schemes have been proposed as yet. The first scheme was proposed in 1999 and is called Compressed RTP (CRTP) [1]. It is the simplest one, yet its performance over relatively error-free links with low round trip delay is remarkable. This scheme was extended and Enhanced CRTP (ECRTP) was proposed in [2]. This compression scheme builds upon the insights from the preceding research efforts with regard to CRTP and aims to ameliorate the performance of that compression scheme over links with high delay and packet error rate. A major step forward regarding header compression schemes was made when the Robust Header Compression (ROHC) scheme was introduced by the Internet Engineering Task force (IETF) [3]. This scheme shows very good characteristics, which have been demonstrated in studies [4–6]. However, the advantages of this compression algorithm come at a cost of increased complexity. A general comparison of the aforementioned schemes is provided in [7].

Although these compression schemes have been the subject of significant research efforts in the context of terrestrial wireless systems, their performance over satellite links has not been investigated. The focus of this paper is on the evaluation of CRTP and ECRTP header compression schemes for VoIP traffic in the context of DVB-RCS satellite systems [8]. The role of header compression schemes is pivotal in this kind of systems, since bandwidth constitutes a commodity at a premium. However, the characteristics of the satellite channel may impair the performance of these schemes. Aside from the long propagation delay, the satellite channel also exhibits increased bit error rate. Although some of the header compression schemes provide mechanisms to cope with the problems that arise in this occasion, all of them have been developed having terrestrial links in mind. Thus, it is not obvious which scheme is apt for satellite systems. In this study the performance of CRTP and ECRTP is assessed by means of a testbed that models the detrimental effects of satellite channel and processes real voice traffic generated by softphone applications in real time.

The remainder of the paper is structured as follows. In Section 2, the CRTP compression scheme is delineated and the details of ECRTP are spelt out. The description of the testbed that was used for the needs of our experiments is provided in Section 3, while results are presented and discussed in Section 4. Finally, concluding remarks are drawn in Section 5.

Overview of CRTP and ECRTP

CRTP

The CRTP compression scheme draws on the mechanisms proposed in the Van Jacobson header compression scheme for TCP. This means that it is predicated upon the notion of context, i.e. information that is shared between the two peers, compressor and decompressor. Both the compressor and the decompressor store all the fields of the full header but the ones that are included "as is". Each packet transmitted over a link should carry a tag (CID), which allows the decompressor to identify the context that must be used to decompress the packet, since each context is associated only with a particular session per link.

The development of CRTP was motivated primarily by the problem of sending audio over low speed dial-up modems. Therefore, it performs best on links with low round-trip delay. CRTP hinges upon the fact that most of the fields in the IP/UDP/RTP header train remain constant or seldom change during the session. Thus, the uncompressed header is transmitted only to establish or refresh the context at the decompressor side. For this purpose the FULL_HEADER packet type is used. Except for the uncompressed headers, this packet also carries a 4-bit sequence number to achieve synchronization between the compressor and the decompressor. This sequence number should not be confused with the RTP sequence number, since it is of local importance only, used to maintain synchronization between the compressor and the decompressor and detect packet losses. It is worth pointing out that the CID and the 4-bit sequence number are inserted into the length fields of the IP and UDP headers since the length of the packet can be inferred by the link layer.

Besides the FULL_HEADER packet, CRTP also defines three other packet types. The COMPRESSED_UDP packet is used to communicate the compressed IP and UDP fields, as well as the full UDP payload (which contains an uncompressed RTP header). Thus, this packet is used when there exist differences in the fields of the RTP header that are expected to remain constant, such as the payload type field, while the IP and UDP headers do not need to be updated. In addition to this packet type, the COMPRESSED_RTP packet communicates the compressed IP/UDP/RTP headers to the decompressor. The minimum size of this packet type is 2 bytes and is used to transmit the CID and the 4-bit sequence number and occasionally the first-order difference, i.e., the difference between two consecutive values, with respect to some header fields. These two packet types rely on differential encoding, namely, for fields that change often but predictably no absolute values are sent, rather the difference from the previous value is communicated; when this first order difference remains constant, no information is exchanged regarding the corresponding field but it is assumed that the previously established deltas apply. The last type of packet is called CONTEXT_STATE and is used by the decompressor when synchronization with the compressor is lost. This packet triggers the compressor to refresh the context at the decompressor. In order to avoid flooding the reverse channel, the decompressor does not send such a packet for every compressed packet that is received in the interim.

When working over error-free links with low round trip delay, the performance of CRTP is outstanding since it compresses the IP/UDP/RTP headers down to 2 octets (or 4 when the UDP checksum is enabled). Nonetheless, when it comes to wireless links, which are characterized by high bit error rates, its performance degrades [9–11], since the loss of synchronization between compressor and decompressor is frequent. Even when the TWICE algorithm is employed to ameliorate the performance by trying to repair the context without having to wait for a round-trip over the link, the performance does not improve significantly. Moreover, the TWICE algorithm necessitates the UDP checksum to be enabled, thus resulting in a reduction in compression efficiency.

ECRTP

Enhanced CRTP has been developed with the aim of making CRTP work well over links with higher bit error rates and long round-trip delay. Moreover, ECRTP provides mechanisms to deal with packet reordering. ECRTP relies on a hybrid open- and closed-loop error recovery/avoidance approach. The closed-loop mechanism is similar to the one that CRTP uses, that is, the transmission of CONTEXT_STATE packets when the decompressor loses synchronization with the compressor. Therefore, the major difference between ECRTP and CRTP lies in the open-loop mechanism, which essentially translates into the transmission of context updates more than once. Specifically, when a change in an absolute or delta value occurs, the compressor incorporates this change in the next N packets. Thus, using the TWICE algorithm the decompressor is able to keep in sync with the compressor provided that no more than N packets are lost.

In addition to the open-loop mechanism, ECRTP also specifies extensions to the COMPRESSED_UDP packet to allow sending the absolute values rather than the first-order differences for some volatile fields. By doing so, ECRTP manages to partially refresh the decompressor state.

Another problem that ECRTP aims to overcome is the inability of the decompressor to verify that the context is still in sync after a packet loss when the UDP checksum is not enabled. To this end, the compressor adds a "header checksum," similar to UDP checksum, which is removed by the decompressor before the packet is passed to upper layers.

Testbed Description

In order to assess the performance of the aforementioned header compression schemes in the hostile environment of satellite systems, we implemented a testbed that models the impairments of the satellite channel and performs header compression on real VoIP traffic. For the implementation of this testbed, we relied upon the netgraph framework that is indigenous to the FreeBSD operating system [12]. The netgraph framework presented itself as the best candidate for our implementation because it allows to divert or inject packets at the data link layer, namely at the layer where header compression takes place.

The architecture of the implementation is depicted in Fig. 21.1. An open-source softphone application is used to generate RTP traffic, which flows downstream and is diverted by the netgraph framework to the modules that we have developed, shaded in dark gray in this figure. The first module on the left-hand side of the figure, *delay*, accounts for the delay and packet loss that voice packets experience before reaching the DVB-RCS terminal. This module adds a random amount of delay to all packets traveling upstream or downstream, based on a user defined distribution with configurable mean value and standard deviation. Moreover, it discards packets according to a binomial distribution. The *CRTP/ECRTP* module implements packet compression and decompression according to the CRTP or ECRTP rules. The last module, namely the module named *sat_delay* models the satellite channel, that is, it adds delay and discards packets.

As far as delay is concerned, the whole delay from the time instant of the transmission from the DVB-RCS terminal till the moment that the packet is received by the

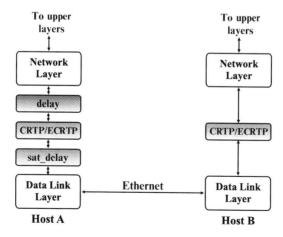

Fig. 21.1 Architecture of the implementation

Table 21.1 Distributions for different error intensities

	Time between Bursts		Burst length	
Error intensity	Min	Max	Min	Max
None	Link without any errors			
Low	10 s	30 s	60 ms	60 ms
Medium	5 s	15 s	100 ms	300 ms
High	2 s	15 s	100 ms	300 ms
High	0 s	10 s	100 ms	300 ms

gateway is modeled by a user defined distribution and accounts for both the propagation and queuing delays. Regarding packet loss, the channel is modeled by a two-state Markovian process [13]. When the channel is in good state, all the packets are received correctly. On the other hand, all the packets that are received while the channel is in bad state are discarded. The transitions between these two states take place after a random amount of time, which follows a uniform distribution. The characteristics of these distributions are described in Table 21.1.

The scenario that we used in our experiments is as follows. Host A establishes a session with Host B using a VoIP softphone. The netgraph framework diverts all the packets that are associated with this session to the module delay, which adds delay to all packets that pass through it according to a normal distribution with mean value 30 ms and standard deviation 3 ms (wireline Internet part), whereas the packet error rate is 0.5%. The CRTP/ECRTP module compresses all the packets associated with the session and pass them to the sat_delay module (satellite segment). The latter adds some delay according to a uniform distribution with mean value 290 ms and standard deviation 5 ms. Moreover, it discards packets according to the model described in Table 21.1. When the packets arrive at host B, the netgraph framework diverts them to the CRTP/ECRTP module which performs decompression and all the packets that are decompressed successfully are passed to the upper layers. One advantage of this testbed is that allows the assessment of speech quality in the context of satellite systems using either objective or subjective test methods, though this issue is beyond the scope of our study.

Performance Evaluation

This section presents the results that were obtained using the testbed described above. These results constitute average values over 60 minutes long VoIP sessions. The voice codec that was used was Speex, generating a frame every 20 ms. VAD was not enabled, consequently the stream to be processed enters the module graph with well-behaved characteristics. However, this is not the case regarding the input of the header compressing module, since the packet loss already inflicted on the module above has somewhat damaged the predictability between consecutive header contents. The UDP checksum was enabled in order to allow the employment of the TWICE algorithm (a realistic approach, since in most OSs there is no provision for disabling the UDP checksum, while in IPv6 the UDP checksum is mandatory). The objective of our experiments is twofold: to compare the performance of the two compression schemes in terms of (1) compression ratio and (2) packet loss. The former performance indicator reveals the ability of the compression scheme to save bandwidth, whereas the latter metric evaluates its ability to keep the compressor and the decompressor in sync even during adverse fading conditions.

However, before commenting on these two issues, we first examine each compression scheme with regard to the percentage of the different packet types that it uses. The results are tabulated in Table 21.2. In this table are presented the FULL_HEADER, COMPRESSED_UDP and COMPRESSED_RTP packets that are received and the CONTEXT_STATE packets that are sent by Host B, normalized by their sum for easier comparison. An interesting observation is that regarding ECRTP the percentage of FULL_HEADER packets is hardly affected by the intensity of fading, while when it comes to CRTP, the more severe error intensity is, the higher the percentage of these packets. Moreover, the experiments confirmed that CRTP does not rely on COMPRESSED_UDP packets to update the context at the decompressor. At the other extreme, the extension made by ECRTP to this packet type allows using it in many occasions for updating the state at the decompressor, which also explains the lower percentage of FULL_HEADER packets in the case of ECRTP.

Concerning the percentage of CONTEXT_STATE packets, it is lower for ECRTP, as expected. This fact reveals the ability of this scheme to keep the compressor and decompressor in sync. Figure 21.2 depicts the average size of the compressed header for different error intensities. It becomes apparent from this figure that both CRTP and ECRTP perform very close to their nominal maximum compression, though CRTP performs slightly better than ECRTP, since the latter scheme advertises the changes in the stream characteristics multiple times, even when there is no need to do so (open-loop approach). This is also evident from the results presented in Table 21.2, which revealed that the percentage of COMPRESSED_RTP packets is lower when ECRTP is applied. Moreover, this figure portrays another interesting

Table 21.2 Distribution of packet types

Error Intensity	Compression Scheme	Packet Type			
		FULL HEADER	COMPRESSED RTP	COMPRESSED UDP	CONTEXT STATE
None	CRTP	$4.45 \cdot 10^{-3}$	99.9955	0	0
	ECRTP	$3.35 \cdot 10^{-4}$	97.3566	2.64	0
Low	CRTP	$8.43 \cdot 10^{-3}$	99.9876	0	$3.93 \cdot 10^{-3}$
	ECRTP	$3.36 \cdot 10^{-3}$	97.3466	2.65	0
Medium	CRTP	$3.92 \cdot 10^{-2}$	99.9255	0	$3.53 \cdot 10^{-2}$
	ECRTP	$3.42 \cdot 10^{-3}$	97.2372	2.74	$1.94 \cdot 10^{-2}$
High	CRTP	$6.59 \cdot 10^{-2}$	99.8722	0	$6.19 \cdot 10^{-2}$
	ECRTP	$3.46 \cdot 10^{3}$	97.1896	2.77	$3.69 \cdot 10^{2}$
Huge	CRTP	0.1034	99.7943	0	0.1023
	ECRTP	$3.50 \cdot 10^{3}$	97.0535	2.86	$8.3 \cdot 10^{2}$

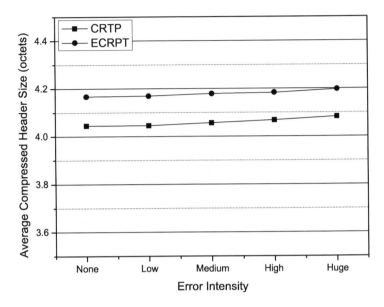

Fig. 21.2 Average size of compressed header versuserror intensity

behavior. The average size of the compressed header increases as channel conditions worsen, since the number of times that the decompressor invalidates its context increases. Nonetheless, the increase in the case of ECRTP is slightly lower than in the case of CRTP, which is an indication that ECRTP accepts more gracefully the higher packet error rates.

Our next objective is to assess these compression schemes in terms of packet loss, expecting from the results thus far that ECRTP will outperform CRTP. Figure 21.3, which presents the percentage of packets that are lost due to desynchronization between the compressor and the decompressor substantiates our expectations. It must be noted that packets that are lost due to adverse fading conditions are not taken into consideration. Instead, we are interested only in the number of packets that are received correctly but cannot be decompressed successfully. ECRTP attains much lower packet loss by virtue of the open-loop error recovery mechanism it uses. For medium and high error intensities, the packet loss of ECRTP is half the packet loss of CRTP, while for low error intensities it is zero. In addition to Fig. 21.3, Fig. 21.4 is a histogram of the number of consecutive packets that are lost when the compressor and the decompressor are out of sync. It should be pointed here that because of space and presentation concerns, the whole histogram is not presented but only its most significant portion. The occurrences that are presented in the depicted interval of the x-axis correspond to more than 95% of the total recorded occurrences.

A first observation reveals that CRTP is more prone to packet loss compared to ECRTP. Under further scrutiny, we observe that for low error intensities ECRTP does not suffer from packet loss due to desync. Further, according to this figure, for

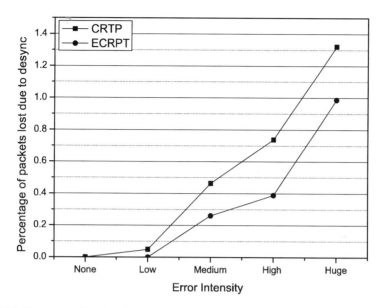

Fig. 21.3 Percentage of packets lost due to desync

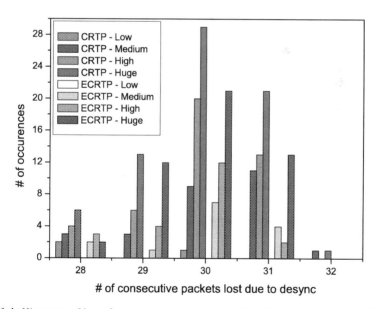

Fig. 21.4 Histogram of loss of sync occurrences versus number of packets lost per loss of sync

both CRTP and ECRTP, when synchronization is lost, usually 30 or more packets will be lost until the decompressor and compressor regain sync. This observation highlights the detrimental effects of the long round-trip delay of satellite links on compression schemes. Taking into account that a packet is generated every 20 ms, it takes about 600ms to recover from the loss of synchronization (equal to a round-trip time). Nonetheless, when ECRTP is applied, loss of synchronization is less frequent. The histogram of Fig. 21.4 shows in the most direct way that the combination of long round-trip times and closed loop control have a particularly destructive effect in speech quality. Even though the percentage of lost packets due to desync remains in reasonable levels (well below the 1% threshold for most cases), the fact that a long string of frames is lost has a heavy impact in perceived voice quality, since the decompressor cannot mask this loss of information. On the other hand however, the fact that the packet losses are concentrated in time leads to the conclusion that such occurrences are separated by substantial intervals. For CRTP, the average time between loss of synchronization occurrences is slightly less than a minute in the worst case, while in the case of ECRTP the corresponding inter-arrival time lies between 1 and 1.5 minute.

Conclusions

In this work the performance of two well-known IP/UDP/RTP header compression schemes, i.e. CRTP and Enhanced CRTP, was evaluated in the context of satellite systems. For the needs of the comparison, a testbed was developed in the FreeBSD operating system that models the behavior of the satellite channel and allows the assessment of these schemes on real voice traffic. The results obtained manifested the superiority of ECRTP to CRTP. In particular, ECRTP presented only a small increase in the average size of the compressed header, which was compensated by the decreased number of packets lost due to the loss of synchronization between the compressor and the decompressor.

References

[1] Casner S, Jacobson V (1999) Compressing IP/UDP/RTP Headers for Low-Speed Serial Links, RFC 2508, Internet Engineering Task Force, Feb. 1999.
[2] Koren T et al. (2003) Enhanced Compressed RTP (CRTP) for Links with High Delay, Packet Loss and Reordering, RFC 3545, Internet Engineering Task Force, Jul. 2003.
[3] Bormann C et al. (2001) RObust Header Compression (ROHC): Framework and Four Profiles: RTP, UDP, ESP, and ncompressed, RFC 3095, Internet Engineering Task Force, Jul. 2001.
[4] Rein S et al (2005) Voice Quality Evaluation in Wireless Packet Communication Systems: A Tutorial and Performance Results for ROHC, in Proc. of the IEEE Wireless Communications, vol. 12, no. 1, New Orleans, LA, USA, Mar 13–15 2005, pp. 60–67.
[5] Fitzek FHP et al (2005) "Robust Header Compression (ROHC) Performance for Multimedia Transmission over 3G/4G Wireless Networks," Wireless Personal Communications, vol. 32, no. 1, pp. 23–41, 2005.
[6] Jin H et al. (2004) Performance Comparison of Header Compression Schemes for RTP/UPD/IP Packets, in Proc. of the IEEE Wireless Communications and Net-working Conference (WCNC) 2004, vol. 3, Atlanta, Georgia, USA, March 21–25 2004, pp. 1691–1696.

[7] Ishac JA (2001) Survey of Header Compression Techniques," NASA/TM-2001–211154, [Available Online: http://gltrs.grc.nasa.gov/cgi-bin/GLTRS/browse.pl?2001/TM-2001–211154.html], Sep. 2001.

[8] ETSI (2005) EN 301 790 - Digital Video Broadcasting (DVB); Interaction channel for satellite distribution systems, vol.4.1, Sep. 2005.

[9] Degermark M et al (2000) Evaluation of CRTP Performance over Cellular Radio Links, IEEE Personal Commun. Mag., vol. 7, no. 4, pp. 20–25, Aug. 2000.

[10] Mamais G et al (1998), Evaluation of the Casner-Jacobson Algorithm for Compressing the RTP/UDP/IP Headers, in Proc. of the 3rd IEEE Symposium on Computers and Communications (ISCC) 1998, Athens, Greece, 29 June–2 July 1998, pp. 543–548.

[11] Cellatoglu A et al (2001) Robust Header Compression for Real-Time Services in Cellular Networks, in Proc. of the 2nd International Conference on 3G Mobile Communication Technologies, London, UK, Mar. 2001, pp. 124–128.

[12] Elischer J, Cobbs A (1999) The netgraph networking system, [Available Online: http://www.elischer.org/netgraph], Jan. 1999.

[13] Astuti D, Kojo M (2004) TCP and Link Layer Enhancements in DVBS/DVB-RCS Satellite Systems, Berkeley - Helsinki Ph.D. Student Workshop on Telecommunication Software Architectures, Jun. 2004.

Chapter 22
Secure Multicast in the Broadband Satellite Multimedia Networks

H. Cruickshank, S. Iyengar, L. Fan, Z. Sun, R.J. Mort, and M. Mezzalla

Abstract Satellites are expected to play an essential role in providing broadband and multicast services to regions that cannot be economically reached by terrestrial networks, in particular the more remote regions of Europe and the rest of the world. Security can be a problem for such global services. Security of multicast data is becoming an important issue to operators and users of satellite networks.

This paper presents a proposed secure multicast architecture that fits well into the ETSI BSM overall architectural concepts. Multicast service scenarios with detailed security architecture are presented using the BSM standard interface between satellite dependent and independent layers. This architecture accommodates link and network layer security (Internet) security systems and interworking with non- satellite entities such as AAA, PEPs and COPs entities.

Introduction

Current broadband satellite services are regarded as a niche market due to the high cost of launching a satellite system, and the relatively limited available bandwidth compared to terrestrial counterparts. To improve take-up of broadband satellite, it is essential to provide cost-effective solutions, to efficiently accommodate new multimedia applications, and to integrate satellites into next generation networks. These issues are being addressed within an IST project called SATSIX (Satellite-based communications systems within IPv6). SATSIX [1] implements innovative concepts and cost-effective solutions for broadband satellite systems and services. Security is a critical issue for the success of next generation satellite networks. SATSIX Project is defining a set of security requirements and solutions for such networks. This work is directly related to the on-going standardisation work the standardisation activities for satellite systems within ETSI Broadband Satellite Multimedia (satellite) group [2] and the MSEC group (multicast security) in the IETF [3].

L. Fan et al. (eds.), *IP Networking over Next-Generation Satellite Systems*
© Springer 2008

The ETSI Satellite Architectural Concepts

The ETSI Broadband Satellite Multimedia (BSM) working group aims for the development of broadband satellite systems providing services based on the Internet Protocol (IP) by means of common standards. These standards will allow building blocks and services for such satellite systems to become more readily available. This work is focussed on the efficient transport of IP data streams and on how to interoperate resulting satellite networks with terrestrial IP networks. The satellite standards are being designed to use existing standards such as DVB-RCS while remaining open to emerging standards and other available technologies [4, 5].

Satellite architecture considers all satellite networks that can provide IP multimedia service by using a common interface between the satellite specific and non-specific layers. This is called Satellite Independent Service Access Point interface or SI-SAP interface in BSM terminology. The SI-SAP defines the Satellite Independent to Satellite Dependent (SI-SD) interface. This interface also corresponds to the endpoints of the BSM satellite bearer services as illustrated in Fig. 22.1.

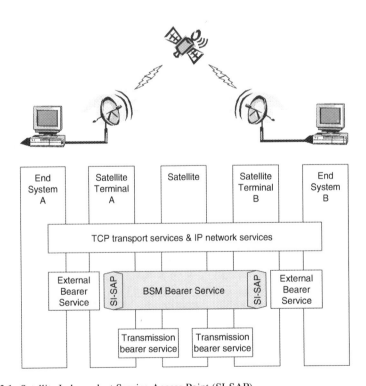

Fig. 22.1 Satellite Independent Service Access Point (SI-SAP)

The SI-SAP services are accessed via the SI-SAP interface and they can be used to transport standard IP-based network services. These SI-SAP services are applicable to any IP-based satellite network, including both transparent and regenerative satellites, and including both star and mesh topologies. For example, in the case of a star network topology, one SI-SAP would be located in the remote Satellite Terminal (ST) and the other in the hub ST; whereas in the case of a mesh network topology, the SI-SAPs would be located in the source and destination STs.

The present specification divides the services into user plane (U-plane), control plane (C-plane) and management plane (M-plane) services. This is a logical (functional) division but these different planes may also be physically separated. Figure 22.2 illustrates the concept of the SI-SAP corresponds as an interface between the Satellite Independent Layers and the Satellite Dependent Layers. Two adaptation functions are defined to adapt these layers to/from the SI-SAP:

- Satellite Independent Adaptation Functions (SIAF) to adapt between the IP network layers and the SI-SAP;
- Satellite Dependent Adaptation Functions (SDAF) to adapt between the SI-SAP and the native services of the satellite dependent SLC layer.

The functionality of the SIAF and the SDAF may be NULL for some of the SI-SAP services.

The SI-SAP is logically divided into 3 separate interfaces (SAP) as shown in Fig. 22.3:

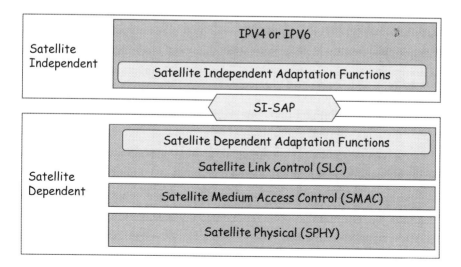

Fig. 22.2 Satellite Independent Service Access Point (SI-SAP) IP reference model

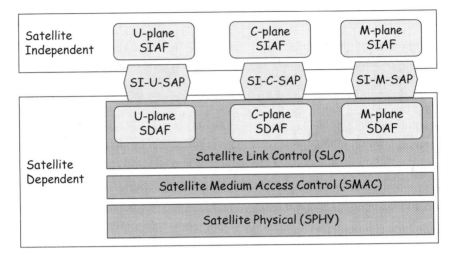

Fig. 22.3 Satellite Independent Service Access Point (SI-SAP) Expanded reference model

User Plane SAP (SI-U-SAP)

The SI-U-SAP is concerned with the services that are used to transport IP packets between STs. IP packets are interworked into the SI-U-SAP endpoint at the source ST, transported transparently to the corresponding SI-U-SAP endpoints at the destination STs where they are interworked into the destination ports. The U-plane services can be used to transport IP packets to/from internal ports as well as external ports. For example, ST management traffic based on SNMP can be transported via the U-plane.

Control Plane SAP (SI-C-SAP)

The SI-C-SAP is concerned with control and signalling related to the user plane services:

- **Control** refers to local services, that are concerned with control of the behaviour of the lower layers of the local stack;
- **Signalling** refers to remote services that involve interaction with the Network Control Centre (NCC) or with the peer ST.

Management Plane (SI-M-SAP)

The SI-M-SAP is concerned with satellite system management services.

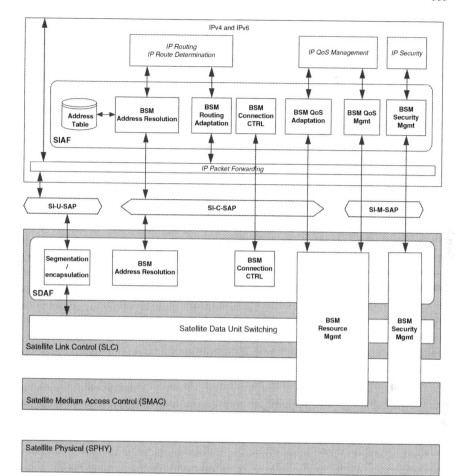

Fig. 22.4 satellite Protocol Stack for unicast/multicast services

Figure 22.4 shows that there are a small number of generic functions that need to cross the SI-SAP and those are related to connection/session management, resource management or security [6]. The architecture is covers IPv4/IPv6 aspects for unicast and multicast services. The BSM protocols are based on the OSI layered protocol stack. For the IP services most of the work has concentrated on the network layers with links to the underlying data link and MAC layers. The reason for this is simple: the developed protocols for IP over satellite should primarily be located in the satellite independent part of the BSM stack to be applicable to a range of different satellite dependent lower layers such as for example DVB-RCS.

Secure IP Multicasting Concept

IP multicast makes efficient use of bandwidth by setting up a mid-point between unicast traffic (one - to-one) and broadcast IP traffic (one-to-all in a network). This is well suited for one-to-many or many-to-many bulk data transfer or multimedia (audio/video) streaming transmission to a large number of heterogeneous receivers. Any host in a subnet can join a multicast group without its subnet router passing identification information about the host to other routers upstream in the distribution tree. This allows IP Multicast to scale to a large number of participating hosts.

However, from the perspective of security, additional mechanisms and services must be built atop the basic IP Multicast model. This decoupling of security from the IP Multicast model is advantageous, since it allows differing security models and architectures to be deployed, without affecting the multicast distribution tree which delivers the multicast data end-to-end. This decoupling is also important from the application's perspective, since each application requires different forms of host information and other security parameters, and may deploy differing user-identification and user-authentication mechanisms [7]. From the security prospective, the group management and routing should be transparent to security.

As shown in Fig. 22.5, there are several factors or aspects of IP Multicast that influence the approaches and mechanisms used to secure it. The most relevant factors include:

- Multicast application type: one-to-one or many-to-many.
- Group dynamics and Scalability issues: satellite groups can very large and dynamic (i.e. membership can change frequently).
- Underlying trust model: This relates to the trust relationship between senders, receivers and security managers.

The combinations of these factors will be the deciding factor on the group security policy parameters such as:

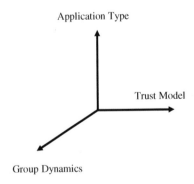

Fig. 22.5 Factors affecting secure multicast system design

- Procedures for establishing, running and terminating the group.
- Key distribution methods such as the use of Logical Key Hierarchy (LKH). For one-to many applications with large and dynamic groups, the rekeying policy must be defined carefully in order to reduce the bandwidth need to key management traffic.
- The strength of the encryption algorithms and digital signature needed, such as using Digital Encryption Standard (DES), triple DES or Advanced Encryption Standard (AES) for privacy and RSA or DSS signature systems.
- What to do when things go wrong: Such as network failure, Denial of Service (DoS) attacks, problems with entity authentication and authorisation. The procedures are usually defined in the security policy. There must be a clear way for creation, dissemination and enforcement of security policies.

It is expected that satellite multicast groups can be large, dynamic and span several administrative domains (such as satellite and fixed/mobile network domains) with varying security policy rules. Key management is a complex issue for large multicast groups. Here we adopt the IETF MSEC group approach [8–10]. Examples of MSEC multicast key management protocols are Group Secure Association Key Management Protocol (GSAKMP), Group Domain of Interpretation (GDOI), Multimedia Internet KEYing (MIKEY) and Flat Multicast Key Exchange (FMKE) [3].

Threats to Satellite Networks and Counter Measure Requirements

Threats and potential attacks in satellite networks can be categorized into network, software, hardware and Human threat types [7].

The first type is software threat. Many systems fail because of mistakes in software implementation. Some systems use temporary files to protect against data loss during a system crash, or virtual memory to increase the available memory; these features can accidentally leave plaintext accessible to un-authorized people. Moreover confidential information of a company or clients should be stored securely at the provider's site in order to prevent misuse of confidential information of clients. Thus, security requirements for software threats are: Protection against software viruses; good software design that prevents unauthorized access and good security software design with strong encryption and digital signature algorithms, good random number generators and secure storage of keys and internal data.

The second type is hardware threats. All hardware systems including hosts (e.g. client stations), satellite terminals and network equipment (e.g. routers and firewalls) can provide a way of attack if not properly configured, since they will become the entry point of attack. Unauthorized access to these machines also poses a threat since it means access to the system. In addition, if all the major hardware systems are not backed up in case of emergency like power outage or denial of service attacks then it poses a serious threat as the data stored in these systems as well as the availability of the service as a whole is disrupted. Thus, security requirements for hardware threats are: Provide secure and robust backups to prevent loss of data and protect against hardware theft.

The third type is human threats (attacks). There are two types of such threats: Insider and outsider attacks. Insider attacks occur when legitimate users of a system behave in unintended or unauthorized ways. Most known computer crime has involved insider attacks that compromised the security of the system. Outsider attacks may use techniques such as wiretapping (active), intercepting, replaying, modifying messages or disrupting services using denial of service attacks etc, in order to carry out the network attacks as mentioned earlier. Thus, security requirements for human threats are: Users and Administrators - authentication, accounting and traceability of actions; the ability to track and report on security events through proper logging and event correlation; Protection of satellite systems from unauthorized people; proper training of users and administrators regarding using good security practices for choosing passwords and controlling access to computers and buildings.

Finally the last type is network threats which are the focus for this work. The simplest type of network threat is a passive threat. Passive attacks include eavesdropping or monitoring of transmissions, with a goal to obtain information that is being transmitted. In broadcast networks (especially those utilizing low-cost physical layer interfaces, such as DVB) counter measures must be provided for passive threats. An example of such threat is an intruder monitoring the satellite transmissions and being able to extract traffic communicated between IP hosts. Active attacks are more complex that passive attacks. They are more difficult to implement successfully than passive attacks, and usually require more sophisticated resources. Thus, the security requirements for network threats are: Source authentication; Confidentiality and integrity of data; Protect the management of the infrastructure from unauthorized people; traceability (such as using intrusion detection systems) to monitor their network and log files to record the activities on the network and protection against denial of service attacks.

Multicast Threat Analysis

In addition, there is further analysis of threats to IP transmission over DVB networks in ??draft-cruickshank-ipdvb-sec-req-04.txt (see bibliography). For the purpose of the present document, three threat satellite examples have been identified:

EXAMPLE 1: Monitoring: The intruder monitors (passively) the satellite broadcasts in order to gain information about data and/or tracking the communicating parties.

EXAMPLE 2: Local high jacking of the satellite transmission: Here it is assumed that the intruder is sophisticated and able to block the original transmission from the satellite system and deliver a modified version of the MPEG-TS transmission to a single satellite Receiver or a small group of Receivers (e.g. in a single company site). The global satellite system might not be aware of such attacks.

EXAMPLE 3: Global high jacking of the satellite transmission: Again it is assumed that the intruder is very sophisticated and able to high jack the whole satellite transmission to all Receivers.

The above analysis shows the need for satellite security services such as data confidentiality, integrity, sender authentication/authorization and efficient key management system.

Satellite Security Services Definition

Examining the above threats show that eavesdropping (passive attacks) can be considered as a major threat to satellite networks, especially for broadcast services. Software, hardware and human threats will need some general measures such as good software and hardware design and maintenance, proper satellite equipment testing, and proper training of satellite personnel and customers regarding basic security issues.

However it is likely that the major active threats will be network threats to satellite networks such as impersonation, message modifications and denial of service attacks. These threats will require appropriate security counter measures. In order to counter these major threats there is a need to define the security services required such as:

- **Confidentiality (or privacy) service** is used to create a private session. Data encryption is typically used to provide this service. Confidentiality can be used as a countermeasure against eavesdropping, masquerading, traffic analysis and leakage of sensitive information.
- **Integrity service** guarantees that the messages are received with no modification by unauthorized entities. In order to provide this service the mechanisms used are encryption, Message Authentication Codes (MAC) or digital signatures. Examples of digital signature schemes are the Digital Signature Standard (DSS); Rivest, Shamir and Adleman (RSA) which are based on public key technology. Certificates signed by a trusted Certification Authority (CA) are used to bind the identity of an entity to its public key. Integrity service prevents manipulation of messages such as messages may be deliberately modified, inserted, replayed, or deleted by an intruder.
- **Authentication** This is similar to the integrity services, however the purpose is different. It is used to verify the identity of entities involved in a communication (e.g. users, STs and IP routers). The simplest technique is user ID and password. More sophisticated authentication mechanisms are encryption, Message Authentication Codes (MAC) and digital signatures, similar to integrity service. Authentication can be mutual (both communicating entities) or one way (only the originator).

- **Authorization and Access control** is a service where each individual user privileges are verified. This service is normally needed in conjunction with authentication in order to provide access control. This prevents an unauthorized use of a resource such as intruders can access services by masquerading as users or network entities (including insider attacks). Also it prevents denial of service attacks such as disturbing, misusing network services or resource exhaustion and overloading.
- **Non-repudiation** is a service that prevents a sender or receiver from denying its acts. The main mechanism used for this service is digital signatures.
- **Availability** Ensuring that all legitimate entities should experience correct access to services and facilities of the satellite network. It prevents the intruder from disturbing or misusing the network services leading to a denial of service attack.
- **Key management** denotes the procedures in which security keys are securely conveyed to the appropriate parties. There are two types of key management may be used, manual and automatic. Manual procedures are typically handled by system administrator and automatic procedures are handled by key management protocols. Key management is one of the most difficult problems in globe communication systems such as satellite.

Summary of Multicast Security Service Requirements

The threat analysis shows the need for the following security requirements:

- Security key management is scalable in order to cater for large and dynamic multicast groups.
- Data confidentiality is the major requirement against passive threats to satellite links.
- In a multiple senders scenario (such as conferencing), strict control is required for all ingress STs in a multicast session, in order to prevent Denial of Service attacks.
- Decoupling of satellite multicast key management functions from satellite data encryption. This will allow the re-use of existing security management systems (e.g. GDOI RFC 3547 and GSAKMP RFC 4535 [11], plus other systems such as DVB-RCS (see EN 301 790) and/or the development of new systems, as required.
- For end-to-end secure multicast, satellite role is confined only to access control to the satellite resources.
- Optional security for QoS and other management control messages (e.g. bandwidth requests).

Further analysis of the multicast key management requirements are presented in later sections.

Overview of DVB-RCS Security

The BSM security architecture focus on using network level security such as and IPsec. It also allows the use of satellite link layer security such as DVB-RCS security. This section presents an overview of the DVB-RCS security. It is assumed that the reader is familiar with IPsec concepts. Therefore IPsec details are not presented in this paper.

The DVB-RCS specification [11] defines the return (or "interaction") channel for communication between a Return Channel Satellite Terminal (RCST) and a Gateway/hub ground station. The term RCST simply refers to a satellite terminal, the term gateway refer to a large ground station that is connected to other networks such as the Internet, and the hub is the entity that is responsible for multiplexing data destined to multiple RCSTs onto the satellite broadcast channels. The gateway and hub station are typically co-located in satellite systems. The satellite network is monitored and controlled by the NCC. Signalling is transmitted between the NCC and the RCSTs over the forward link and the return link.

The DVB-RCS security specification currently supports the authentication of each RCST to the NCC, and the encryption of both forward and return link traffic, and these functions are described in the following sub-sections.

DVB-RCS Authentication

Each RCST holds a shared secret key, called a cookie, known only to the given RCST and the NCC. This cookie is used during key exchanges, as will now be described.

A logon is initiated by a RCST, for example when the first user of the RCST wishes to use the satellite link for data transfer. This is followed by an initial handshake between the NCC and the RCST to agree the security profile (i.e. the cryptographic algorithms and key sizes to be used): this is performed by the Security Sign-On and Security Sign-On Response messages (Fig. 22.6). The current DVB-RCS specification supports a single session key per RCST, this key being used to encrypt data traffic in both directions on the satellite link. In the process of authentication, the specification then allows one of three key exchange mechanisms to occur. The objectives of these key exchange messages are firstly to authenticate the RCST and secondly for the RCST and NCC to agree the session key to be used. The three key exchanges and their principal features are as follows:

- Main Key Exchange (MKE): this uses the Diffie-Hellman algorithm to develop a shared secret between the NCC and RCST, known only to these two entities; it also uses the cookie (secret key) held in the RCST to authenticate the RCST to the NCC; optionally it can use the newly developed shared secret to update the cookie; and finally it derives a session key from the newly developed shared secret.

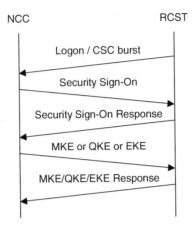

NCC   RCST

Logon / CSC burst

Security Sign-On

Security Sign-On Response

MKE or QKE or EKE

MKE/QKE/EKE Response

Fig. 22.6 DVB-RCS security establishment

- Quick Key Exchange (QKE): this uses the cookie to authenticate the RCST to the NCC; and derives a session key from the cookie.
- Explicit Key Exchange (EKE): this transmits (encrypted) a key from the NCC to the RCST; this key is then used as the session key.

Following logon, the NCC can initiate further key exchanges as required to update the session key.

DVB-RCS Encryption

The session key obtained during authentication is used in DVB-RCS to encrypt IP datagrams in both forward and reverse directions, and this is shown in Fig. 22.7 as "individual user scrambling". This allows data destined for different RCSTs to be encrypted with different keys. The Figure also shows an alternative encryption approach, "Common Scrambling", used to encrypt traffic in DVB Conditional Access (CA) systems. CA is typically used for encryption of broadcast traffic such as TV transmissions. Although it could be used for IP traffic, CA relies on a single key to encrypt all IP datagrams and thus encrypted traffic destined for one RCST could be decrypted by any other RCST.

The DVB satellite specifications support two data link layer platforms: DVB/MPEG2 and ATM. Encryption is performed differently in the two cases:

- In DVB Multi Protocol Encapsulation (MPE), two scrambling control fields are defined: these are the payload_scrambling_control field and the address_scrambling_control field. MPE is placed within the DSM-CC - Digital Storage Media Command and Control(DSM-CC) section in Fig. 22.7.

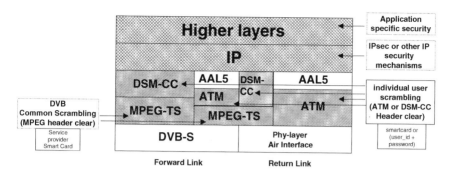

Fig. 22.7 Security layers for satellite interactive network (example), from [11]

- The DVB-RCS specification requires that the payload and/or address may be scrambled, but the MAC address must not be encrypted. The payload includes the IP datagram.
- In an ATM stream, the 48-byte cell payload is encrypted using the session key, but the 5-byte ATM header is not encrypted. If a single ATM VPI/VCI is used to carry traffic on the forward or return link then no information is given away to an eavesdropper by the ATM address.

The principal advantages of individual user scrambling compared to MPEG-TS common scrambling are:

- Encrypted traffic sent to or from one RCST can not be decrypted by any other RCST;
- There are no fixed relationships between IP addresses, the MPEG-TS program identifier (PID) and session key used for encryption, thus making eavesdropping more difficult;
- Data decryption at the RCST does not have to be performed on all the packets in MPEG-TS stream, but only on the individual unicast or multicast packets that are required by the given RCST, identified by the MAC address in each MPE Section header. This simplifies the RCST design.

DVB-RCS Multicast Extensions

The current DVB-RCS standard supports the following security requirements:

- Enables RCST authentication to NCC, by means of a logon phase.
- Maintains the forward and return uplinks and downlinks secure from eavesdropping, either by unauthorized persons or by other users of the satellite system, provided "individual user scrambling" is implemented.
- Supports separate keys for unicast traffic to each RCST so that no RCST can decrypt unicast traffic intended for a different RCST.

- Supports periodic rekeying of unicast channels using the EKE messages.
- Supports logout.

The security procedure in the current DVB-RCS standard (as presented above) has several gaps regarding multicast security. The standard currently mentions the use of Explicit Key Exchange (EKE) for multicast. However it is not clear how a particular key exchanged with EKE can be linked to multicast in general or to a particular multicast service, since only one key is used per session and this key needs to support all unicast and multicast traffic through the RCST. In particular, in order to support multicast security the following additional requirements can be stated:

- Support separate keys for each multicast channel.
- Transmit multicast keys efficiently to RCSTs.

Support different security profiles (i.e. a specific set of cryptographic algorithms and parameters) for each channel's keys, both unicast and multicast.

- Support periodic rekeying of multicast channels - this is usually performed at regular intervals to reduce the probability of successful cryptanalysis of the encrypted traffic.
- Support rekeying to perform ejection of a compromised member of a multicast group.

In order to support multiple keys per RCST, the Main Key Exchange (MKE), Quick Key Exchange (QKE) and Explicit Key Exchange (EKE) messages need to both support multiple keys and also identify which unicast or multicast channel uses a given key. These messages therefore need extending to allow this. One of the following two mechanisms can be adopted to enable the NCC to send the keys for each multicast group to an RCST:

- At logon: keys for all multicast groups are distributed to each RCST, at the same time as the RCST's individual (unicast) session key. This mechanism is suitable where there are a small number of multicast groups. The advantage of this mechanism is its simplicity, but the disadvantage is that if there are a large number of multicast keys and each RCST is only expected to join a small number of groups then a large amount of network capacity is wasted in sending unwanted keys.
- On demand: keys for multicast groups are issued on demand, when the RCST joins a multicast group. This mechanism is scalable and suitable for systems with a large number of multicast groups, and requires new message types to enable a RCST to request the key(s).

In addition, in order to provide periodic rekeying and existing member ejection, the MKE/QKE/EKE messages can be used. However, key management architectures exist that are highly scalable to large multicast groups; one particularly promising mechanism which is receiving a high degree of interest is Logical Key Hierarchy (LKH) RFC 2627. LKH requires that multiple keys be unicast to each

RCST when it joins a multicast group, and that some further keys be multicast to the group when rekeying takes place. LKH therefore requires two new extensions to the EKE message, which we call Extended EKE and Rekey EKE.

Therefore four sets of DVB-RCS security specification amendments, can be considered to support secure multicast:

- Support for multiple keys per RCST, to enable unicast and multicast traffic to use separate keys.
- Transmission of multicast keys at logon.
- RCST on-demand request for keys, to provide a scalable mechanism for multicast group joining.
- Extensions to EKE to support multicast rekeying, to enable key updates in case an RCST is compromised (Extended EKE and Rekey EKE).

Multicast Service Scenarios

This section presents some high level scenarios that highlight key management issue for secure multicast services in satellite networks. Each scenario should counter all of the threats identified above. Security policies play an important role in defining the rules that govern a secure multicast session and the privileges of each ingress ST/Gateway (multicast sender) and egress STs (multicast receiver). All satellite security entities will enforce the rules of these policies.

Scenario 1: End-to-end Secure Multicast - External Multicast Source to Satellite Network

This scenario is transparent to the satellite network. The satellite network plays no part in the secure multicast service setup or management. The multicast source (IP host/server), receivers (IP host) and key management are outside the satellite security administrative domain. Figure 22.8 shows this scenario, where the access to satellite network is controlled and managed by the satellite multicast source management functions [12].

Scenario 2: Multicast Group with Static Membership

This scenario covers the secure multicast services, where the group membership is fixed. In other words, the satellite security manager has a fixed list of ingress/egress STs. The egress STs may join the secure session before or during the multicast session. One practical example of such scenario is a Virtual Private Network (VPN) over satellites.

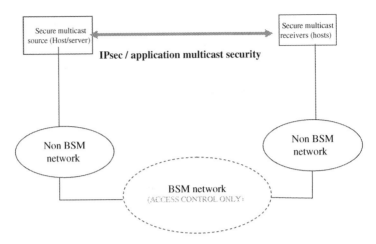

Fig. 22.8 Scenario 1: End-to-end multicast group - satellite transparent

The key management procedures here are simple: Initially, the secure multicast group members are authenticated by the key management server. The data encryption key and key management messages (generated by the satellite key management server) are distributed to all ingress and egress STs (by unicast or multicast transmissions). These keys can either be fixed for the whole duration of multicast session, or updated periodically (depending on the group policy). Figure 22.9 shows this scenario, where the security policy defines the rules for the secure group and the data security is the actual encryption/integrity of the multicast data.

Scenario 3: Multicast Group with Dynamic Membership

As shown in Fig. 22.10, this scenario covers the satellite secure multicast services, where the secure multicast group membership can vary with time. Assuming that there is a single ingress ST/gateway, the number of egress STs in a secure multicast session is dynamic. In other words, the satellite key management server has to manage a variable number egress STs. One example of such scenario an Internet Service Provider (ISP) providing a real time streaming service to a group of customers.

The key management procedures are more complex than scenario 2: The satellite key management server may perform the following tasks:

- Authentication of the secure multicast group members.
- Distribute keys to STs that are members of this session at the time of their joining the group.
- Re-key when an ST joins and/or leaves (depending on the rules in the security policy for this group).
- Periodic rekeying if required.

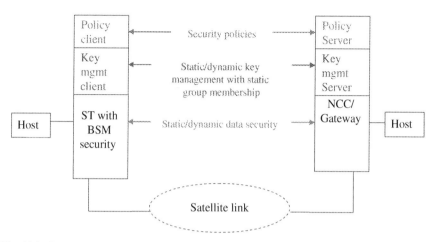

Fig. 22.9 Static multicast key management

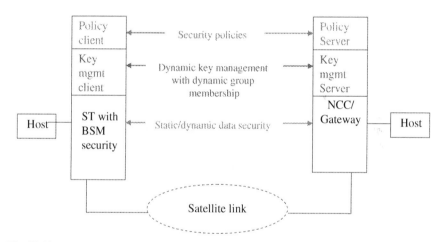

Fig. 22.10 Dynamic multicast key management

Scenario 4: Multiple Senders

The group management here can be either dynamic or static. However let us assume it is a dynamic group. One example of such scenario is a multimedia conferencing over satellites with multiple ingress STs. Similar to scenario 3, the satellite key management server has to perform key distribution and re-keying tasks. In addition, there can be two types of security association for ingress STs:

- Scenario 4a: All ingress STs (satellite multicast source) share one key and one security association.
- Scenario 4b: Each ingress ST has its own key and security association.

Scenario 4a is the simpler of the two, where one key and one security association is used for the whole group (all ingress and egress STs). However, protection against replay attacks is not possible because sequence numbering of secure packets is not possible in the presence of multiple senders (ingress STs).

For scenario 4b, each egress ST needs to handle multiple decryption keys in a multicast session (one key per ingress ST). Protection against replay attacks is possible with the use of one sequence number set per ingress ST.

In general, multiple source groups have special requirements for protection against Denial of Service (DoS) and for minimizing state needed for sender authentication. There can be two access control levels: The satellite Local and Network security managers to allow or deny ingress STs sending to the group. Therefore strict sender rules and a well defined security policy is critical to prevent DoS attacks.

Multicast Security Reference Framework

Figure 22.11, shows the main entities and functions relating to multicast security, and depicts the inter-relations among them [3, 7]. It also expresses the complex multicast security from the perspective of architecture (centralized and distributed), multicast group types (1-to-N and M-to-N), and classes of protocols (the exchanged messages) needed to secure multicast packets. Regarding the scenarios presented in "Overview of DVB-RCS Security", a single sender corresponds to 1-to-N type and multiple senders corresponds to M-to-N type. The satellite Network security manager correspond to the Domain Controller and Key Server (DCKS) in Fig. 22.11. The sender and receivers can be interpreted as end server/hosts or satellite ingress/egress STs. They show the place where encryption/decryption is applied. The following subsection presents the 3 functional areas in generic terms, which is applicable to satellite networks.

The Reference Framework can be viewed horizontally and vertically. Horizontally, it displays both the entities and functions as singular boxes, expressing each of the three broad problem areas (i.e. multicast security policies, management of keying material and multicast data handling). Vertically, it expresses the basic architecture designs for solutions; namely, a centralized architecture and a distributed architecture. In short, a distributed design is a superset of the centralized design which involves more than one group controllers and policy servers. This helps to prevent a single point of failure and performance bottleneck in the centralized design. In clause 4, all scenarios presented multicast services in a centralized architecture. Distributed key management is out of scope for this paper.

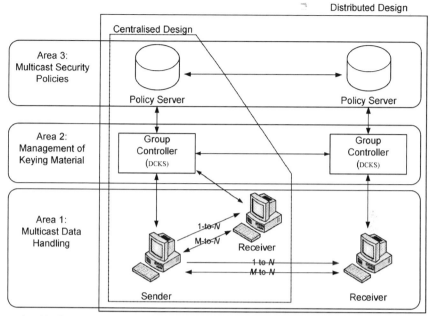

1-to-*N* : Only a single sender is allowed to transmit data to the group
M-to-*N* : Many group members can transmit data to the group

Fig. 22.11 Secure multicast reference framework

Multicast Data Handling (Privacy and Integrity)

This corresponds to functional area 1 in Fig. 22.11. In a secure multicast group, the
multicast data typically need to be:

- Encrypted using a group key, mainly for access control or confidentiality.
- Join authentication: New members are authenticated using the registration protocol specified later in this section.
- Authenticated, for verifying the source and integrity of the data. Authentication
  takes two flavours:

  - *Source authentication:* This functionality guarantees that the multicast
    data originated by the claimed source and was not modified en route
    (either by a group member or an external attacker). Typically, examples
    of source authentication algorithms are: TESLA RFC 4082 or RSA digital
    signature RFC 4359 [3].
  - *Group authentication*: This type of authentication only guarantees that the
    data generated (or last modified) by some group members. It does not guar-
    antee data integrity unless all group members are trusted. Typically, Message
    Authentication Code (MAC) is used to provide group authentication.

For data secrecy, the sender needs to encrypt the data stream with a secret key which is known by all group members that are authorized to receive multicast data stream. When the group becomes large, scalable distribution and rekeying of a group key can be a complex problem. The ETSI technical report TR 102 287 [9] presents the a scalable solution for keying to large multicast groups. The solution called Logical Key Hierarchy.

Group Security Association (GSA)

This also corresponds to functional area 1 in Fig. 22.11. In the context of unicast, the two-party Security Association (SA) management model is used to secure the communication between both parties. For example an SA in the Internet Protocol Security (IPSec) is identified by a Security Parameter Index (SPI). However, the unicast SA is simply not mapable to groups in the case of IP multicast, and the wider field of group communications, as there are many group members (senders and receivers). A GSA contains all of the SA attributes in point-to-point key management such as attributes include cryptographic keys, algorithm, identifier and other related attributes used to associate with the security material; as well as some additional attributes pertaining to the group.

The GSA is an aggregate of three categories of SAs. The first one is a "pull" SA between the group member and the DCKS, second is a "push" SA between the DCKS and all the group members, and the third is an SA to protect application data from the sender-members to receiver-members. In fact, each sender to the group may use a unique key for their data and use a separate SA. These three categories of SAs, which correspond to three different kinds of communications commonly required for group communications, are shown in Fig. 22.12. These categories are elaborated further in the following subsections.

Key Management

This corresponds to functional area 2 in Fig. 22.11. It is concerned with the security of distribution and refreshment of keying material. The term "keying material" refers to the cryptographic key belonging to a multicast group, the state associated with the keys and the other security parameters related to the keys.

A group key management protocol supports protected communication between members of a secure group. As group membership may vary over time, a group key management protocol can ensure that only members of a secure group can gain access to group data (by gaining access to group keys) and authenticate group data. The goal of a group key management protocol is to provide legitimate group members with the up-to-date cryptographic state they need for secrecy and authentication.

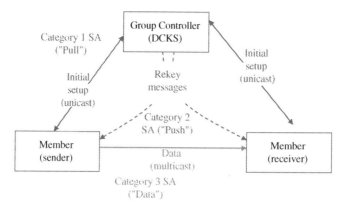

Fig. 22.12 Group Security Association (GSA) definition

Registration Protocol

In this protocol, the GC and a joining member mutually authenticate each other. If the authentication succeeds and the GC finds that the joining member is authorized, then the GC supplies the joining member with the following information:

- Sufficient information to initialize the Data SA within the joining member. This information is given only if the group security policy calls for initializing the Data SA at registration.
- Sufficient information to initialize a Rekey SA within the joining member. This information is given if the group security policy calls for a rekey protocol.

Some registration protocols need to tunnel through a data-signalling protocol to take advantage of existing security functionality, and/or to optimize the total session setup time. It may be advantageous to tunnel the key exchange procedure inside call establishment (MIKEY RFC 2830) [3] so that both can complete at the same time. The registration protocol ensures that the transfer of information from GC to member is done in an authenticated and confidential manner over a registration SA. A complementary deregistration protocol serves to explicitly remove Registration SA state.

Rekey Protocol

The purpose of the group rekey protocol is for transport of keys and SAs between a GC and the members of a secure group. The GC may periodically update or change the Data SA, by sending rekeying information to the group members. Rekey messages may result from group membership changes from:

- Changes in group security policy.
- Creation of new traffic keys or key distribution keys.

- Key expiry.
  Generally, the goals of the rekey protocol are:
- To synchronize a GSA.
- To provide privacy and (symmetric and asymmetric) authentication, replay protection and DoS protection.
- Efficient rekeying after changes in group membership or when keys expire.
- Reliable delivery of rekey messages.
- Member recovery from an out-of-sync GSA.
- Support multicast or multi-unicast.

Rekey messages are protected by the Rekey SA, which is initialized in the registration protocol. They contain information for updating the Rekey SA and/or the Data SA and can be sent via multicast to the group members or via unicast from the GC to a particular group member. There are two methods of authenticating rekey messages: group-based and source authentication.

The rekey protocol ensures that all members receive the rekey information in a timely manner. In addition, the rekey protocol specifies mechanisms for the parties to contact the GC and re-synch if their keys expired and an updated key has not been received. The rekey protocol for large-scale groups offers mechanism to avoid implosion problems and to ensure reliability in its delivery of keying material. A scalable solution for keying to large multicast groups called Logical Key Hierarchy (LKH) is presented in [9].

Although the Rekey SA is established by the registration protocol, it is updated using a rekey protocol. When a member departs the group, it destroys its local copy of the GSA. Using a de-registration message may be an efficient way for a member to inform the GC that it has destroyed, or is about to destroy, the SAs. Such a message may prompt the GC to cryptographically remove the member from the group (i.e. to prevent the member from having access to future group communication). In large-scale multicast applications, however, de-registration can potentially cause implosion at the GC.

Data Security Protocol

The data security protocol uses Traffic Encryption Keys (TEKs) to protect data streams sent and received by the data security protocol. Thus the registration protocol and/or the rekey protocol establish the Key Encryption Keys (KEKs) and TEKs. Regardless of the data security protocol used, the GC is responsible for supplying the TEKs, or information to derive the TEKs for traffic protection.

Figure 22.13 depicts the design of a group key management protocol. Each group member, sender or receiver, uses the registration protocol to get authorized and authenticated access to a particular group, its policies, and its keys. The two types of key used are the KEKs and TEKs. For group authentication of rekey or data, key integrity or traffic integrity keys may be used, as well. The KEK may be a single key that protect the rekey message, typically containing a new Rekey SA (containing a KEK) and/or Data SA (containing a TEK).

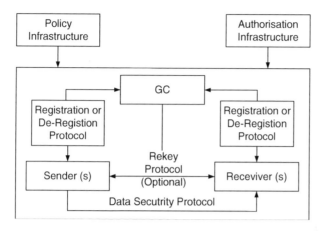

Fig. 22.13 Design of a group key management model

There are a few distinct outcomes to a successful registration protocol exchange:

- If the GC uses rekey messages, then the admitted member receives the Rekey SA. The Rekey SA contains the group's rekey policy, and at least one group KEK. In addition, the GC sends a group integrity key for integrity protection of rekey messages. If a group key management algorithm is used for efficient rekeying, the GC also sends one or more KEKs as specified by the key distribution policy of the group key management.
- If rekey messages are not used for the group, then the admitted member receives TEKs (as part of the Data Security SAs) that are passed to the member's data security protocol.
- The GC may pass the KEKs or TEKs to the member even if rekey messages are used, for efficiency reasons and according to group policy.

It is also worth noting that the rekey protocol is primarily responsible for scalability of the group key management architecture. Hence, it is imperative that the above listed properties are provided in a scalable manner. For instance, the rekey properties may use a scalable group key management algorithm to reduce the number of keys sent in a rekey message.

Security Policy Establishment and Enforcement

The general framework for using security policies is described in TS 102 465 [8]. Satellite security policy provides the rules for operation for all the elements of the multicast reference framework. Security policy design and detailed description is out of scope of the present document. The communication protocols between the Policy Server and the Key Server can be realized using various mechanisms:

- Using standard policy infrastructure such as a COPS Policy Decision Point (PDP) and Policy Enforcement Point architecture (PEP) [3].
- Using the key management protocol to transfer the security policy such as the GSAKMP protocol.
- Using other protocols such as Session Initiation Protocol (SIP) to transfer the security policy or even through web services.

At minimum, however, this security service will be realized in a set of policy definitions, such as every session security conditions and actions.

Summary of Multicast key Management Requirements

In general, multicast services have the following key-management requirements (these requirements however are not intended to be an exhaustive list nor applicable to all applications and services):

- Group members receive Security Associations (SAs) which including encryption keys, authentication/integrity keys, cryptographic policy that describes the keys, and attributes such as an index for referencing the SA or particular objects contained in the SA.
- In addition to the policy associated with group keys, the Group Controller (GC) may define and enforce group membership, key management, data security, and other policies that may or may not be communicated to the entire membership.
- Keys will have a predetermined lifetime and may be periodically refreshed.
- Key materials should be delivered securely to members of the group so that they are secret, integrity-protected, and verifiably obtained from an authorized source.
- The key management protocol should be secure against replay attacks and Denial of Service (DoS) attacks.
- The protocol should facilitate addition and removal of group members. Members who are added may optionally be denied access to the key material used before they joined the group, and removed members should lose access to the key material following their departure.
- The protocol should support a scalable group rekeying operation without unicast exchanges between members and a GC, to avoid overwhelming a GC managing a large group.
- The key management protocol should offer a framework for replacing or renewing security transforms, authorization infrastructure and authentication systems.
- The key management protocol should be secure against collusions among excluded members and non-members. In other words, combining the knowledge of the colluding entities should not result in revealing additional group secrets.

- The key management protocol should provide a mechanism to securely recover from a compromise of some or all of the key material.
- The key management protocol may need to address real-world deployment issues such as Network Address Translation (NAT)-traversal and interfacing with legacy authentication mechanisms. It should be flexible for use at any layer of the protocol stack such as satellite link layer, IP layer or above. Example satellite security cases with satellite link layer security and IPsec are presented in clause 6.

Some of the above requirements are illustrated in the secure multicast scenarios in clause 4 such the enforcement of group rules using policies, dynamic group membership and protection against DoS and replay attacks.

For scenarios 2 and 3 (static or dynamic) if IPsec is used, then the satellite secure multicast management system requires that the IPsec subsystem to support unidirectional Group Security Associations (GSA). As such only one group member authorized to transmit (Ingress ST or Hub) can use this type of group security association to enforce that group policy. In the inverse direction (e.g. egress STs), the GSA does not have a SAD entry, and the SPD configuration is optionally setup to discard unauthorized attempts to transmit unicast or multicast packets to the group. Similar procedures can be used for link layer security systems such as DVB-RCS.

Regarding scenario 4 (multiple senders), all (or a large subset) of the Group Members are authorized as multicast ingress STs. In such service model, creating a distinct SA with anti-replay state for every potential source ST does not scale to large groups. The group may share one SA for all of its ingress STs. In this case, the SA should not use the anti-replay protection service for the multicast data flow to the Group Receivers (egress STs).

Detailed Satellite Security Functional Architecture

The BSM general security architecture document (TS 102 465, [8]) defines the architecture elements and interactions across the SI-SAP interface. The same architecture applies to secure multicast over satellite. If a unicast session is present in combination with multicast (such as unicast satellite return channel or a multicast ingress ST forwarding data to ingress Hub) the security procedures presented in the satellite general security architecture document (TS 102 465) is applied. The focus here is the multicast key management exchanges (registration, re-keying and data key distribution as presented in "Multicast Services Scenarios"). The correct relationship between security entities: satellite **Network** security manager (the security key server as presented in "Overview of DVB-RCS Security" scenarios), ST **Local** security manager (the security client), ingress and egress STs are enforced by the correct use of security policies, with the strict enforcement of the roles for each entity (e.g. the permissions to send in the multiple senders scenario).

Case 1: Secure Multicast in the Satellite Dependent (SD) Layer (Below SI-SAP)

As shown in Fig. 22.14 all data security features (data privacy and integrity) are implement at the link layer in this scenario. Typical examples of such system are DVB-RCS with Multi Protocol Encapsulation (MPE) [11]or Unidirectional Lightweight Encapsulation (ULE) RFC 4326 IP encapsulation [3]. This is transparent to the satellite network. However, the satellite **Local** and **Network** security managers are able to enforce the satellite security multicast policy rules. The multicast security policy is distributed through the SI-M-SAP interface. Using link layer security has the major advantage of authenticating satellite terminals (BSM STs and gateways), which is not possible using security methods above the SI-SAP.

Case 2: Secure Multicast with Network Layer Security (Above SI-SAP)

As shown in Fig. 22.15, this case illustrates the use of network layer security (such as IPsec) for secure multicast over satellite network in a security gateway-to-gateway configuration such as VPN over satellites scenario. IPsec protocol operates above the SI-SAP. Security is provided between a security gateways (that can be co-located with a BSM ST or Gateway). The security gateway consists of three functional entities:

1) Secure data handling entity (privacy/integrity engine): IPsec operates here in tunnel mode. The data security protocol was described in "Multicast Service Scenarios".
2) Key management entity: In a star topology, there is a **Network** security manager for the whole satellite network (co-located with satellite gateway/hub). In addition there is a **Local** security manager in each ST. The key management protocols such as registration and rekeying were described in "Multicast Service Scenarios".
3) Security policy client and server and distribution mechanism as described in "Multicast Service Scenarios".

Figure 22.15 shows that the SI-U-SAP (the user interface) ONLY is used to communicate all secure information (user data key management messages and multicast security policies). Any of the key multicast key management systems described in "Multicast Service Scenarios" can used here. The Security Association Identity SID is used in all security management message exchanges. The client/server authentication/authorization processes is not shown here.

Case 3: Mixed Secure Multicast (Security Manager Above SI-SAP and Security Engine Below SI-SAP)

As shown in Fig. 22.16, this case illustrates the use of link layer security (below SI-SAP) for data security. The security manager is above the SI-SAP for the multicast key management. This can be a star or mesh topology with centralized security

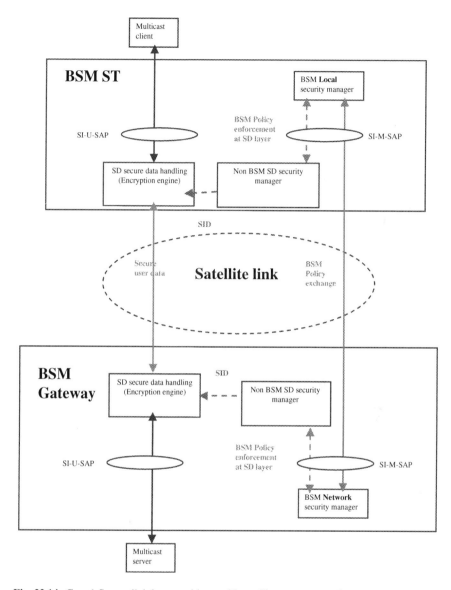

Fig. 22.14 Case 1 Secure link layer multicast with satellite access control

key management. Typical examples of such system are DVB-RCS with MPE or ULE. Like case 2, security is provided between security gateways (that can be co-located with satellite ST or Gateway).

Figure 22.16 shows that the SI-U-SAP (the user interface) is used to communi-cate secure user data. While the key management secure information is passed through the SI-C-SAP interface. The SI-M-SAP (management interface) is used to

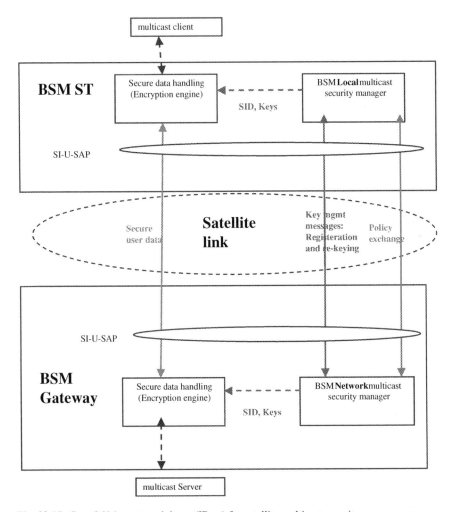

Fig. 22.15 Case 2 Using network layer (IPsec) for satellite multicast security

pass the multicast security policies between the satellite Network and STs Local security manager.

The specific satellite systems (SD layer) security is used. For example, for DVB-RCS satellite systems, the logon and key exchanges procedures of DVB-RCS recommendations (see "Overview of DVB-RCS Security" for more details on logon) is used to establish all security associations. For satellite systems operating with ULE, then the ULE specific key management procedures are used [3]. This will ensure the mutual authentication between all security entities, establishing the keys used subsequently to secure the user data. Using link layer security will also authenticate satellite terminals (STs

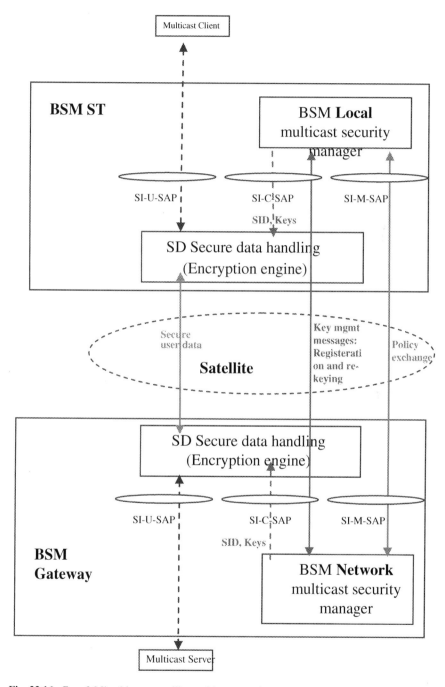

Fig. 22.16 Case 3 Mixed layers satellite multicast security

and gateways), which is not possible with using IPsec (case 2). The Security association identity SID is used in all security management message exchanges.

Case 4: End-to-End Secure Multicast

This case is used to transport non-satellite secure multicast traffic over the satellite network. It is transparent to satellite security system, except applying access control on such traffic. As shown in Fig. 22.17, the multicast security policy is used to enforce the security rules in the satellite ST and Gateway. Again the SI-M-SAP is used for policy distribution.

Interactions Between Multicast Security and Other Satellite Network Entities

Interactions Between Multicast Security and QoS Satellite Entities

QoS Provisioning for Key Management Messages

The satellite general QoS architecture and interworking with RSVP and Diffserv are presented in detailed in TS 102 462 [13], TS 102 463[14] and TS 102 464 [15] respectively. Reliability of key management messages is important for the efficient working of secure multicast in satellite networks. If RSVP or DiffServ are used for QoS provisioning, then key management messages should be transport in a better than best effort flows. For example, Fig. 22.18 shows the transport of these messages in a Guaranteed Service (GS) flow.

In the case of DiffServ the security policies are defined in the SLA. The Common Open Policy Service (COPS) protocol can used to carry QoS or security information between satellite management entities and satellite terminals (gateways/ST) (RFC 2748 [3]) and to update them dynamically (even in the case of DiffServ). In addition, if COPS is used for QoS provisioning, then COPS Policy Provisioning protocol (COPS-PR) can be used for security policy transfer (RFC 3084, [3]). The transport of security message flows is defined in the QoS policy at the time of service provisioning setup.

Securing RSVP and Diffserv Message Exchanges

Another interaction between security and QoS in a satellite network is the security and authenticity of QoS signalling between NCC, ingress and egress STs. TS 102 463 [14]describes a centralized and distributed RSVP architectures. Figure 22.19 shows the centralized architecture as an example.

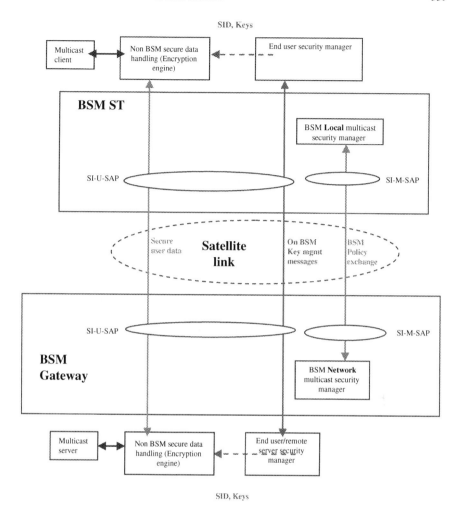

SID, Keys

Fig. 22.17 Case 4 Secure end-to-end multicast with satellite access control

Securing RSVP messaging using IPsec has several complications as presented in RFC 4230 [3]. Therefore, the use of IPsec (above the SI-SAP) is not recommended. Optionally security can be applied to the QoS control message exchanges (below the SI-SAP) between the NCC and ingress ST (Fig. 22.19). If security is required, then case 1 or 3 ("Multicast Security Reference Framework") can used, where the SD layer security (e.g. using authentication and/or encryption) is applied.

The security policy can define the security services needed for QoS signalling at the time of service provisioning setup. The choices for QoS signalling messages are:

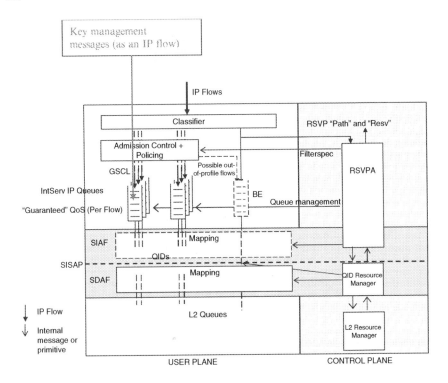

Fig. 22.18 Example Security messages mapping in ingress QoS Architecture (Dynamic SD Resources)

- No security.
- Authentication only.
- Authentication and encryption.

In case of static DiffServ, or dynamic DiffServ without explicit signalling (such as RSVP or NSIS) QoS signalling is not provided at IP layer. The IP queues are limited to a small number (maximum number is 64), and they are usually static or slowly-changing (on a time scale of hours or days). So the control plane operation does not need any satellite specific security considerations. The following way of operating (described in TS 102 464 [15]) shows that, even in case of dynamic DiffServ, messages in the control-plane are only exchanged below the SI-SAP (Fig. 22.20):

1) The IP resource manager interacts with the IP classifier and queuing module in the user-plane and with external IP signalling (optional) to understand

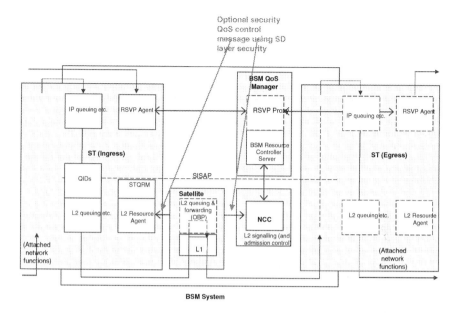

Fig. 22.19 Secure QoS signalling in satellite centralized RSVP Architecture (Dynamic SD Resources)

when and whether new resources are needed (or not needed anymore). The IP DiffServ queues may remain static, but the traffic situation might change.
MESSAGES: (1a) terminal defined, (1b) standard IP signalling, e.g. RSVP, NSIS.

2) So if the situation changes this module should notify the STQRM and take appropriate actions to allocate/release resources at the SD layers.
MESSAGES: (2) SI-SAP primitives.

3) This triggers an exchange of messages along the red lines, down to the SD layer and the NCC, since the resources at the SD layer will be, most likely, centrally managed.
MESSAGES: (3a) terminal defined, (3b) satellite defined.

4) The reply of the NCC will follow the green lines up to the IP resource manager.
MESSAGES: (4a) satellite defined, (4b) terminal defined, (4c) SI-SAP primitives.

5) These replies will enable the (re)configuration of the queue structure and of the mapping shown in TS 102 463.
MESSAGES: (5a) terminal defined, (5b) terminal defined, (5c) terminal defined.

6) In the end these operations might trigger responses at IP signalling level (optional).
MESSAGES: (6) standard IP signalling e.g. RSVP, NSIS.

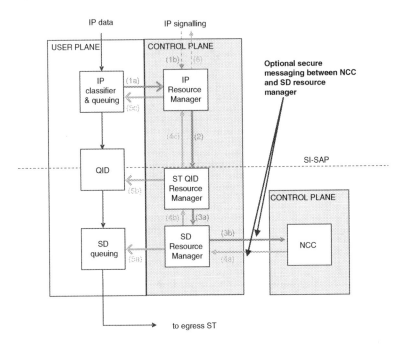

Fig. 22.20 Control-Plane operations of a DiffServ-aware ST - satellite security interactions

Since these messages exchanges between the ST and NCC in the in SD layer, Optional security service may be applied. If security is required, then case 1 or 3 ("Multicast Security Reference Framework") can used, where the SD layer security (e.g. using authentication and/or encryption) is applied to the messages 3b and 4a in Fig. 22.20.

Interactions with IPv6 Related Entities

In an all IPv6 satellite network, secure multicast issues are similar to an all IPv4 satellite network. However, if IPv6 and IPv4 combinations are used, then additional security issues related to the use of NATs are considered.

A straightforward and standards-based architecture that effectively avoids the satellite security manager interaction with NAT gateways (draft-ietf-msec-ipsec-extensions-04.txt, [3]) is the IPv6 over IPv4 transition mechanism RFC 2529 [3]. In IPv6 over IPv4 (a.k.a. "6over4"), the underlying IPv4 network is treated as a virtual multicast-capable Local Area Network. The IPv6 traffic tunnels over that IPv4 virtual link layer.

Applying satellite security system (with IPsec) in a 6over4 architecture leverages the fact that an administrative domain deploying satellite security system would already be planning to deploy IPv4 multicast router(s). The group's IPv6 multicast routing can execute in parallel to IPv4 multicast routing on that same physical router infrastructure. In particular, IPv6 multicast routers operating with 6over4 mode enabled on their network interfaces replaces the NAT gateways at administrative domain public/private boundaries.

Within the satellite security system, all references to IP addresses are IPv6 addresses for all security association endpoints and these addresses do not change over the group's lifetime. This yields a substantial reduction in complexity and error cases over the NAT-based approaches. This reduction in complexity can translate into better security.

Using COPS for Security Policy Provisioning

The IETF has defined the Common Open Policy Service (COPS) protocol (RFC 2748) [3] as a scalable protocol that allows policy servers (Policy Decision Point, PDPs) to communicate policy decisions to network devices. COPS is designed to support multiple types of policy clients. In satellite network, COPS can used to carry QoS or security information between satellite management entities and satellite terminals (gateways/ST).

RFC 3084 [3], describes the use of the COPS protocol for support of policy provisioning (COPS-PR). This specification is independent of the type of policy being provisioned (QoS, Security, etc.). The data model is based on the concept of Policy Information Bases (PIBs) that define the policy data. In order to support a model that includes multiple Policy-PDPs controlling non-overlapping areas of policy on a single Policy Enforcement Point (Policy-PEP), the client-type specified by Policy-PEP to the Policy-PDP is unique for the area of policy being managed. A single client-type for a given area of policy (e.g. security) will be used for all PIBs that exist in that area. The client should treat all the COPS-PR client-types it supports as non-overlapping and independent namespaces where instances MUST NOT be shared.

Radius/ Diameter

Authentication, Authorization and Accounting (AAA) protocols such as and RADIUS (RFC 2865) [3] was initially deployed to provide dial-up PPP and terminal server access. This can be achieved using the Remote Authentication Dial-In User Service (RADIUS) protocol. The RADIUS client is responsible for passing user information to designated RADIUS servers, and then acting on the response which is returned. RADIUS servers are responsible for receiving user connection requests, authenticating the user, and then returning all configuration information

Fig. 22.21 Host/machine authentication process

necessary for the client to deliver service to the user. A RADIUS server can act as a proxy client to other RADIUS servers or other kinds of authentication servers. To do so, the client creates an "Access-Request" containing such Attributes as the user's name, the user's password, the ID of the client and the Port ID which the user is accessing. When a password is present, it is hidden using a method based on the RSA Message Digest Algorithm MD5.

RADIUS is widely implemented and used. Experience has shown that it can suffer degraded performance and lost data when used in large scale systems, in part because it does not include provisions for congestion control. As a result, DIAMETER (RFC 3588) [3] should be considered as an alternative protocol to RADIUS. IPsec can be used with both RADIUS and DIAMETER. For example in RFC 3162 [3], RADIUS support for IPsec is not required. However, IPsec support is mandatory in DIAMETER, and TLS support is optional.

In satellite networks, communications between DIAMETER client and server are transparent to satellite security. However if RADIUS is used then either IPsec or link layer security must be used to carry such authentication/authorization messages.

For the purpose of the present document, the RADIUS/ DIAMETER concepts are abstracted. Therefore, three authentication entities are defined below and the authentication process is illustrated in Fig. 22.21:

- Supplicant: The client or machine requesting access to the network.
- Authenticator: The second component of the architecture is the access device or gateway, which is typically a switch or an access-point or a hub. The device in an authentication system that physically allows or blocks access to the network.
- Authentication server: This is typically a RADIUS/DIAMETER server or others against which the users will authenticate and from which they can even receive their authorization rules.

Interactions Between Satellite Security and Network Address Translation (NAT)

Perhaps the most common use of IPsec (RFC 4301) [3] is in providing Virtual Private Networking (VPN) capabilities. One popular use of VPNs is to provide telecommuter access to the corporate Intranet. Today, Network Address Translations (NATs) as described in (RFC 3022 [3]), are widely deployed in home gateways, as well as in other locations likely to be used by telecommuters, such as hotels. However, IPsec-NAT compatibility issue is a transitional problem and is related to the limited address space in IPv4. In IPv6 will address the address scarcity is not a problem. Therefore, to be useful, an IPsec-NAT compatibility solution MUST be deployable on a shorter time scale than IPv6. Here are some example compatibility issues between IPsec and NAT (3715 [3]):

- Incompatibility between IPsec AH (RFC 4302 [3]) and NAT. Since the AH header incorporates the IP source and destination addresses in the keyed message integrity check.
- Incompatibility between checksums and NAT. TCP and UDP checksums have a dependency on the IP source and destination addresses through inclusion of the "pseudo-header" in the calculation. As a result, where checksums are calculated and checked upon receipt, they will be invalidated by passage through a NAT or reverse NAT device.

Therefore if IPsec is used in satellite networks then NAT issues should be addressed for remote access,
terminal-to-terminal and end-to-end scenarios. One solution is using the Realm Specific IP (RSIP, RFC 3103,
RFC 3104) [3]. This solution will work for only a single NAT and does not work with multiple NATs.
A more generic solution is to adopt the IETF recommendations in satellite networks which resolves the compatibility issues, by implementing the following:

- UDP encapsulation of IPSec ESP packets as specified in RFC 3948 [3].
- IPSec key management and NAT traversal as specified in RFC 3947 [3].
- IPSec AH mode should not be used.

Interactions Between Security and Performance Enhancing Proxies (PEP)

The Transmission Control Protocol (RFC 0793) [[3] (TCP) is used as the transport layer protocol by many Internet and intranet applications. However, in certain environments, TCP and other higher layer protocol performance is limited by the link characteristics of the environment. Performance Enhancing Proxy (PEP) can perform mitigation techniques (RFC 3135) [3]. PEP is used to improve the performance of the Internet protocols on network paths where native performance suffers due to characteristics of a link (such as satellite links) or sub network on the path. A large spectrum of PEP devices exists (RFC 3449) [3], ranging from simple devices (e.g. ACK filtering) to more sophisticated devices (e.g. stateful devices that split a TCP connection into two separate parts).

However there are some security implications for using PEP is satellite environment. The most detrimental negative implication of breaking the end-to-end semantics of a connection is that it disables end-to-end use of IPsec. In general, a user or network administrator must choose between using PEPs and using IPsec. If IPsec is employed end-to-end, PEPs that are implemented on intermediate nodes in the network cannot examine the transport or application headers of IP packets because encryption of IP packets via IPsec's ESP header (in either transport or tunnel mode) renders the TCP header and payload unintelligible to the PEPs. Without being able to examine the transport or application headers, a PEP may not function optimally or at all.

If a PEP implementation is non-transparent to the users and the users trust the PEP in the middle, IPsec can be used separately between each end system and PEP. However, in most cases this is an undesirable or unacceptable alternative as the end systems cannot trust PEPs in general. With a transparent PEP implementation, it is difficult for the end systems to trust the PEP because they may not be aware of its existence. Even if the user is aware of the PEP, setting up acceptable security associations with the PEP while maintaining the PEPs transparent nature is problematic (if not impossible).

There are some steps which can be taken to allow the use of IPsec and PEPs to coexist. If an end user can select the use of IPsec for some traffic and not for other traffic, PEP processing can be applied to the traffic sent without IPsec. Another alternative is to implement IPsec between the two PEPs of a distributed PEP implementation. This at least protects the traffic between the two PEPs. (The issue of trusting the PEPs does not change).

Thus the requirement is that security must be implemented in such away that allows PEP entity access to the transport protocol headers (such as TCP). In addition,

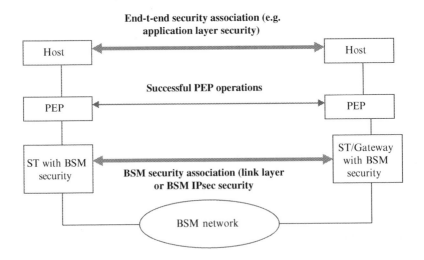

Fig. 22.22 Suitable Security associations for interworking with PEPs

link and application layer security are transparent to PEPs. If IPsec is used, then PEP operations must be performed outside the IPsec processing as shown in Fig. 22.22. IABG final report on an ESA project [16] provides further information about PEPs and security issues over satellites.

In summary, and as shown in Fig. 22.22, PEPs can be used successfully in satellite network with the following configurations:

- With Link layer security which operates on the satellite link only (such as DVB-RCS security).
- With IPsec provided that IPsec is used between the satellite ST/Gateway, where IPsec encryption is performed on incoming traffic (ingress ST/Gateway) after the PEP operations and decryption is performed on outgoing traffic (egress ST/Gateway) before the PEP operations.

Conclusions

Whilst TV broadcasting is probably the best-known application of satellite technology, satellite service providers are now expanding their services to include Internet data transmission. Also satellites are expected to play an essential role in providing broadband and multicast services to regions that cannot be economically reached by terrestrial networks, in particular the more remote regions of Europe and the rest of the world. Security can be a problem for such global services. This has focused on security of multicast data over satellite networks.

The paper started with and introduction to the ETSI BSM architecture and concepts. BSM work is aims for efficient transport of IP data streams and on how to interoperate resulting satellite networks with terrestrial IP networks. The BSM standards are being designed to use existing satellite standards such as DVB-RCS and IETF standards such as IPsec and it will remain open to emerging standards and other available technologies. This paper has presented a security solution that conforms to the BSM standard.

"Threats to Satellite Networks and Counter Measure Requirements" presented a threat analysis to multicast satellite services and it show the need for security requirements such as scalable multicast key management, Data confidentiality to protect satellite broadcasts; data origin authentication and reliable security message transport through the use of satellite specific QoS mechanisms.

The security architecture in this paper focuses on using network level security such as and IPsec. It is assumed that most reader is familiar with IPsec concepts and therefore IPsec details are not presented. The solution also allows for the use of satellite link layer security such as DVB-RCS security and an overview of the DVB-RCS security has been presented ("Overview of DVB-RCS Security") for the unfamiliar reader.

"Multicast Services Scenarios" has presented four high level scenarios that highlight key management issues for secure multicast services in satellite networks and addresses the threats mentioned in "Threats to Satellite Networks and Counter

Measure Requirements". In addition, it shows the important role of security policies in defining the rules that govern a secure multicast session and the privileges
of each ingress ST/Gateway (multicast sender) and egress STs (multicast receiver).
Then the paper introduced a detailed reference framework ("Multicast Security
Reference Framework") for key management highlighting three functional area for
secure multicast: multicast security policies, management of keying material and
multicast data handling. The framework is based on the IETF MSEC group work.

Based on the satellite scenarios and key management functional areas, detailed
security architecture is shown in "Multicast Services Scenarios". Four security
cases are presented: satellite independent, satellite dependent, mixed and end-to-
end architectures. Interworking through the BSM standard interface (Satellite
Independent Service Access Point, SI-SAP) is shown is all these cases. To complete
this architecture, the paper has presented interactions with other entities in satellite
networks in "Interactions Between Multicast Security and Other Satellite Network
Entities". Example interaction with QoS provisioning entities is presented in order
to provide reliability for security messages; and also providing security for QoS
request/response messages. In addition, the paper presents other interaction with
non-satellite entities such as policy provisioning with COPS, authorization with
AAA and interworking Performance Enhancing Proxies. Again interworking
through the SI-SAP has been highlighted.

The architecture proposed in this paper show clearly satellite security management service can be provided at various layers of the protocol stack (such as IP and
link layers) to accommodate various services and applications. This architecture
facilitates good interworking with satellite (QoS) and non satellite entities (COPS,
AAA and PEPs). Finally such solution allows closer integration of satellite security
system with the terrestrial Internet security solutions.

Acknowledgements The authors gratefully acknowledge the support of the EU Information
Society Technologies SATSIX Project [1], the European Commission [17]. Also the authors
acknowledge support of the satellite group at the European Telecommunications Standards
Institute (ETSI) [2].

References

[1] SATSIX web: http://www.ist-satsix.org
[2] ETSI home page: http://portal.etsi.org/portal_common/home.asp?TbId = 540
[3] IETF home page: www.ietf.org
[4] ETSI TS 102 292, "Satellite Equipment and Systems (SES); Broadband Satellite Multimedia;
    IP Interworking over Satellite; Satellite Functional Architecture".
[5] ETSI TR 101 984: "Satellite Earth Stations and Systems (SES); Broadband Satellite
    Multimedia; Services and Architectures".
[6] ETSI TS 102 294: "Satellite Earth Stations and Systems (SES); Broadband Satellite
    Multimedia (BSM) Services and Architectures; IP Interworking via Satellite; Multicast
    Functional Architecture".
[7] ETSI TS 102 466, "Satellite Equipment and Systems (SES); Broadband Satellite Multimedia;
    IP Interworking over Satellite; Multicast Security Functional Architecture".

[8] ETSI TS 102 465: "Satellite Earth Stations and Systems (SES); Broadband Satellite Multimedia (BSM); General Security Architecture".

[9] ETSI TR 102 287, "Satellite Equipment and Systems (SES); Broadband Satellite Multimedia; IP Interworking over Satellite; Security Aspects."

[10] T. Hardjono and B. Weis, "The Multicast Group Security Architecturel", RFC 3740, March 2005. www.ietf.org

[11] EN 301 790, "Digital Video Broadcasting (DVB); Interaction Channel for Satellite Distribution Systems", European Telecommunications Standards Institute (ETSI).

[12] ETSI TS 102 461: "Satellite Earth Stations and Systems (SES); Broadband Satellite Multimedia (BSM); Multicast Source Management".

[13] ETSI TS 102 462: "Satellite Earth Stations and Systems (SES); Broadband Satellite Multimedia (BSM); QoS Functional Architecture".

[14] ETSI TS 102 463: "Satellite Earth Stations and Systems (SES); Broadband Satellite Multimedia (BSM); Interworking with IntServ QoS".

[15] ETSI TS 102 464: "Satellite Earth Stations and Systems (SES); Broadband Satellite Multimedia (BSM); Interworking with DiffServ QoS".

[16] IABG final report: ESA project "IP Security over Satellites". Contract No. 15555/01/NL/US. 2002

[17] European Commission home page: http://www.cordis.lu/

About the Editors

Dr. Linghang Fan is a research fellow at the Centre for Communication Systems Research in the University of Surrey, UK. He received his B. Eng. in Automatic Control from Southeast University, China, and his MSc and Ph.D. in Telecommunications from the University of Bradford, UK. From 1998 to 2000, he was a researcher at the University of Bradford and worked on EU projects SINUS and SUMO. In 2003, he joined the University of Surrey and worked on EU projects STRIKE, Ambient Networks, MAESTRO and SATNEX. Currently, he is working on the EU projects SATSIX and ECGIN. He has published more than forty papers in international journals and conferences. His research interests include mobile/wireless communications, mobile Internet, and communication networking.

Dr. Haitham Cruickshank is a lecturer at the University of Surrey. He has worked there since January of 1996 on several European research projects in the ACTS, ESPRIT, TEN-TELECOM, and IST programs. His main research interests are network security, satellite network architectures, VoIP, and IP conferencing over satellites. He also teaches Data and Internet Networking and Satellite Communication courses at the University of Surrey. He is a member of the Satellite and Space Communications Committee of the IEEE ComSoc and a chartered engineer and corporate member of the IEEE in UK. He is an active member of the IETF and ETSI BSM working group in the security area.

Professor Zhili Sun is the Chair of Communication Networking in the Centre for Communication Systems Research, Department of Electronic Engineering, University of Surrey, UK. He received his BSc in Mathematics from Nanjing University in China and Ph.D. in Computing Science from Lancaster University, UK. He was a Postdoctoral Researcher, from 1989 to 1993, in the Telecommunications Group, Queen Mary, University of London before coming to Surrey. He has been a principal investigator and technical coordinator in many European projects including the ESPRIT BISANTE, VIP-TEN, GEOCAST, ICEBERGS, SatLife, SATSIX, and Euro-NGI projects. He has also been a principal investigator in UK EPSRC, the European Space Agency (ESA) and industrial projects on IP multicast security. He has supervised many Ph.D.

and research fellows. He published a book entitled *Satellite Networking: Principles and Protocols* (Wiley, 2005) and over 120 papers in International journals and conferences. He has also been a member of technical committees for international conferences and a member of reviewers for EU and UK research proposals.

Index